Lecture Notes in Electrical Engineering

Volume 1192

The book series *Lecture Notes in Electrical Engineering* (LNEE) publishes the latest developments in Electrical Engineering—quickly, informally and in high quality. While original research reported in proceedings and monographs has traditionally formed the core of LNEE, we also encourage authors to submit books devoted to supporting student education and professional training in the various fields and applications areas of electrical engineering. The series cover classical and emerging topics concerning:

- Communication Engineering, Information Theory and Networks
- Electronics Engineering and Microelectronics
- Signal, Image and Speech Processing
- Wireless and Mobile Communication
- Circuits and Systems
- Energy Systems, Power Electronics and Electrical Machines
- Electro-optical Engineering
- Instrumentation Engineering
- Avionics Engineering
- Control Systems
- Internet-of-Things and Cybersecurity
- Biomedical Devices, MEMS and NEMS

For general information about this book series, comments or suggestions, please contact leontina.dicecco@springer.com.

To submit a proposal or request further information, please contact the Publishing Editor in your country:

China

Jasmine Dou, Editor (jasmine.dou@springer.com)

India, Japan, Rest of Asia

Swati Meherishi, Editorial Director (Swati.Meherishi@springer.com)

Southeast Asia, Australia, New Zealand

Ramesh Nath Premnath, Editor (ramesh.premnath@springernature.com)

USA, Canada

Michael Luby, Senior Editor (michael.luby@springer.com)

All other Countries

Leontina Di Cecco, Senior Editor (leontina.dicecco@springer.com)

**** This series is indexed by EI Compendex and Scopus databases. ****

Jitendra Kumar · S. N. Singh · Om P. Malik
Editors

Recent Advances in Power Systems

Select Proceedings of EPREC 2023, Volume 1

Springer

Editors
Jitendra Kumar
Department of Electrical Engineering
National Institute of Technology
Jamshedpur
Jamshedpur, Jharkhand, India

S. N. Singh
Atal Bihari Vajpayee Indian Institute
of Information Technology
and Management
Gwalior, Madhya Pradesh, India

Om P. Malik
Department of Electrical and Computer
Engineering
University of Calgary
Calgary, Canada

ISSN 1876-1100 ISSN 1876-1119 (electronic)
Lecture Notes in Electrical Engineering
ISBN 978-981-97-2787-2 ISBN 978-981-97-2788-9 (eBook)
https://doi.org/10.1007/978-981-97-2788-9

This Springer imprint is published by the registered company Springer Nature Singapore Pte Ltd.
The registered company address is: 152 Beach Road, #21-01/04 Gateway East, Singapore 189721, Singapore

Preface

This book presents select proceedings of the Electric Power and Renewable Energy Conference-2023 (EPREC-2023). It provides rigorous discussions, case studies, and recent developments in the emerging areas of control application in power system, especially protection strategies, electric vehicle, microgrid operation, protection issues, operation of FACTS and HVDC, smart grid, etc. The readers would be benefited in enhancing their knowledge and skills in these domain areas. The book will be a valuable reference for beginners, researchers, and professionals interested in control applications in power system.

Jamshedpur, India — Jitendra Kumar
Gwalior, India — S. N. Singh
Calgary, Canada — Om P. Malik

Contents

About the Editors

Dr. Jitendra Kumar is currently an Assistant Professor at the Department of Electrical Engineering, National Institute of Technology Jamshedpur, India. He received his B.Tech. from IMS Engineering College, Ghaziabad, India, in 2009, his M.Tech. from the National Institute of Technology Kurukshetra, Kurukshetra, in 2011, and the Ph.D. degree in electrical engineering from the Indian Institute of Technology Roorkee, Roorkee, India, in 2017. He has over four years of teaching experience and taught subjects like power systems, advanced power systems, control systems, control and instrumentation, power system operation and control, signal and system, power system protection, and circuit and network theory. He has many papers in reputed journals. His major areas of research interests include power system protection and restructuring, Protection algorithms design in smart grid and microgrid environments, and Design of protection algorithms in FACTS environment. His research area also includes protection strategies during power swings.

Dr. S. N. Singh (Fellow, IEEE) is a Professor of Electrical Engineering at IIT Kanpur and has been the Vice-Chancellor of MMMUT Gorakhpur. He has obtained his M.Tech. and Ph.D. degrees in Electrical Engineering from IIT Kanpur. Apart from a receiver of various awards including the Young Engineer Award of the Indian National Academy of Engineering, the Khosla Research Award of IIT Roorkee, the Young Engineer Award of CBIP, New Delhi (India), Humboldt Fellowship of Germany and Otto Mønsted Fellowship of Denmark. Sir is also a fellow of the prestigious Institute of Electrical and Electronics Engineers (IEEE), USA, Indian National Academy of Engineering (INAE), Institution of Engineering and Technology (IET), UK, Institution of Engineers (India), Institution of Electronics and Telecom. Engineers (India), and Alexander von Humboldt (AvH), Germany. Professor Singh has authored more than 500 publications in reputed Journals, Conferences, Seminars, Short Papers, Invited Papers, and Technical Reports. He has supervised more than 40 Ph.Ds. and written two Books on Electric Power Generation, Transmission and Distribution, and Basic Electrical Engineering published by PHI. He has also completed more than 30 sponsored and consultancy projects.

Dr. Om P. Malik (Life Fellow, IEEE) graduated in electrical engineering from Delhi Polytechnic in 1952 and obtained a master's degree in electrical machine design from Roorkee University in 1962. He received a Ph.D. degree in electrical engineering from the University of London and a D.I.C. from the Imperial College (now the University) of Science and Technology, London in 1965. Dr. Malik worked in electric utilities in India from 1952 to 1961 and spent one year, 1959–1960, as a Confederation of British Industries Senior Scholar with manufacturing companies in the UK. He taught at the University of Windsor, Ontario, Canada, from 1966 to 1968 and joined the University of Calgary, Alberta, Canada, in 1968. He was acting Dean of the Faculty of Engineering in 1981, Associate Dean (Academic) from July 1979 to December 1990, and Associate Dean (Student Affairs) from July 1995 to December 1998. Presently, he is a Professor Emeritus in the Department of Electrical and Computer Engineering at the University of Calgary.

An Extensive Critique on Performance Analysis of Natural Ester-Based Edible Oils

Duraisamy Sarathkumar, Antony Raj Raymon, S. Sukitha, J. Sumithra, P. Sujeet, and R. Muralidharan

Abstract Nowadays, oil adulteration represents a big problem. The nutritional value and benefits of edible vegetable oils can be lost as a result of adulteration. Deterioration in quality may aid in the creation of hazardous and reactive oxidation products, which ultimately pose health hazards like cancer and inflammation. The identification of fatty acids using a variety of techniques is often followed by the computation of appropriate ratios in order to identify superfluous plant oils or visceral fats. However, since mineral oil has been added to edible oil, the detection of extraneous oil has become more challenging. The dielectric qualities of the mineral oils (breakdown voltage (kV) assessment), which are thermo-physical and electrochemical properties, are frequently what control the shelf lifespan of the oils recycled in transformers. This characteristic existed found in certain edible oils, and it can be used to detect oil adulteration with ease. The goal of this study is to identify an alternate insulating fluid for oil-cooled power transformer applications that has better dielectric characteristics. Additionally, the research targets the extension of the proposed fluid to power cables, reactors, circuit breakers, and capacitors. Another purpose of this paper is to find the oil adulteration easily. Electric property, thermal property, physical property and chemical property of pure edible oil, pure mineral oil and the mixture of edible oil and the mineral oil with different ratios such as 50% mineral oil and 50% edible

D. Sarathkumar (✉) · A. R. Raymon · S. Sukitha · J. Sumithra · P. Sujeet · R. Muralidharan
Department of Electrical and Electronics Engineering, Kongu Engineering College, Perundurai, Erode 638060, Tamil Nadu, India
e-mail: sarath.eee@kongu.ac.in

A. R. Raymon
e-mail: raymonhve@gmail.com

S. Sukitha
e-mail: sukithas.19eee@kongu.edu

J. Sumithra
e-mail: sumithraj.19eee@kongu.edu

P. Sujeet
e-mail: sujeetp.19eee@kongu.edu

R. Muralidharan
e-mail: muralidharanr.19eee@kongu.edu

J. Kumar et al. (eds.), *Recent Advances in Power Systems*, Lecture Notes in Electrical Engineering 1192, https://doi.org/10.1007/978-981-97-2788-9_1

oil, 70% mineral oil and 30% edible oil, the experiment is done and the result is analysed in this paper.

Keywords Breakdown voltage · Flash and fire point · Colour · Viscosity · Quality · Free fatty acid (FFA) · Cloud and pour point · Acid value (AV) · Iodine value (IV) · Edible oil

1 Introduction

Power transformer is the heart of any power system networks that integrates primary generation and distribution sides. Increasing day-by-day world's power demand and the expectation is to be doubled in 2030. Future growth of the transformer market will primarily be influenced by factors such as rising electricity demand in developing nations, increased demand for high-voltage power transformers, replacement of outdated electric power equipment in developed economies and global investments in the power sector. Such consumptions not only increase the usage of insulating fluids in power equipment's but affects the environment with hazardous by-products such as furans and particular matters. Hence in industrial settings where the environment is a concern, edible oils work well as lubricants or lubricant additives in place of petroleum oils.

The growth and regulation of numerous body functions are considerably aided by the consumption of oil and fat, which are significant dietary components. Oil serves as a major source of energy and a vehicle for the transportation of nutrients required for growth and metabolism, protecting brain cells, reducing the risk of heart disease, etc. Coconut, sunflower, olive, sesame, soy and groundnut oils are a few of the primary sources of edible oil. As the population of our country slowly grows, so must the production of cooking oil. Nowadays, the oil adulteration represents a big problem. Adding less quality or inexpensive vegetable oils, animal fat or mineral oil to the quality oil is the main objective. The adulteration is found by using various types of physical, chemical, thermal and dielectric properties such as viscosity, colour, free fatty acid value, flash and fire point, cloud and pour point, breakdown voltage, etc.

In industrial settings where the environment is a concern, edible oils are excellent substitutes for petroleum oils as lubricants or lubricant additives. With the high price of petroleum, the development of alternative industrial lubricants that are economically viable is urgently desired. Vegetable oils increasingly being used in lubricants for industrial and transportation uses as environmental concerns rise. They are biodegradable and renewable. Low-temperature viscosities, oxidative instability and hydrolytic instability are linked to vegetable oils or triglycerides. Blending several oils can enhance the lubricating qualities of vegetable oils in addition to functionalization. Vegetable oils that have been blended typically have better pour, flash and fire points.

The article [1] describes how to spot milk fat adulteration. The dielectric properties of mineral oils, which include electrochemical and thermo-physical features, are typically utilized to predict the shelf lifetime of the oils castoff in transformers (breakdown voltage (kV) test). This property was assessed in a range of milk fats, margarine oil, vegetable oils and some animal fats to see if there was a notable variation that could be used to separate these oils and fats from one another [2].

The research [3] examines the steps and evaluates the outcomes of the free fatty test for different kinds of cooking oils. This study shows that used frying oil can produce a chemical that is extremely dangerous. The process flow for determining the quality of cooking oil, calculating the quantity of free fatty acids, designing a detector, determining the quality and developing a classification system utilizing an artificial neural network is well explained in this study. It also demonstrates that doing a cooking oil quality inspection does not necessitate any chemical expertise [4].

In paper [5], the methods for blending crude oil, the research into technology, the tools required, and the mathematical models for calculating the best rates of blending blended oil are all observed. The goal of the investigation of density and viscosity in this paper was to ascertain how these properties changed with the mixing ratio and to compare the findings to those of current numerical models. This article examines the changes in refinery product management, including the amount of white products, level of oil fuel, and variations in delivery price, as well as changes in physical qualities including density, viscosity, freezing point and flow point [6].

The paper [7] gives the information about the use of edible oils in transformer for alternative of mineral oil. Petroleum oils can be replaced with edible oils as lubricants. With the high price of petroleum, the development of alternative industrial lubricants that are economically viable is urgently desired. Vegetable oils are increasingly being used in lubricants for industrial and transportation uses as environmental concerns rise [8].

The following lists the sections of the paper. Section 2 presents the block diagram of the proposed system. The hardware component description for the suggested scheme is described in Sect. 3. Section 4 gives the performance analysis of various oil samples. Section 5 describes the results and discussion. In Sect. 6, the conclusion of the suggested system was explained.

2 Block Diagram of the Suggested Scheme

The general block diagram provides the electrical, physical, thermal and chemical property and the test done in particular property (Fig. 1).

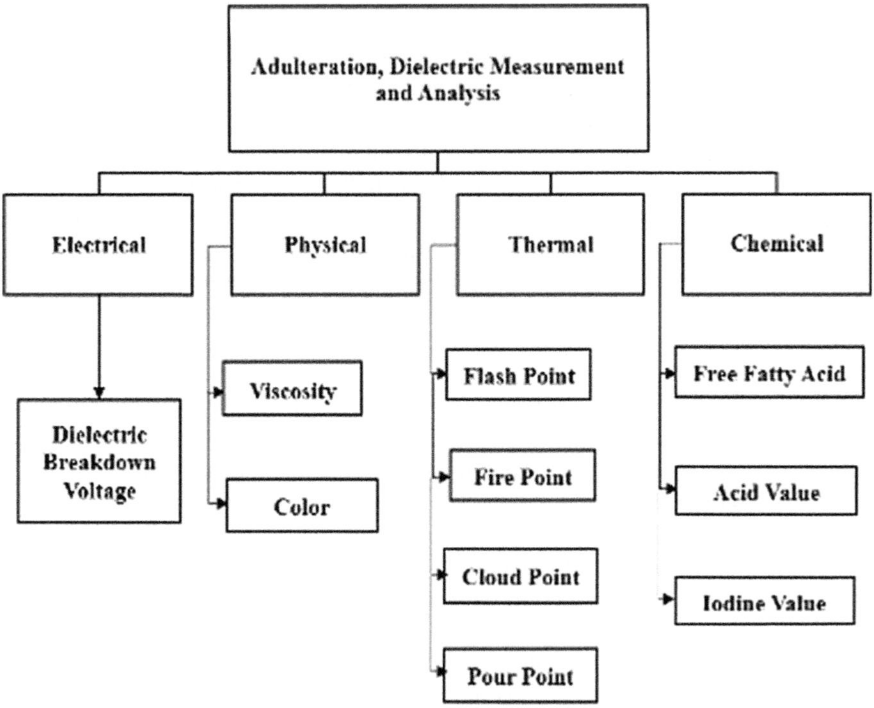

Fig. 1 Block diagram of the proposed scheme

3 Hardware Component Description for Proposed System

Several hardware constituents are used in our proposed method to analyse the performance of various oil samples. It consists of several components such as breakdown voltage transformer oil kit, redwood viscometer with digital temperature indicator, colour spectrophotometer, Cleveland's open-cup flashy and fire point kit with digital temperature indicator conical flask, graduated cylinder, cloud and pour point kit, burette, burette stand and burette clamp hot plate, conical flask, iodine flask, and pipette used to find adulteration [9].

3.1 Electrical Property

A. **Breakdown Voltage Transformer Oil Kit**

The threshold voltage at which the beginning of breakdown takes place is known as breakdown voltage. The open gap voltage rises, though, before current may flow, until it has made a channel for ions to travel through the dielectric [10]. Voltage

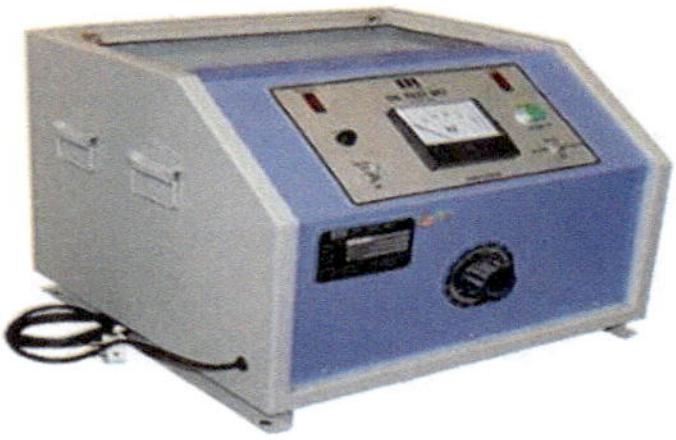

Fig. 2 Breakdown voltage transformer oil equipment

decreases and stabilities at the working gap level once the current begins to flow. To assess its insulating power, a sample of dielectric transformer oil is collected, and its breakdown voltage is measured. The transformer oil is of poor grade the lower the consequent breakdown voltage. Figure 2 represents physical appearance of breakdown voltage transformer oil kit.

3.2 Physical Properties

A. **Redwood Viscometer with Digital Temperature Indicator**

The term viscosity refers to how a fluid or gas flows. It is the oil's most important physical characteristic since it has an impact on both wear rate and fuel efficiency. It displays the degree of internal particle friction in a liquid. One of a liquid's most crucial physical properties is its viscosity [11]. Viscosity and temperature are inversely proportional to each other. Figure 3 shows the physical appearance of redwood viscometer with digital temperature indicator.

$$\text{Kinematic Viscosity} = \text{At} - (\text{B}/\text{t}),\ \text{in centistoke}$$

where

A = 0.284,

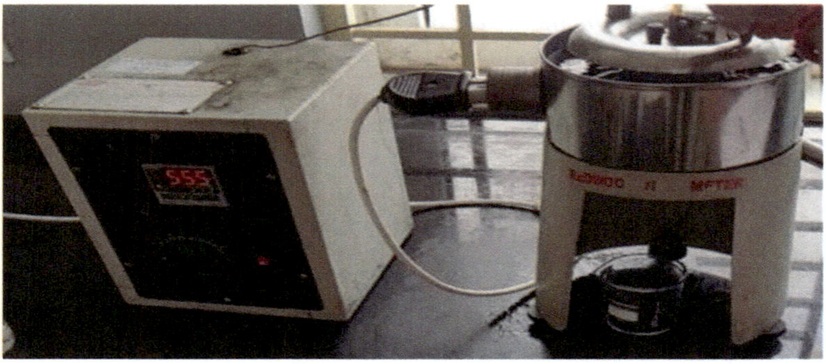

Fig. 3 Redwood viscometer with digital temperature indicator

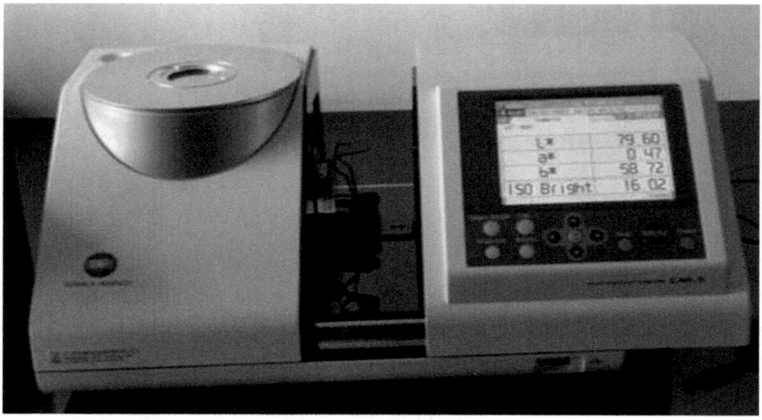

Fig. 4 Colour spectrophotometer

t = Time in seconds,
B = 190.

B. **Colour Spectrophotometer**

The conventional and affordable method of measuring light absorption or the concentration of substances in a solution is the spectrophotometer. When light of a specific wavelength is sent through the sample using a light beam, each substance in the solution either absorbs or transmits the light of that wavelength [3]. Figure 4 shows the hardware setup of colour spectrophotometer.

3.3 Thermal Properties

A. **Cleveland's Open Cup Fire and Flash Point Kit with Digital Temperature Indicator**

The Cleveland open-cup method, also referred to as the Cleveland open-cup tester, is one of three primary chemistry methods for evaluating a petroleum product's flash point. The flash point is the heat at which the arc or flash point is created by the oil, whereas the greatest temperature at which the oil has completely burned is known as the fire point. In situations where spills, leaks or ruptured lines may come into touch with an ignition source, a lubricant's high flash point is a crucial feature [12]. Figure 5 shows the hardware setup of Cleveland's open-cup fire and flash point kit with digital temperature indicator.

B. **Cloud and Pour Point Kit**

When oil is chilled to a low temperature, wax (paraffin) starts to separate at a certain point, known as the cloud point. This point is a crucial indicator of the practical performance of an oil in low-temperature automotive applications. The lowest temperature

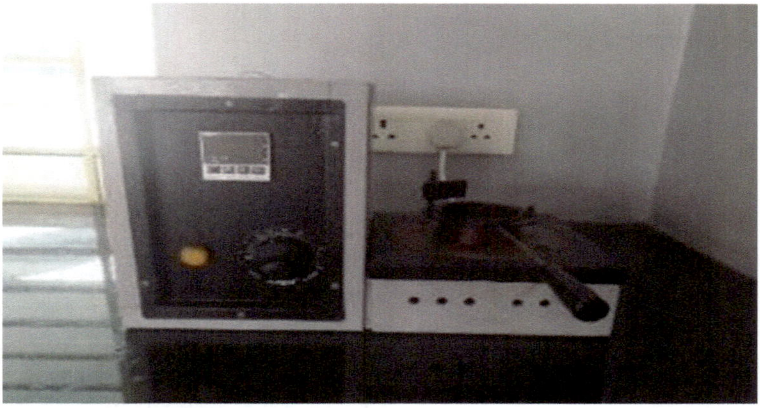

Fig. 5 Cleveland's open-cup fire and flash point kit with digital temperature indicator

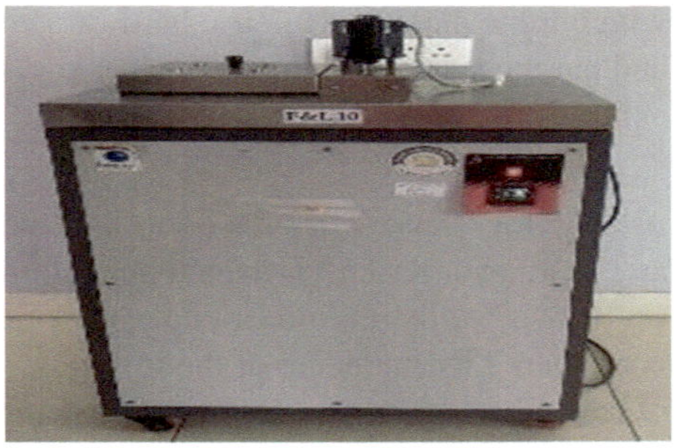

Fig. 6 Cloud and pour point apparatus

for an oil is known as its pour point at which it has been demonstrated in a particular scientific test to flow by gravity. Specifically, the pour point is 5 °F (3 °C) above the temperature at which oil in a lab sample container does not move while held horizontally for 5 s [12]. Figure 6 represents the hardware setup of cloud and pour point apparatus.

3.4 Chemical Properties (Free Fatty Acid Test, Acid Test and Iodine Test)

A. **Burette Stand, Burette Clamp and Burette**

In volumetric analysis, a burette is a lab tool used. A valve at one end of a long glass used in the device controls the flow of liquid. The purpose of a burette clamp is to support and maintain a burette on a stand so that it is fixed and more practical for the experiment. A burette stand, also known as a titration stand or retort stand, is a piece of laboratory apparatus widely used for titration research and can be equipped

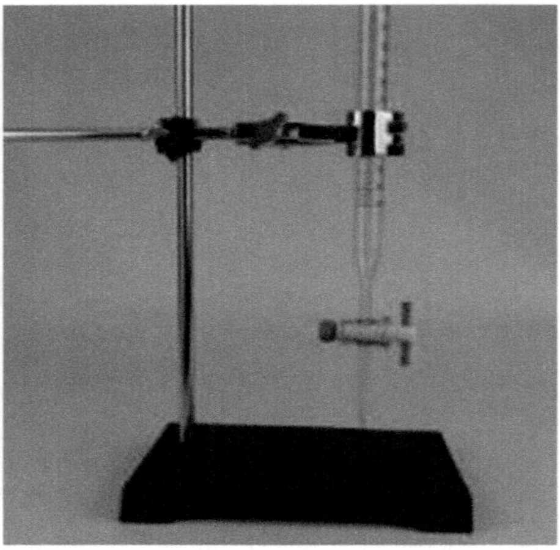

Fig. 7 Burette, burette clamp and burette stand

with burette clamps [10]. The hardware configuration of the burette, burette clamp and burette stand used in our suggested system is shown in Fig. 7.

B. Conical Flask

A laboratory flask with a flat bottom, conical body and cylindrical neck is known as an Erlenmeyer flask. It is sometimes referred to as a titration flask or a conical flask (in British English). Emil Erlenmeyer, a German scientist who developed it in 1860, is honoured by the name of the invention. Figure 8 shows the physical appearance of conical flask used for testing of oil samples.

C. Iodine Flask

An iodine flask is frequently used to measure iodine. Each flask has a ground socket and a cup-shaped top. These flasks are made of glass borosilicate. It is immune to thermal shocks due to the borosilicate glass's incredibly low thermal expansion coefficients. The ground glass joint on the iodine flask prevents the vapours from

Fig. 8 Conical flask

Fig. 9 Iodine flask

leaking out, improving the accuracy of your inquiry [12]. Iodine flask hardware for our suggested system is shown in Fig. 9.

D. **Graduated Cylinder**

Labs use a small, graduated cylinder-shaped container to precisely measure liquids. It is more accurate than a beaker because of the permanently stated incremental graduations incorporated into the clear cylinder. Long, thin containers called graduated cylinders are used to quantify the volume of liquids. The graduated cylinder configuration for evaluating oil samples is shown in Fig. 10.

E. **Sodium Hydroxide**

NaOH, or sodium hydroxide, has no smell. It is made of impenetrable white crystals that absorb moisture from the air. It is extremely corrosive and irritating to use sodium hydroxide. Whatever tissue it comes into contact with could be severely burned and permanently harmed. Figure 11 depicts the sodium hydroxide's physical characteristics as they are utilized to analyse oil samples.

F. **Potassium Iodide**

Potassium iodide is a compound made up of two different elements. Potassium and iodine were the main components that went into making this specific chemical. To stop radioactive iodine (I-131), one type of radioactive substance, from being

Fig. 10 Graduated cylinder

Fig. 11 Sodium hydroxide

absorbed by the thyroid, potassium iodide (KI), a non-radioactive form of iodine, can be utilized. The physical characteristics of the potassium iodide used to analyse oil samples are depicted in Fig. 12.

G. **Ethanol**

Ethanol is a renewable fuel made from corn and other plant materials (C_2H_5OH). Ethanol is used in the production of medicines, polymers, lacquers, polishes, plasticisers and cosmetics. The physical characteristics of the ethanol test samples are depicted in Fig. 13.

H. **Phenolphthalein Indicator**

A chemical molecule from the phthalein family called phenolphthalein ($C_{20}H_{14}O_4$) is widely employed as an acid–base indicator. Phenolphthalein is often used as a chemical indicator. The physical characteristics of the phenolphthalein indicator used for testing are shown in Fig. 14.

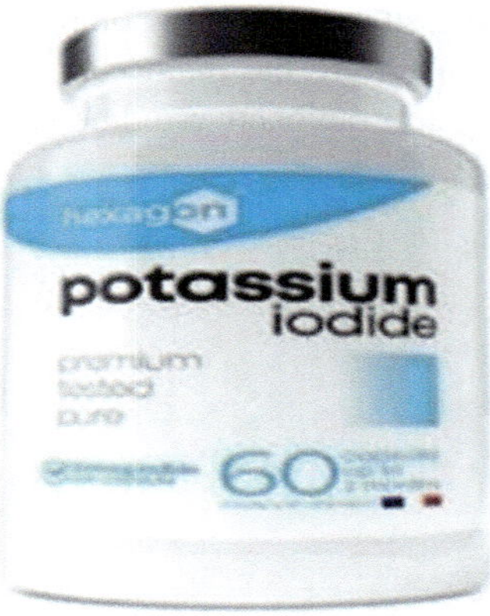

Fig. 12 Potassium iodide

Fig. 13 Ethanol

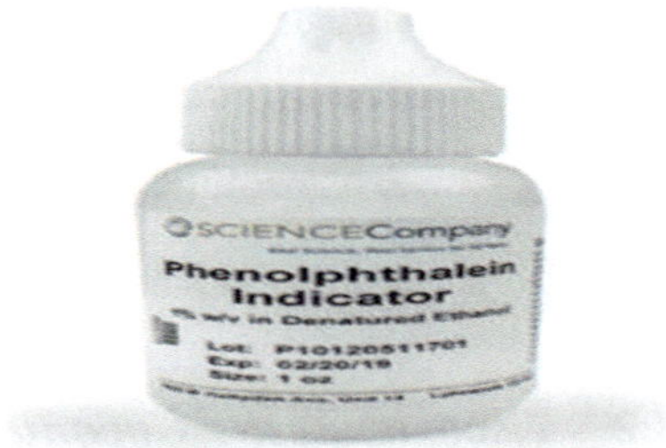

Fig. 14 Phenolphthalein indicator

4 Performance Analysis of Various Oil Samples

4.1 Breakdown Voltage Test

Figure 15 represents the performance analysis of breakdown voltage testing of edible oil and mineral oil. By the analysis of the experiments done, the sesame oil has the maximum breakdown voltage for 30% edible oil and 70% mineral oil.

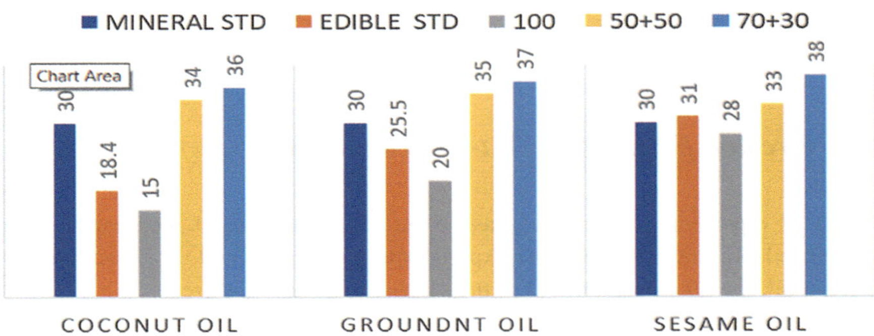

Fig. 15 Results of testing of breakdown voltage of edible oil and mineral oil

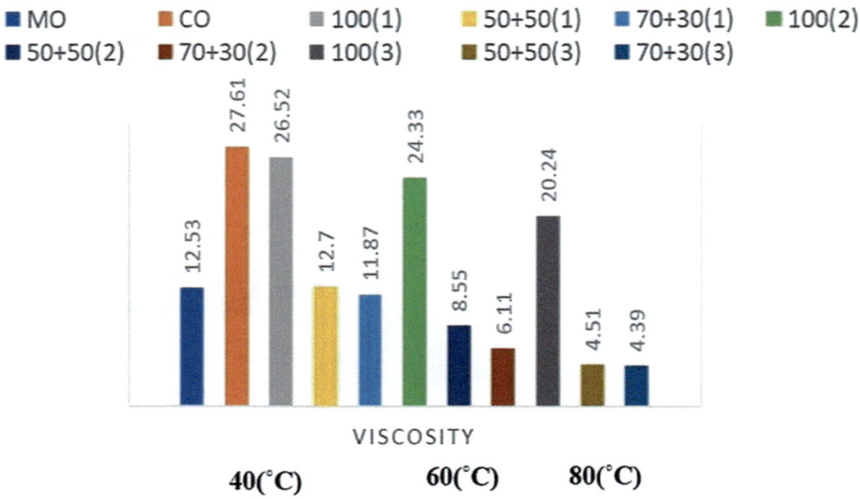

Fig. 16 Result of viscosity testing of coconut oil and mineral oil

4.2 Viscosity Test—Coconut Oil

Figure 16 shows the performance analysis of viscosity testing of mineral oil and coconut oil. By the analysis of the experiments done, the pure coconut oil has the highest viscosity at 40 °C.

4.3 Colour Test

Figure 17 represents the colour test analysis of edible oils. By the analysis of the experiments done in colour test, the pure coconut oil, groundnut oil, and sesame oil have highest lightness index.

4.4 Fire and Flash Point Test—Coconut Oil

Figure 18 represents the fire and flash point result analysis of inorganic oil and coconut oil. By the analysis of the experiments done in fire and flash point test for coconut oil, the pure coconut oil has the highest withstand for fire.

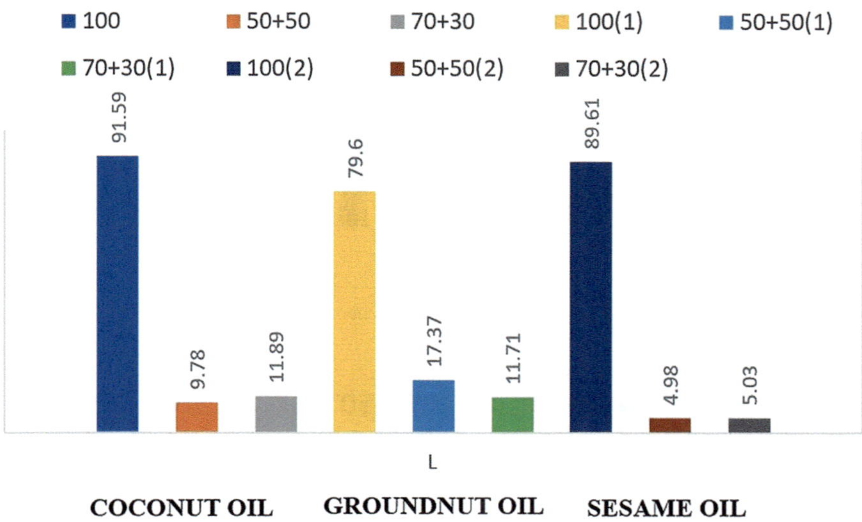

Fig. 17 Lightness index (L*) of edible oils

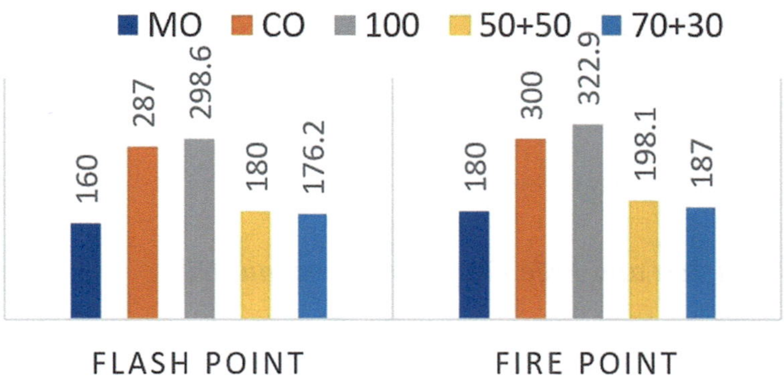

Fig. 18 Fire and flash point of mineral oil and coconut oil

4.5 Cloud and Pour Point Test—Coconut Oil

By the analysis of the experiments done of cloud and pour point for coconut oil, the coconut oil can also withstand to lower temperature. Figure 19 shows the performance analysis of pour point and cloud point of mineral oil and coconut oil.

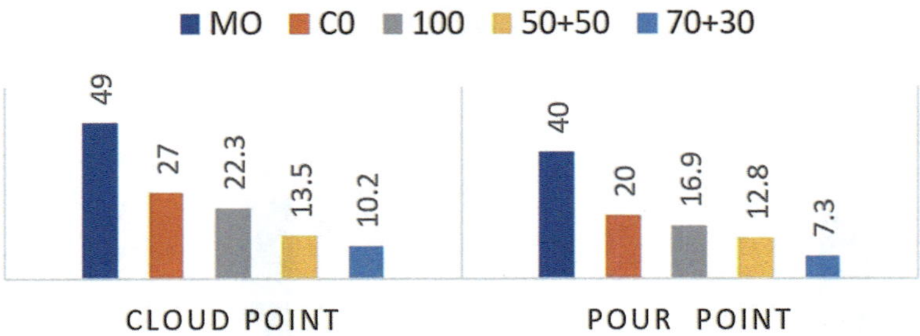

Fig. 19 Pour point and cloud point of mineral oil and coconut oil

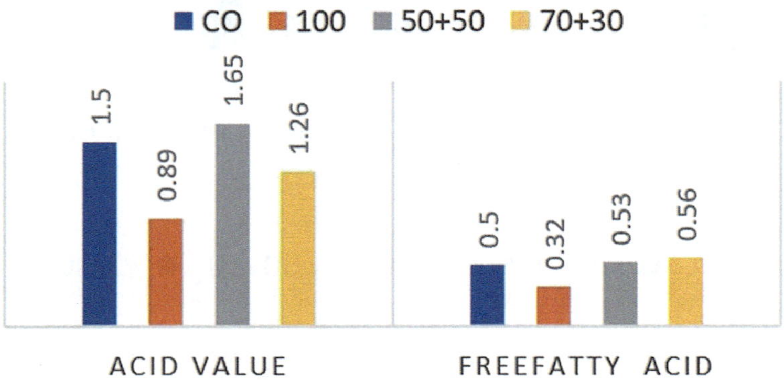

Fig. 20 Acid and free fatty value of coconut oil and mineral oil

4.6 Acid Test and Free Fatty Test—Coconut Oil

Figure 20 shows the performance analysis of acid and free fatty value of coconut oil and mineral oil. By the analysis of the experiments done of acid and free fatty acid value of the 50% mineral oil and 50% coconut oil has the highest acid value and free fatty acid rate.

5 Results and Discussions of Various Oil Samples

5.1 Breakdown Voltage Test

Table 1 represents the test results for breakdown voltage test. By the experiments done, the performance of breakdown voltage of pure mineral oil, pure edible oil, mineral oil and edible oil mixed with different ratios is between the standards.

Table 1 Results for breakdown voltage test

Oil name	Breakdown voltage (kV)
Mineral oil	≤30
Coconut oil	9.3–18.4
Groundnut oil	20.2–25.5
Sesame oil	29.5–31.8

Table 2 Results for viscosity test

Oil name	Viscosity (Centistoke)
Mineral oil	12.53 at 40 (°C)
Coconut oil	27.61 at 40 (°C)
Groundnut oil	30.32 at 40 (°C)
Sesame oil	**34.98 at 40 (°C)**

Table 3 Results for colour test

Oil type	L*	a*	b*
Pure mineral oil	1.76	1.28	1.00
Pure edible oil	1.73	1.39	1.86
Mineral oil	1.61	1.17	1.56
Edible oil	1.14	1.57	1.68

5.2 *Viscosity Test*

Table 2 shows the results for viscosity test. By the experiments done, the performance of viscosity value of pure mineral oil, pure edible oil, mineral oil and edible oil mixed with different ratios is between the standards.

5.3 *Colour Test*

Table 3 shows the results for colour test. By the experiments done, the performance of Lightness Index (L*), Chromaticity of Green and Red (a*), Chromaticity of Green and Blue (b*) of pure mineral oil, pure edible oil, mineral oil and edible oil is noted.

5.4 *Flash and Fire Point Test*

Table 4 demonstrates the results for fire and flash point test. By the experiments done, the performance of fire and flash point of pure mineral oil, pure edible oil, mineral oil and edible oil mixed with different ratios is between the standards.

Table 4 Results for flash and fire point test

Oil name	Flash point (°C)	Fire point (°C)
Mineral oil	160	180
Coconut oil	287	300
Groundnut oil	280	290
Sesame oil	300	320

Table 5 Results for cloud and pour point test

Oil name	Cloud point (°C)	Pour point (°C)
Mineral oil	Below 49	≤40
Coconut oil	Below 27	Below 20
Groundnut oil	−3–12	−15–10

Table 6 Results for acid test and free fatty test

Oil name	Acid value	Free fatty acid
Coconut oil	0.5–1.5	0.50
Groundnut oil	1–3	0.70
Sesame oil	1–2	0.25–3

5.5 Cloud and Pour Point Test

Table 5 shows the results for pour point and cloud point test by the experiments done, the performance of cloud and pour point of pure mineral oil, pure edible oil, mineral oil and edible oil mixed with different ratios is between the standards.

5.6 Free Fatty Test and Acid Test

Table 6 represents the results for free fatty test and acid test. By the experiments done, the performance of free fatty value and acid value of pure mineral oil, pure edible oil, mineral oil and edible oil mixed with different ratios is between the standards.

5.7 Iodine Test Result

Table 7 shows the results for iodine test result. By the experiments done, the performance of iodine value of pure mineral oil, pure edible oil, mineral oil and edible oil mixed with different ratios is between the standards.

Table 7 Results for iodine test result

Oil name	Iodine value
Coconut oil	7–10
Groundnut oil	105–115
Sesame oil	100–120

6 Conclusion

Various kinds of testing methods are tested for different kinds of edible oil with different oil ratio and combined with mineral oil in different ratios. The ratio of oil taken is 100% edible oil, 50% mineral oil + 50% edible oil, 70% mineral oil + 30% edible oil. The quality of edible oil can be checked by using breakdown voltage test, viscosity test, colour test and chemical test. The oil properties can be clearly identified by using breakdown voltage test. Based on the research works the mineral oil in addition with mineral oil can be used for cooling medium in transformer. By comparing all the above results, the 50% mineral oil + 50% edible oil can be preferred for cooling medium in transformer. The 50% mineral oil + 50% edible oil has good standard values in comparison to 70% mineral oil + 30% edible oil.

It has been shown that quality detection of cooking oil can be used to improve the process for maintaining oil standard and optimizing it for oil manufacturing at a fair rate. With the help of this strategy, the manufactures may guarantee the grade of the oil in their own area.

References

1. Raymon A, Pakianathan PS, Rajamani MPE, Karthik R (2013) Enhancing the critical characteristics of natural esters with antioxidants for power transformer applications. IEEE Trans Dielectr Electr Insul 20(3):899–912. https://doi.org/10.1109/TDEI.2013.6518959
2. Srinivasan M, Raymon A (2018) Comparison of potential of natural and synthetic multifunctional additives with natural ester and mineral oil for transformer applications. J Electr Eng 18(4):194–201
3. Duraisamy S, Murugesan S, Murugan K, Raj RA (2022) Reclamation of aged transformer oil employing combined adsorbents techniques using response surface for transformer applications. IEEE Trans Dielectr Electr Insul. https://doi.org/10.1109/TDEI.2022.3226162.
4. Sarathkumar D, Stonier AA, Srinivasan M, Senthamil LS (2022) Review on power restoration techniques for smart power distribution systems (Lecture notes in electrical engineering), vol 823. Springer, Singapore. https://doi.org/10.1007/978-981-16-7472-3_6
5. Sarathkumar D, Raj RA, Ganeshraj M, Kangiyaraj P, Kathir K, Kavin P (2022) IOT based password secured residential distribution transformer for technical safety. In: 2022 1st international conference on sustainable technology for power and energy systems (STPES), Srinagar, India, pp 1–6. https://doi.org/10.1109/STPES54845.2022.10006490
6. Sarathkumar D, Srinivasan M, Raj RA (2022) An extensive critique on transformer oil reclamation techniques. In: 2022 IEEE 7th international conference on recent advances and innovations in engineering (ICRAIE), Mangalore, India, pp 438–443. https://doi.org/10.1109/ICRAIE56454.2022.10054184

7. Karthik R, Raymon A (2016) Effect of silicone oxide nano particles on dielectric characteristics of natural ester. In: 2016 IEEE international conference on high voltage engineering and application (ICHVE), Chengdu, China, pp 1–3. https://doi.org/10.1109/ICHVE.2016.7800576
8. Duraisamy S, Murugesan S, Palanichamy M et al (2022) Restoration of critical dielectric properties of waste/aged transformer oil using biodegradable biopolymer-activated clay composite for power and distribution transformers. Biomass Conv Bioref 12:4817–4833. https://doi.org/10.1007/s13399-022-03065-0
9. Raj RA, Murugesan S, Ramanujam S, Stonier AA (2022) Empirical model application to analyze reliability and hazards in Pongamia oil using breakdown voltage characteristics. IEEE Trans Dielectr Electr Insul 29(5):1948–1957. https://doi.org/10.1109/TDEI.2022.3194490
10. Raj RA, Murugesan S (2022) Optimization of dielectric properties of Pongamia pinnata methyl ester for power transformers using response surface methodology. IEEE Trans Dielectr Electr Insul 29(5):1931–1939. https://doi.org/10.1109/TDEI.2022.3190257
11. Raymon A, Ravi S, Abid Y, Modisa M (2020) Enhancement of dielectric properties of baobab oil and mongongo oil using cost-effective additive for power transformer insulating fluids. Environ Technol Innov 20:1–13
12. Raj RA, Murugesan S, Sarathkumar D, Venkatachary SK (2022) A practical method for restoring waste cooking and motor oil as possible biofuel energy. In: 2022 2nd international conference on innovative sustainable computational technologies (CISCT), Dehradun, India, pp 1–6. https://doi.org/10.1109/CISCT55310.2022.10046608

Analysis of Distribution System Considering Load Models and Impact of Capacitor Minimizing Cost Using GAMS

Minakshi and Ashwani Kumar

Abstract Present-day distribution system operation has become complex due to load growth and reactive power requirements which leads to higher cost of reactive power sources and increased losses. Reactive power management and increasing demand in power system require the power system network operator to plan their generation to meet load demand maintaining the voltages and improving the overall efficiency of the system. The generation utilities are shifting their generation from conventional to non-conventional resources due to environmental concerns. In this paper, analysis of the distribution system has been carried out to minimize the power loss and cost of the power loss along with capacitor cost. The impact of different load models and reactive support from the capacitor bank has been considered for the analysis. Through load modeling, the planning and control of a particular power system can be done. In emergency condition, there will be chances of better control of power system by capturing the load variation and guide the power system operator to avoid further critical circumstances. The load growth has been also considered in analysis. The main objective of the work is: (i) minimizing cost of loss, cost of capacitors; (ii) load model impact on the cost and losses; and (iii) impact of load growth on losses and cost. Results on IEEE-33 radial distribution system are obtained using GAMS.

Keywords Load model · Capacitor cost · Power loss cost · Reactive power support · Voltage profile · Radial distribution system · General algebraic modeling system · Constant power-constant current-constant impedance load model · Residential-industrial-commercial load model

---

Minakshi (✉) · A. Kumar
National Institute of Technology, Kurukshetra 136119, India
e-mail: MINAKSHI_32114325@nitkkr.ac.in

A. Kumar
e-mail: ashwani.k.sharma@nitkkr.ac.in

J. Kumar et al. (eds.), *Recent Advances in Power Systems*, Lecture Notes in Electrical Engineering 1192, https://doi.org/10.1007/978-981-97-2788-9_2

1 Introduction

With the increase in power demand around the globe, there is a rapid increment in power loss and voltage deviation. Load modeling is done to improve the accuracy, planning, controlling, and analysis of the power system. Discussion on characteristics of load and composition of load is important and done [1, 2]. Review on conventional and new trends of load modeling, when the demand of consumer side differs there is variation in load consumption which is time dependent. Various research progress, industrial development, and new trends in load modeling are discussed in [3]. The study of load models is done using different approaches. In [4] two methods are used for the static model using conic optimal power flow solver. With objective of accuracy of different approach, when there is an undervoltage. According to observation, linear regression approach need to be done because it provides more accuracy whenever voltage is not in its operating range. A least square algorithm is used for analysis of the ZIP (constant impedance, constant current, and constant power) parameters. Prediction for active power demand profile of residential devices is obtained for the case study of Istanbul Technical University smart home. The accuracy of ZIP model is more when there is increment of 10 V as per the result [5]. Consumer usage of electrical power demand is handled by MCMC (Markov Chain Monte Carlo), concluded that if there is an increase in supply voltage, then there will be increase in power demand [6]. An optimization method for unbalanced distribution system is proposed for the day-ahead planning by integrating solar PV (photovoltaic) generation and CVR (conservation of voltage reduction) technique is used for load modeling parameter. The objective of MINLP (mixed integer non-linear programming) problem is to minimize the consumption of energy at substation, results are obtained for system in which energy saving and power demand reduced with the help of optimization technique [7]. For the power system dynamic stability main issue is accuracy of load model. Artificial Neural Network (ANN)-based approach load model is proposed which gave more accuracy. Artificial neural network-based model using load composition as its input and ZIP parameters. To check the accuracy, the result of proposed model has compared available measurement data where mean absolute percentage error is 1.7%. The main thing is that it must be updated periodically whenever new datasets are available [8]. In distribution system network line loss varies with different load models in network. Every year active power loss and reactive power increase when load growth is taken into consideration. These can be resolved by optimal placement of distributed generation [9]. Model that consists of both ZIP and RIC (residential, industrial, and commercial) model called composite load model was implemented by the WECC (Western system coordination council) after the 1996 blackout. Composite load model is defined as the combination of static and dynamic load models. WECC composite load model represents the quality of new obtained model that is identification of electrical distance between consumer and transmission bus and representation of dynamic characteristic of various consumers. This composite model determines the impact of new load demand on dynamic performance of power system [10]. Due to large number of parameter and complexity, new challenges arise for composite load model. A cutting-edge parameter reduction approach is based on active sub-space method. This method has better accuracy and it needs less size of sample as compared to Monte Carlo sampling-based methods [11]. Loads such as speaker, fans, pumps, refrigerator, and air conditioner are examples of residential, commercial, and industrial load. Due to these inductive loads, there is an increase in total active and reactive power losses in lines, decrement in voltage profile, and rise in energy cost. I^2R loss can affect the power transfer which results in low voltage profile. Installation of capacitor in distribution system can be used to improve voltage profile and reduce the power loss. In recent years, different metaheuristic algorithms are used for the placement of capacitor. Optimal planning of capacitor is done by using genetic algorithm. The benefits are power loss minimization and the settlement of cost [12]. Considering transient overvoltage for improving the convergence time and increase the problem-solving process. Placement of shunt capacitor the approach used that is two stages in first NSGA-II (non-dominated sorting genetic algorithm) and then decision-making technique. Second stage is used to find best result obtained from Pareto front. With the objective of minimizing power loss and total harmonic distortion which maintains the power quality issues [13]. The placement and sizing of shunt capacitor for reactive power compensation and minimizing voltage fluctuation. Power factor correction can provide better power with minimum losses by results obtained from MATLAB software [14].

In this paper, analysis of the IEEE-33 radial distribution system has been carried out for the minimization of the total power loss and cost of the total power loss along with capacitor cost. The analysis of different load models is covered in Sect. 2. In Sect. 3, the load growth has been considered for analysis. In Sect. 4, problem formulation is presented. In Sect. 5, methodology of GAMS (general algebraic modeling system) is discussed. In Sect. 6, results of the multi-objective function study are shown along with a detailed comparison of the proposed methods with some benchmark algorithms. Section 7 draws the conclusion.

2 Load Modeling

Practically it is not possible to show the load composition of each and every component as there are innumerous devices connected in a power system network. Load models are non-linear in nature that's why load modeling is important for the planning, operation, design, and controlling of a power system. To prevent any system collapse and to secure the stability of power system a proper and accurate load model should be used. With the use of un-appropriate load models, there will be many unwanted situations which can lead problematic results such as voltage instability. To maintain system stability voltage security monitoring system were developed. In that various approaches are figured to achieve processing at regular interval of every one minute [15].

For the classification of load model, it is categorized in two: static models and dynamic models. Steps for load modeling are to select a load model structure and

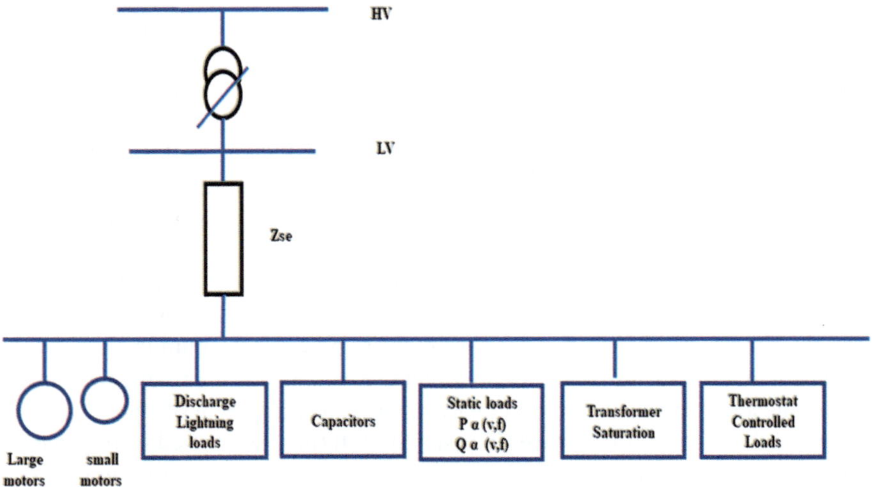

Fig. 1 Static and dynamic load model

then identify the load model parameter. Figure 1 shows the different types of loads connected to the substation.

2.1 *Static Load Model*

These load models do not vary with time. New active and reactive power demands are expressed in terms of voltage in the form of exponential and polynomial. The static studies are for steady-state analysis of the system.

Exponential Load Model

The relation of active and reactive power at load bus with voltage in an exponential equation form is represented below:

$$P_{dn} = P_d * \left(\frac{V}{V_o}\right)^a \tag{1}$$

$$Q_{dn} = Q_d * \left(\frac{V}{V_o}\right)^b \tag{2}$$

where P_d, Q_d are nominal active power and reactive power demand at load bus,

P_{dn} and Q_{dn} are new demand, and a and b are the load coefficient and the value of these exponents can be 0, 1, 2. By the value of these coefficients, different models can be designed. In each iteration of load studies power remains constant.

Polynomial Load Model

It is a mixed model combination of all three different models, also called the ZIP load model, here Z signifies constant impedance, I as the constant current, and P represents the constant power. Model of a polynomial is given below:

$$P_{dn} = P_d\left[a_p\left(\frac{V}{V_o}\right)^2 + b_p\left(\frac{V}{V_o}\right) + c_p\right] \tag{3}$$

$$Q_{dn} = Q_d\left[a_q\left(\frac{V}{V_o}\right)^2 + b_q\left(\frac{V}{V_o}\right) + c_q\right] \tag{4}$$

where P_d and Q_d are nominal real and reactive powers of load, V_o = nominal voltage of load, and V is actual voltage to calculate the load $a_q+b_q+c_q = 1$ and $a_p+b_p+c_p = 1$. Sum of active and reactive power of load model parameter is 1, cause the total amount of active and reactive is 1 or 100%. Value of fraction for ZIP model parameter taken as $a_p = a_q = b_p = b_q = 0.1$ and $c_p = c_q = 0.8$ [16].

2.2 *Dynamic Load Model*

Model which states the active and reactive powers as function of voltage with respect to time with the availability data from past as well as including present data.

RIC Load Model

Combination of residential, industrial, and commercial with time dependent. α, β, and γ are the percentage loads at each load bus, respectively. Data of summer, load types, and load exponent values of n_{pr}, n_{pc}, n_{pi}, n_{qr}, n_{qc}, n_{qi} provided [17].

$$P_{dn} = P_d\left[\alpha\left(\frac{V}{Vo}\right)^{n_{pr}} + \beta\left(\frac{V}{Vo}\right)^{n_{pc}} + \gamma\left(\frac{V}{Vo}\right)^{n_{pi}}\right] \tag{5}$$

$$Q_{dn} = Q_d\left[\alpha\left(\frac{V}{Vo}\right)^{n_{qr}} + \beta\left(\frac{V}{Vo}\right)^{n_{qc}} + \gamma\left(\frac{V}{Vo}\right)^{n_{qi}}\right] \tag{6}$$

3 Load Growth

Active and reactive power demand with load growth for any X year, the mathematical equation follows:

$$P_n = P_{d0}(1+g)^X \tag{7}$$

$$Q_n = Q_{d0}(1 + g)^X \quad (8)$$

where P_n, P_{d0} are active load power in the Xth year and in the initial year (base case year).

Q_n, Q_{d0} are reactive load power in the Xth year and in the initial year (base case year).

g represents the annual load growth which is taken as 7.5%.

X is the planning period up to which load growth is considered.

4 Problem Formulation

Objective function (OF): Minimization of Total cost.

$$\text{C1} = \text{Total active power loss} * \text{K1}.$$

$$\text{C2} = \text{Total reactive power supplied from capacitors} * \text{K2}.$$

$$\text{C3} = \text{Number of capacitors} * \text{K3}.$$

$$\text{OF: C1} + \text{C2} + \text{C3}$$

Where C1 is cost of active power loss.
C2 is operation and maintenance cost of the capacitors.
C3 is capital cost for installing capacitors.
K1 is equivalent annual cost per unit of power loss—168 $/(kW year).
K2 is running cost of capacitor—3 $/kVAR.
K3 is installation cost of capacitor—1000 $.
Power loss of active and reactive powers and RIC will vary with respect to time:

$$P_{loss} = \sum_i \left(V_i^2 + V_j^2\right) * g_{ij} - 2(V_i * V_j * g_{ij} * cos\left(\delta_i - \delta_j\right) \quad (9)$$

$$Q_{loss} = \sum_i \left(V_i^2 + V_j^2\right) * b_{ij} \quad (10)$$

Power flow equations:

$$P_i = V_i^2 * g_{ij} - \sum_{\substack{j \neq i \\ j = 1}} V_i * V_j(g_{ij} * \cos(\delta_i - \delta_j) + b_{ij} * \sin(\delta_i - \delta_j)) \tag{11}$$

$$Q_i = -\left[V_i^2 * b_{ij} - \sum_{\substack{j \neq i \\ j = 1}} V_i * V_j(g_{ij} * \sin(\delta_i - \delta_j) + b_{ij} * \cos(\delta_i - \delta_j)) \right] \tag{12}$$

where $\delta_i$and δ_j are the bus angles, g_{ij} is conductance and b_{ij} is susceptance between buses i and j, n is the number of buses, and P_{loss} and Q_{loss} are the active and reactive power losses of line which is present between present bus i and j.

$$P_{TL} = \sum_{i,j} P_{loss} \tag{13}$$

$$Q_{TL} = \sum_{i,j} Q_{loss} \tag{14}$$

P_{TL} is the total active loss of the line.
Q_{TL} is the active and reactive power losses of the line.

Balance equation

$$P_{i,g} - P_{i,dn} = P_{ij} \tag{15}$$

where $P_{i,g}$ and $P_{i,dn}$ are active power generation and new demand after load modeling at bus i.

$$Q_{i,g} + Q_{i,c} - Q_{i,dn} = Q_{ij} \tag{16}$$

where $Q_{i,c}$ is reactive power supplied by capacitor at bus i.
$Q_{i,dn}$ is the new reactive power demand after load modeling at bus i.

Inequality constraints

Inequality constraints are given their boundary limit where active and reactive powers lie between maximum and minimum which is taken from generator data of bus system. Boundary limit of voltage lies between 0.95 and 1.05. Optimization problem solved in GAMS 23.4 [18].

$$P_{gmin} \leq P_g \leq P_{gmax} \tag{17}$$

$$Q_{gmin} \leq Q_g \leq Q_{gmax} \tag{18}$$

$$V_{min} \leq V_i \leq V_{max} \tag{19}$$

$$\sum_{i=1}^{2} Q_{i,c} \leq \sum_{i=1}^{n} Q_{i,d} \tag{20}$$

The cost of capacitor and power losses is dependent on the load demand. From Eqs. (15) and (16), it is evident that $P_{i,g}$ and P_{ij} are dependent on the new load demand which is obtained after load modeling. Equations (5) and (6) represent the new load demands which are dependent on the load coefficients. So, from this we can conclude that the cost of capacitor and the power loss are dependent on the load coefficients.

5 Proposed Methodology

The flow diagram of proposed program is presented in Fig. 2 as the load flow method is used for the calculation of active and reactive power flows. Total active and total reactive power loss in base case is measured. For the reactive power compensation, capacitor is installed at specific bus location on the criteria of comparing the low voltage among all 33 buses to compensate reactive power demand. The mathematical modeling for ZIP is performed then for RIC load model for 24 h duration of summer season. New minimized load demand, power loss, voltage deviation, and change in cost are measured.

5.1 *Proposed Algorithm Using GAMS*

The following are the steps for how the progress is made:

Step 1: GAMS will read line and bus data of the system and consider the value of load line and generator data which are given.
Step 2: Define the type of problem based on objective function expression.
Step 3: Carry out load flow studies.
Step 4: After that perform load flow studies by modifying the demand by using Eqs. (3), (4), (5), and (6) for ZIP load model and RIC load model.
Step 5: Carry out load flow studies with load growth while considering both ZIP and RIC models.
Step 6: Multi-objective functions to calculate voltage profile, active and reactive power flow, active and reactive power loss, and number of iterations from step 2 to step 5.

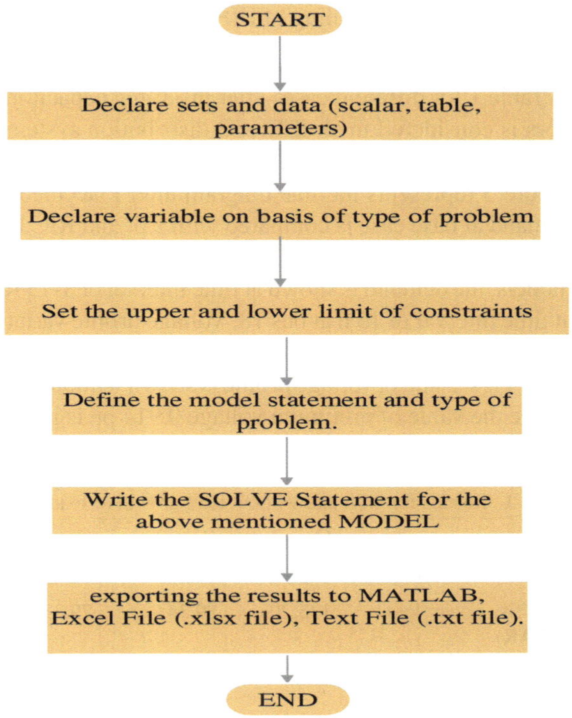

Fig. 2 Flowchart of presented method

Step 7: After getting the results we will compare the results for ZIP and RIC.

6 Results and Discussions

The result brought out shows the impact of load modeling and proves its necessity. It provides a possibility of maintaining better voltage levels within limits for all load models and also reduces the load demand to avoid the emergency situations. Capacitor is important in order to achieve a better voltage profile and to reduce the losses. Installation of multiple capacitors can increase the overall cost of network, so only a couple of capacitors are installed for injecting reactive power at the locations having low voltage levels. Due to all this the total cost can be minimized. The study is evaluated on 33 radial distribution system using the DICOPT solver for this mixed integer non-linear problem in GAMS. Base MVA is taken as 10 MVA.

There are five cases considered for analysis:

Case-1: Base case performed on 33-bus system.
Case-2: Load modeling of RIC without load growth.
Case-3: Load modeling of ZIP without load growth.
Case-4: Load modeling of RIC with load growth.

Case-5: Load modeling of ZIP with load growth.

Results are obtained for power loss, cost, minimum voltage, and maximum voltage in Table 1 for different cases after installing capacitor. Placement of multiple capacitors is considered in an IEEE-33 distribution system at 17 and 32 bus because the voltage at these nodes was found minimum 0.91 pu as compared to other nodes. Figure 3 represents the line diagram after placement of capacitor. In Fig. 4, power demand at base case is compared with ZIP and RIC loads. There is slight decrement in demand for the RIC and ZIP loads due to the variations in voltage. For RIC load models, the demand is shown at time interval of T-15 for node 18 as load is maximum at time interval of 15 h at bus 18. Voltage profile variations and reactive power losses are shown in the subsequent figures for T-15 interval and at bus 18.

In Figs. 5 and 6, voltage profile is shown without load growth and with load growth where the value of minimum voltage 0.914 pu is improved to 0.937 pu at the 18 bus.

Table 1 Results for with and without load growth of load models for IEEE 33-bus system

Parameters	Case-1	Case-2	Case-3	Case-4	Case-5
Active power loss (kW)	202.98	133.61	125.20	274.87	249.9154
Reactive power loss (kVAR)	135.20	89.34	79.33373	184.056	167.6367
Least voltage in pu (bus no.)	0.914 (18)	0.9269 (18)	0.9393 (18)	0.9109 (18)	0.9158 (18)
Max. voltage in pu (bus no.) (except 1 pu)	0.997 (2)	0.9973 (2)	0.9974 (2)	0.9963 (2)	0.9964 (2)
Size of capacitor in kVAR (bus no.)	–	163.05 (17) 509.62 (32)	191.64 (17) 611.81 (32)	715.7286 (17) 236.8785 (32)	288.477 (17) 902.355 (32)
Total cost ($)	33,986.37	26,464.643	25,444.331	1,179,181.1	47,588.49

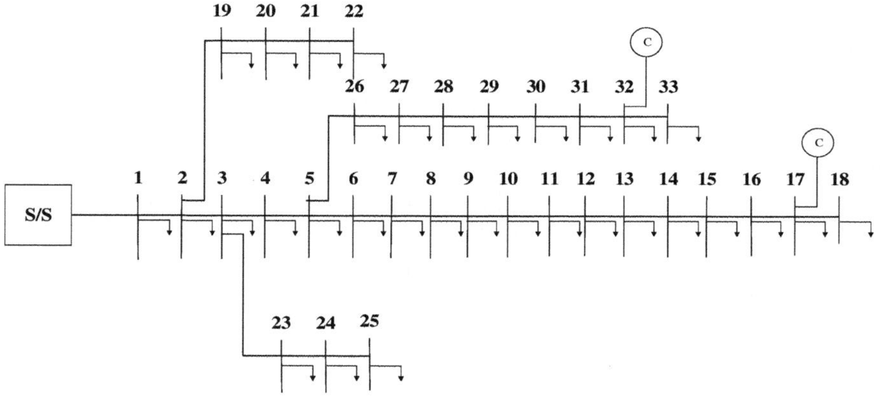

Fig. 3 IEEE 33-bus system line diagram

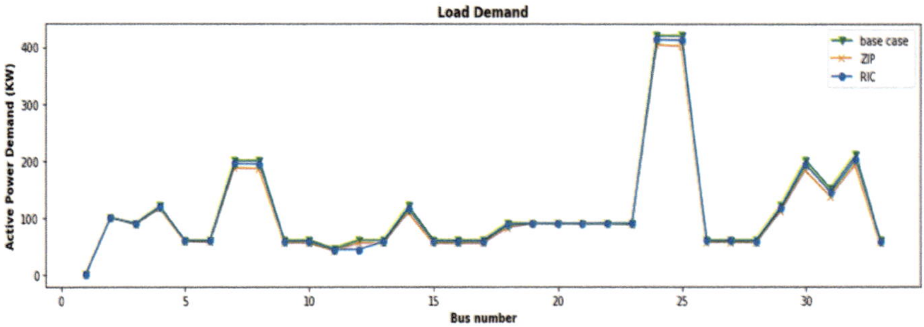

Fig. 4 Active power demand on 33-bus system

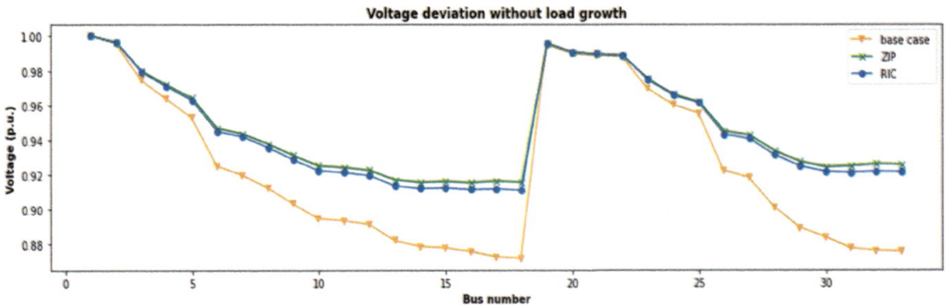

Fig. 5 Voltage profile variation without load growth on 33-bus system

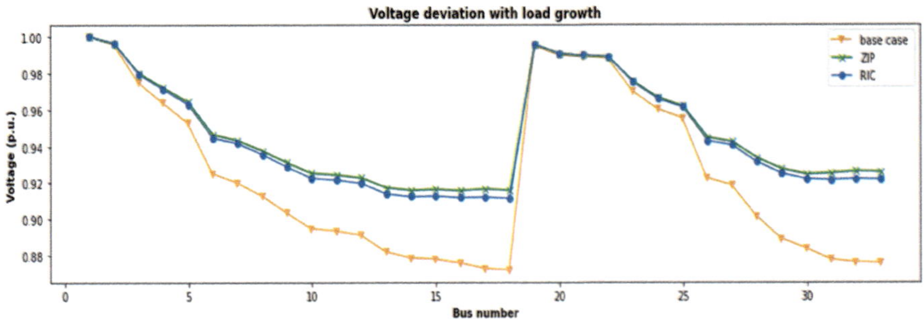

Fig. 6 Voltage profile with load growth on 33-bus system

In Figs. 7 and 8, reactive power loss of base case is improved with the help of load modeling and capacitor installation at buses 17 and 32. Reactive power loss reduction with ZIP modeling is 38% and with RIC modeling is 33.91% without load growth. Furthermore, the reduction of reactive power loss while considering load growth in ZIP load modeling is 44.13% and in RIC load modeling is 38.66%.

In Figs. 9 and 10, active power losses with and without load growth are graphically depicted. The active power losses are reduced by utilizing load modeling and

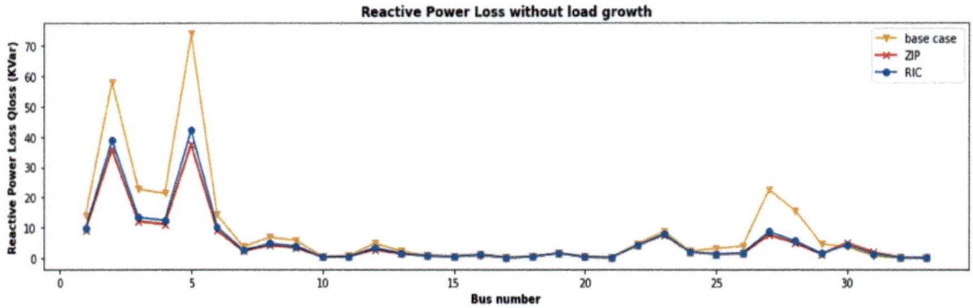

Fig. 7 Reactive power loss without load growth on 33-bus system

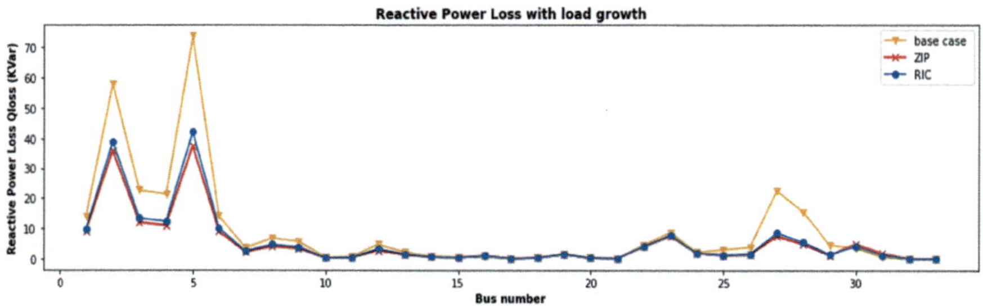

Fig. 8 Reactive power loss with load growth on 33-bus system

capacitor installation. Reduction of active power losses without load growth in ZIP is 38.11% and in RIC is 33.95%. While considering load growth the active power losses are reduced to 44.27% using ZIP model and 38.70% using RIC model.

Total cost of base case is 33,986 $ and the total cost of base case considering load growth after 5 years at a growth rate of 7.5% is 75,343.447 $.

In Fig. 11, new obtained power demand for RIC load model is shown for summer season of 33 RDS.

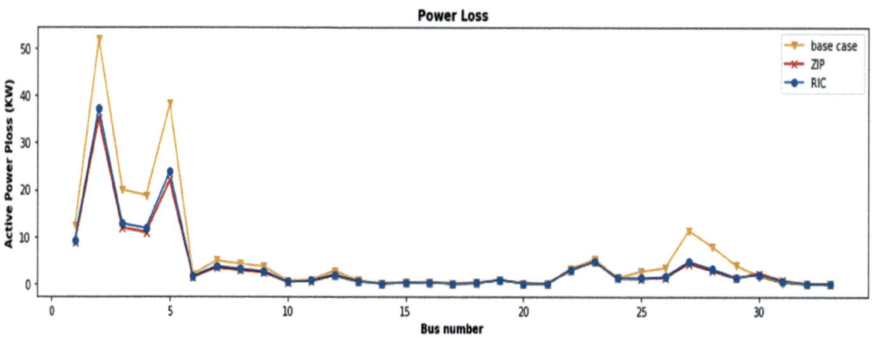

Fig. 9 Active power loss without load growth on 33-bus system

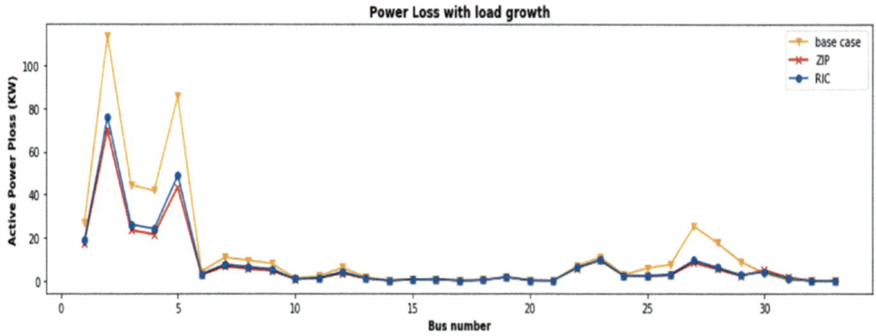

Fig. 10 Active power loss with load growth on 33-bus system

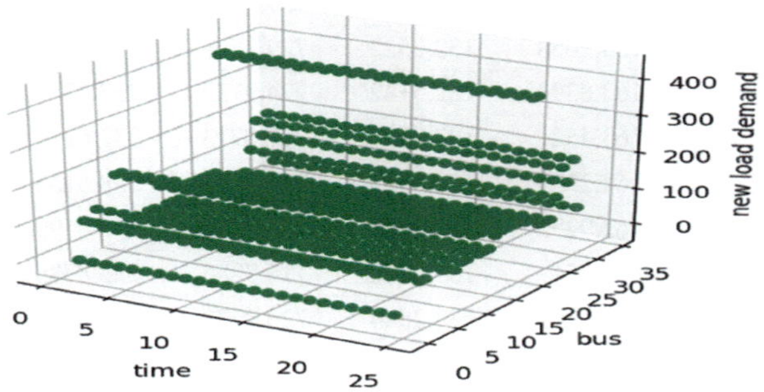

Fig. 11 Power demand of time-varying load on 33-bus system

In Table 2, the hourly active power losses along with sizing of capacitors at the preferred locations with and without consideration of load growth are represented. Table 3 represents the cost of different load models with or without load growth with the integration of capacitors at buses 17 and 32. In Table 4, a detailed comparison of the proposed methods is done with MOGWO (multi-objective gray wolf optimizer), NSGA-II (non-dominated sorting genetic algorithm-II), and FRCGA (fuzzy real coded genetic algorithm). The results show the effectiveness of the proposed methods in terms of reduction of active power loss and reactive power loss in IEEE-33 RDS.

7 Conclusions

In this paper, an analysis is carried out by considering with or without load growth. Different load models ZIP and RIC are developed on IEEE-33 RDS and analyzed to minimize the cost of active power loss, compensate the reactive power, and minimize installation cost of capacitors. Capacitors are placed at bus 17 and 32 for all cases except case-1. Results show that the load demand in RIC model is higher than ZIP

Table 2 RIC load modeling with capacitor installation with and without load growth for 24 h of summer season

Hour	Without load growth			With load growth		
	Qc (1)	Qc (2)	TPL	Qc (1)	Qc (2)	TPL
1	549.6402	177.7049	133.7897	786.4111	262.3774	275.2633
2	548.5317	177.1007	133.6055	784.3631	261.2493	274.6989
3	546.6916	176.345	133.5227	781.0733	259.9018	274.4502
4	544.4877	175.6381	133.6025	777.2195	258.7077	274.7043
5	543.0292	175.5867	134.0301	774.8557	258.8044	276.0312
6	547.055	175.8045	132.9154	781.4329	258.7145	272.5792
7	538.6876	171.0201	131.3077	765.9964	249.806	267.6906
8	518.4943	164.4266	131.9523	730.6157	238.5797	269.7305
9	516.6536	163.9143	132.092	727.4275	237.7435	270.1654
10	517.7546	164.2688	132.0522	729.3523	238.3437	270.0394
11	515.1612	163.8393	132.5142	724.9759	237.7789	271.4668
12	510.3562	163.452	133.7336	717.0773	237.6287	275.2476
13	515.8868	164.2355	132.6355	726.3082	238.5153	271.8368
14	514.412	163.9252	132.8376	723.7989	238.0599	272.4642
15	509.6252	163.051	133.6108	715.7286	236.8785	274.8706
16	514.3958	164.4141	133.2844	723.9775	239.0969	273.8418
17	512.9219	164.3572	133.7131	721.5883	239.1877	275.1731
18	515.505	166.5337	134.8256	726.7252	243.4838	278.6161
19	537.915	174.5405	134.7468	766.2038	257.3191	278.2745
20	540.8694	175.528	132.0533	774.9294	260.3406	272.5494
21	551.1459	179.0237	134.4867	789.4259	264.9762	277.4146
22	554.8103	180.0285	134.2025	795.7522	266.5931	276.5204
23	554.4388	179.8259	134.1408	795.0643	266.214	276.3309
24	552.2235	178.8658	133.9958	791.0771	264.4828	275.8908

Table 3 Cost comparison for all types of load models with and without load growth

Load type	ZIP without load growth	ZIP with load growth	RIC without load growth	RIC with load growth
Cost ($)	25,544.33	47,558.496	5,90,163.7	11,79,181

model. Throughout the day demand changes in RIC model where in ZIP it remains constant. Reduction in the active power losses for ZIP is 44.27% and RIC is 38.70% with a load growth of 5 years at a growth rate of 7.5%. Whereas the active power losses are reduced by 38.11% in ZIP and 33.95% in RIC without considering load growth. Moreover, the cost of active power losses varies every hour as the load continuously varies throughout the 24 h. The study shows that load modeling is effective with

Table 4 The comparison of results with some benchmarked algorithms

Parameters	Case-1	Case-2	Case-3	MOGWO [19]	NSGA-II [19]	FRCGA [20]
V_{min} (pu)	0.914	0.9369	0.9393	0.9456	0.9412	0.9665
V_{max} (pu)	0.997	0.9973	0.9974	0.9977	0.9977	–
Active power loss (kW)	202.98	133.61	125.20	148.78	142.70	148.69
Reactive power loss (kVAR)	135.20	89.34	79.33	102.03	97.56	–

proper management of reactive power and voltage profile. With appropriate sizing of capacitors, the voltage profile improves and the losses in the network decrease. With load growth analysis, an approximate estimation can be made so a proper planning can be carried out.

References

1. Kundur P (1994) Power system stability and control. McGraw-Hill, New York
2. Concordia C, Ihara S (1982) Load representation in power system stability studies. IEEE Trans Power App Syst 969–977
3. Arif A, Wang Z, Wang J, Mather B, Bashualdo H, Zhao D (2018) Load modeling—a review. IEEE Trans Smart Grid 9:5986–5999
4. Nazir FU, Pal BC, Jabr RA (2021) Approximate load models for conic OPF solvers. IEEE Trans Power Syst 36:549–552
5. Bircan M, Durusu A, Kekezoglu B, Elma O, Selamogullari US (2020) Experimental determination of ZIP coefficients for residential appliances and ZIP model based appliance identification: the case of YTU Smart Home. Electr Power Syst Res 179
6. Collin AJ, Tsagarakis G, Kiprakis AE, McLaughlin S (2014) Development of low-voltage load models for the residential load sector. IEEE Trans Power Syst 29
7. Noel da Silva MS, Lima DA (2021) Optimal operation planning of distribution systems under uncertainty in load modeling by using decentralized energy resources. Electr Power Syst Res 196
8. Jun PR, Kyung-Bin S, Kyungsang L (2020) The load model composition method in power systems using artificial neural network. J Electr Eng Technol 519–526
9. Mohanty PK, Lal DK (2021) Effects of load models and load growth in distribution system in presence of distributed generator. In: 2021 1st international conference on power electronics and energy (ICPEE), Bhubaneswar, India
10. Kosterev D, Meklin A, Undrill J, Lesieutre B, Price W, Chassin D, Bravo R, Yang S (2008) Load modeling in power system studies: WECC progress update. In: Proceedings the IEEE PES general meeting, Pittsburgh, pp 1–8
11. Ma Z, Cui B, Wang Z, Zhao D (2021) Parameter reduction of composite load model using active subspace method. IEEE Trans Power Syst 36(6):5441–5452
12. Soma GG (2021) Optimal sizing and placement of capacitor banks in distribution networks using a genetic algorithm. Electricity 2(2):187–204
13. Javadi MS, Nezhad AE, Siano P, Shafie-khah M, Catalão JPS (2017) Shunt capacitor placement in radial distribution networks considering switching transients decision making approach. Int J Electr Power Energy Syst 92:167–180

14. Zaheeruddin, Pande M (2014) Disaster management of distribution line using fixed and switched capacitors. In: IEEE innovative applications of computational intelligence on power, energy and controls with their impact on humanity (CIPECH), Ghaziabad, India, pp 221–225
15. Suzuki M, Wada S, Sato M, Asano T, Kudo Y (1992) Newly developed voltage security monitoring system. IEEE Trans Power Syst 7(3):965–973
16. Murty V, Kumar A (2015) Capacitor allocation in radial distribution system with time varying ZIP load model and energy savings. Procedia Comput Sci 70:377–383
17. Qian K, Zhou C, Allan M, Yuan Y (2011) Effect of load models on assessment of energy losses in distributed generation planning. Int J Electr Power Energy Syst 1243–1250
18. GAMS tutorial homepage. http://www.gams.com/latest/docs/UG_Tutorial.html
19. El-Saeed MAEM, Abdel-Gwaad AF, Farahat MAF (2023) Solving the capacitor placement problem in radial distribution networks. Results Eng 17
20. Abul'Wafa AR (2014) Optimal capacitor placement for enhancing voltage stability in distribution systems using analytical algorithm and Fuzzy-Real Coded GA. Int J Electr. Power Energy Syst 55

Fuzzy Logic Controller-Based DSTATCOM to Improve Power Quality in Distribution System

D. Srinivas, Penta Samyuktha, T. Himabindu, K. Shirisha, S. Ashlesha, and E. Venkatesh

Abstract Power quality (PQ) is gaining a great deal of importance as more sensitive loads are introduced into the utility grid. Because most of the loads are non-linear in utility, the degradation of power quality leads to damage of equipment and temporary shutdowns are the general issues which are associated with PQ problems in utility. Those are like distortion in voltage and current for utility. To solve this problem a modern and important power device is introduced to reduce the distortions in utility grid which is the DSTATCOM. This device is flexible to reduce the distortions. The DSTATCOM is a static synchronous compensator used to mitigate the harmonics in utility grid. A fuzzy logic controller was introduced to improve the performance of DSTATCOM further. The system is designed and simulated using the MATLAB/SIMULINK for different types of faults and simulation studies are presented.

Keywords Power quality · DSTATCOM · Utility grid · Non-linear loads · Fuzzy logic controller

1 Introduction

Requirement of best quality of power with reliability of supply is essential for utility purpose. But these requirements are not meeting the required levels because of introduction of various non-linear loads at the utility end. The essential problems associated with quality of power are reduction in voltage, distortions in currents, and lack of required operating power factor. Reduction in voltage is called as voltage sags [1], which is a main problem in power systems. Whenever voltage sag occurs,

D. Srinivas (✉) · P. Samyuktha
KG Reddy College of Engineering and Technology, Hyderabad, India
e-mail: shrinudepally@gmail.com

T. Himabindu
G. Narayanamma Institute of Science and Technology (For Women), Hyderabad, India

K. Shirisha · S. Ashlesha · E. Venkatesh
Sri Indu College of Engineering and Technology, Hyderabad, India

J. Kumar et al. (eds.), *Recent Advances in Power Systems*, Lecture Notes in Electrical Engineering 1192, https://doi.org/10.1007/978-981-97-2788-9_3

it is measured by different parameters mainly in terms of magnitude and how much time there will be a reduction in that voltage magnitude. Voltage sag is an essential problem at the load end point. If a voltage sag occurs, it will abnormally increase the current, this increased current will cause serious damage to the loads like different motors and transformers at the utility end. It is also considered as essential problem for industries, because most of the industrial loads are motor. Because of presence of various non-linear loads at utility end there is a distortion in currents of distribution systems which are called as harmonics in currents, which leads to disturbances in currents, reduced operating power factor, various losses, and also production of excess heat in operating parts of utility equipment. These harmonics also increase the noise and vibrations in machine drives and also there is a possibility of maloperation of various sensitive equipment present.

With the enhanced development and advancement in operation of solid-state devices called power electronic devices, which are providing the various advanced and improved control strategies in power systems. This improved power electronic device theory is called as FACTS. FACTS is introducing various [2] power electronic controllers with different control strategies adopted for improving the quality of power at utility end, i.e., compensation of reduction in voltage sag, reduction of harmonics in current, and improvement in operating power factor. Among all the FACTS controllers, DSTATCOM is most essential and effective FACTS controller. There are different types of PWM control strategies which are available to control the DSTATCOM. To control the switching devices present [1] in the DSTATCOM, a novel pulse width modulation method has been implemented in this paper. DSTATCOM is a controller which is having a capability to maintain reactive component of current at reduced voltages. It is also used as sources of voltage and frequency by utilizing storage of energy devices like batteries replaced by capacitors. In this paper, design and implementation LCL passive filter based has been discussed using MATLAB/SIMULINK platform. DSTATCOM is parallelly connected to the 11 kV test system. It is designed and designated to increase the quality of power, viz., reduction in voltage, reduced harmonics, and reduced power factor in distribution system [2].

2 Basic Principle of DSTATCOM

STATCOM is basically static synchronous compensators which is used to improve the power quality. DSTATCOM is one device which is used to improve the power quality in utility grid. The main feature of a DSTATCOM is to deliver or sink the reactive power, which can be formed by using a voltage source converter and a DC link capacitor in parallel. DSTATCOM is operated based on the principle of traditional synchronous compensator. The input to the VSC which is used as DSTATCOM is taken from the AC terminal of the distribution line along with inductance at PCC [3]. The inductance may be a filtering inductance or a coupling transformer leakage inductance, as shown in Fig. 1.

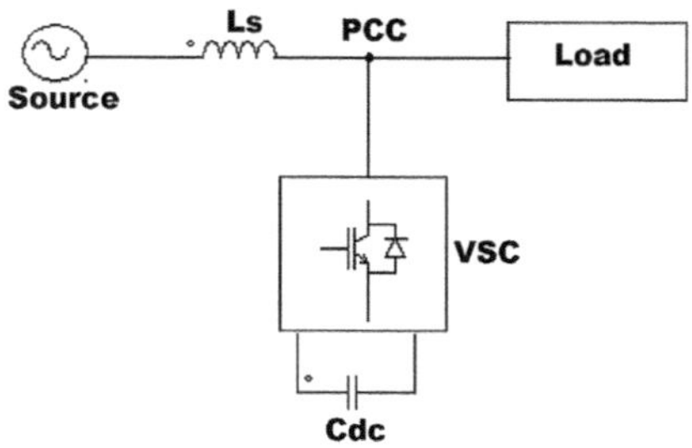

Fig. 1 Basic line diagram of DSTATCOM

DC capacitor which is connected at DC side of the converter is used to reduce the current ripples and it is the main energy-storing element for reactive power storage. Charging of this DC capacitor is done by using a battery source or else we can connect a charged capacitor. When compared with the generated voltage by the voltage source converter with AC line voltage it can be concluded that there will not be any reactive power delivered to the system. The output of the VSC is more than the AC line voltage then DSTATCOM seems to be operating in capacitive mode of operation. The reactive power flow magnitude can be decided by the comparison between the voltages of VSC and AC line voltages. It is difficult to improve the voltage regulation and improvement in power factor at the same time. Better voltage regulation can be achieved at the PCC by DSTATCOM, source currents has to be in phase advance by the source voltages. Whereas improvement in power factor can be achieved by source currents and source voltages in phase with each other. In this paper, improvement in power factor and reduction in harmonics have been studied by using different control strategies for DSTATCOM [4–6].

3 Control Strategies

The control of DSTATCOM can be achieved by applying different PWM control strategies. The PWM control is the main part of DSTATCOM for compensating different quantities in distribution systems. By utilizing PWM control there is much more improvement in response of DSTATCOM and also there is considerable reduction in harmonics and improvement in power factor. In this paper, we are discussing about improvement in power factor and reduction in harmonics by comparing different control strategies like

1. Phase shift control
2. Indirect decoupled current control.

3.1 Indirect Decoupled Current Control

The basic block diagram of indirect coupled control has been shown in Fig. 4. This strategy is based upon instantaneous power theory in which three-phase quantities are transformed into the synchronously rotating frame by using Park's transformation. The required compensation can be done by utilizing direct and quadrature components of currents i_d, i_q, respectively. The error in current can be compensated indirectly by using voltage modulation of DSTATCOM that is why it is called as indirect control strategy [7].

Figure 2 shows the basic block diagram of DSTATCOM, the operation and performance of DSTATCOM can be studied by using MATLAB/SIMULINK platforms.

Phase Shift Control

The schematic diagram of phase shift control is shown in Figs. 3and 4. In this strategy, the required phase shift between VSC and AC line voltage can be generated by comparing the RMS voltage of the line voltages measured with the base voltage there which will give the error between the voltages, and these error voltages can be given to the PI controller, then PI controller gives the angle corresponding phase shift between AC line voltage and DSTATCOM. For generating desired synchronizing signal to drive the PWM circuitry, the generated phase shift angles are added up to the conventional source voltages, which are assumed to be spread equally around 120° rotating frame by using Park's transformation. The required compensation can be done by utilizing direct and quadrature components of currents i_d, i_q, respectively. The error in current can be compensated indirectly by using voltage modulation of DSTATCOM that is why it is called as indirect control strategy:

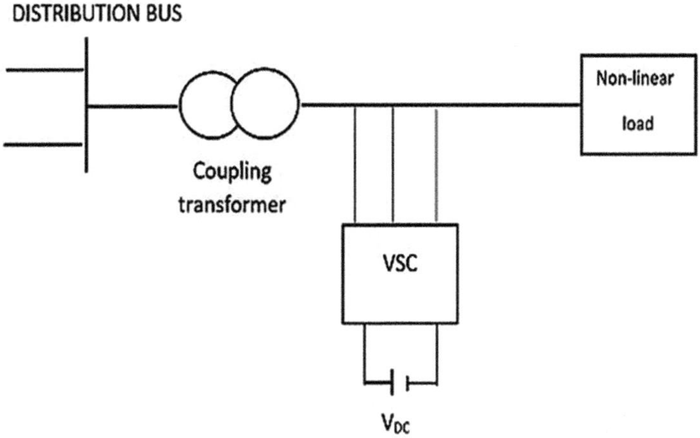

Fig. 2 Basic block diagram of DSTATCOM

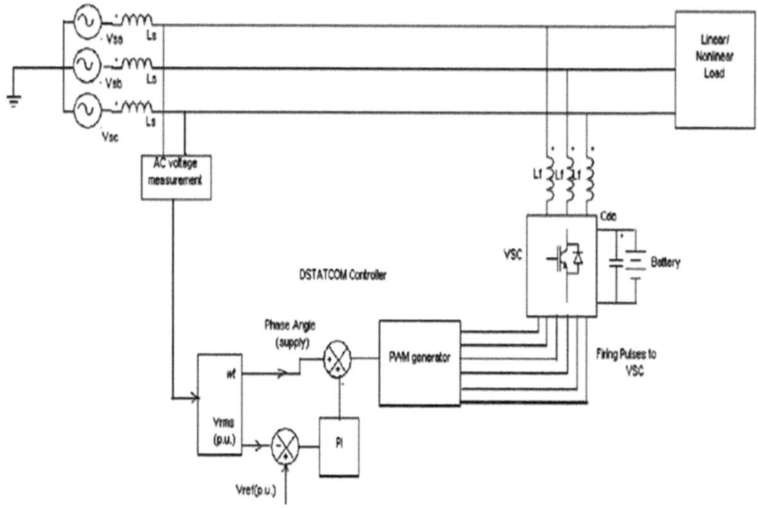

Fig. 3 Control of phase shift of DSTATCOM

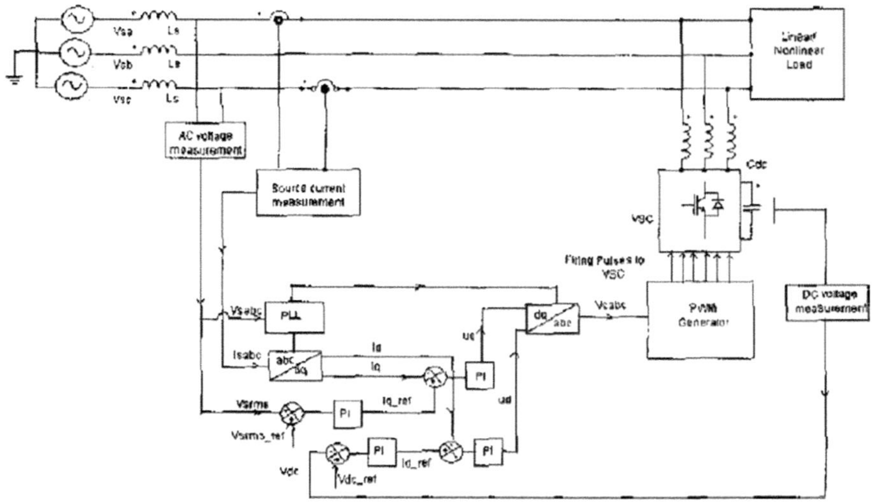

Fig. 4 Block diagram of rotating frame current controller

$$p = V_d i_d + V_q i_q \tag{1}$$

$$q = V_q i_q - V_d i_q \tag{2}$$

There is internal rotating frame current controller in DSTATCOM which is used as PI controller as shown in Fig. 4.

The transformed line currents are compared with base currents which are obtained by open-loop control. The modified signals of rotating synchronous frame are given by

$$U_d = K_{p1}\left[i_{d\_ref} - i_d\right] + K_{i1}\int\left[i_{d\_ref} - i_d\right]dt \tag{3}$$

$$U_q = K_{p2}\left[i_{q\_ref} - i_q\right] + K_{i2}\int\left[i_{q\_ref} - i_q\right]dt \tag{4}$$

4 Proposed Control Strategy

The control block diagram of proposed strategy has been shown in Fig. 5.

In this strategy, the required phase shift between VSC and AC line voltage can be generated by comparing the RMS voltage of the line voltages measured with the base voltage there will give the error between the voltages, these error voltages can be given to the PI controller, then PI controller gives the angle corresponding phase shift between AC line voltage and DSTATCOM. For generating desired synchronizing signal to drive the PWM circuitry, the generated phase shift angles are added up to the conventional source voltages, which are assumed to be spread equally around 120° [8–10] (Table 1).

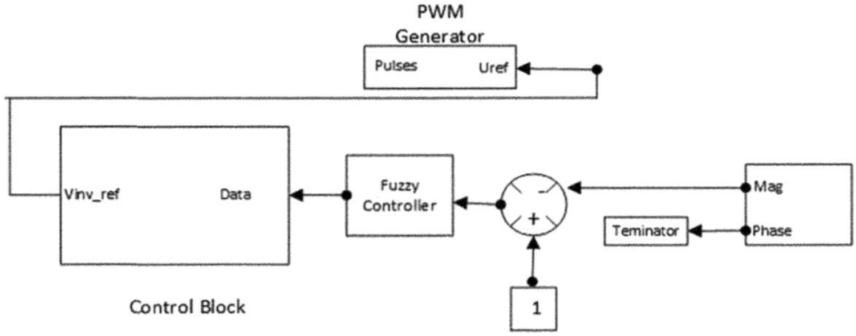

Fig. 5 Control block diagram of proposed strategy

Table 1 Fuzzy rule base

$e(t)$	$\dot{e}(t)$						
	NL	NM	NS	ZO	PS	PM	PL
PL	ZO	PS	PM	PL	PL	PL	PL
PM	NS	ZO	PS	PM	PL	PL	PL
PS	NM	NS	ZO	PS	PM	PL	PL
ZO	NL	NM	NS	ZO	PS	PM	PL
NS	NL	NL	NM	NS	ZO	PS	PM
NM	NL	NL	NL	NM	NS	ZO	PS
NL	NL	NL	NL	NL	NM	NS	ZO

5 MATLAB Design of Case Study Case 1: Test System Without Insertion of DSTATCOM

See Fig. 6.

Results of Test System Without Insertion of DSTATCOM

See Figs. 7, 8, 9 and 10.

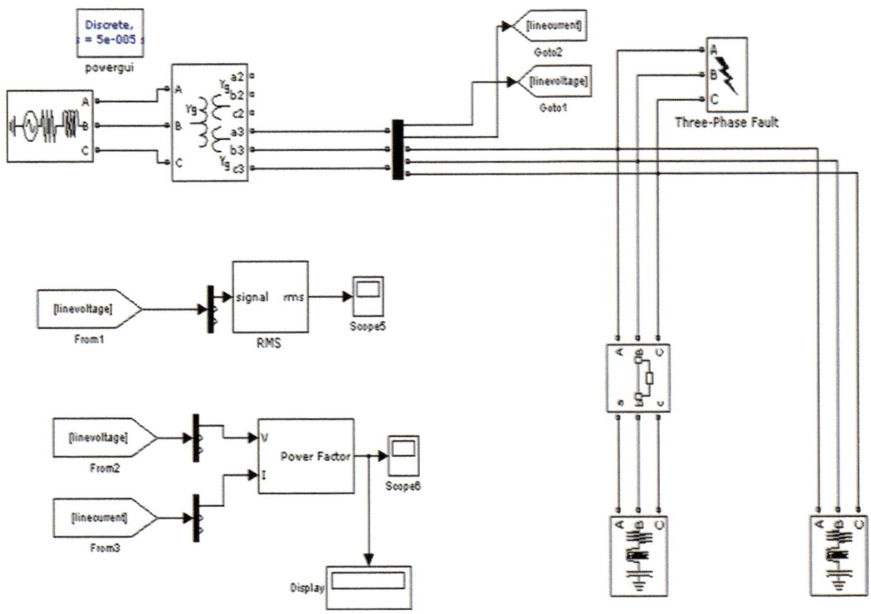

Fig. 6 Simulink model of test system

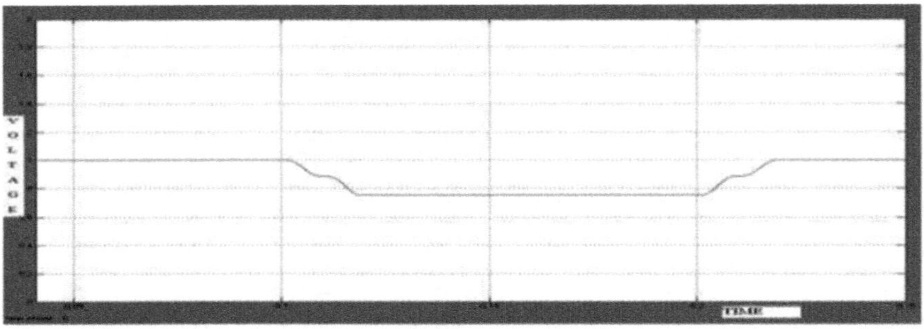

Fig. 7 Voltage at load point for three phase to ground is 0.6600 pu

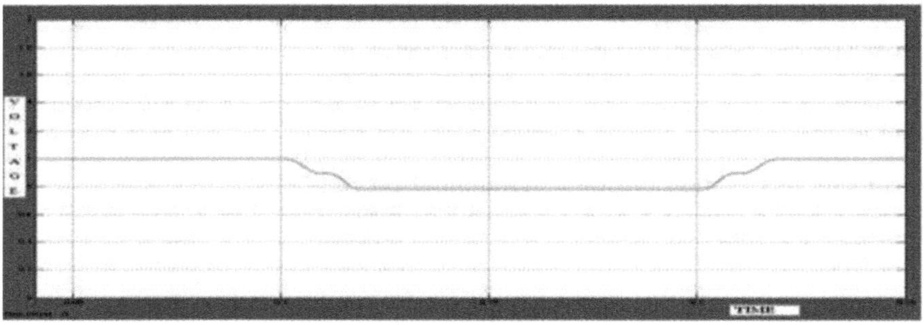

Fig. 8 Voltage at load point for double line to ground is 0.7070 pu

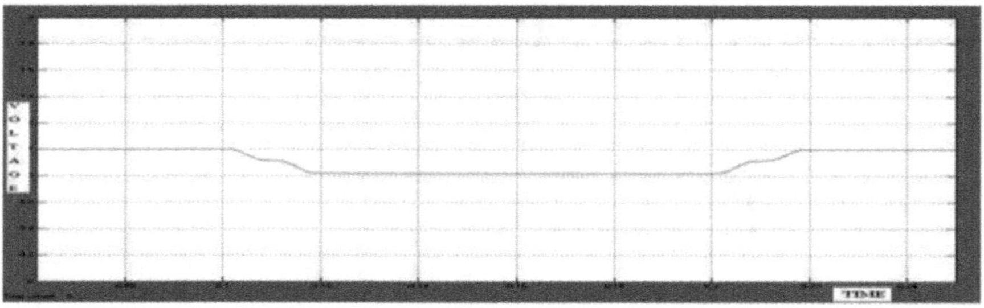

Fig. 9 Voltage at load point for line to line is 0.7585

Case 2: Insertion of DSTATCOM

See Fig. 11.

Results of Insertion of DSTATCOM

See Figs. 12, 13, 14, 15, 16, 17, 18 and 19.

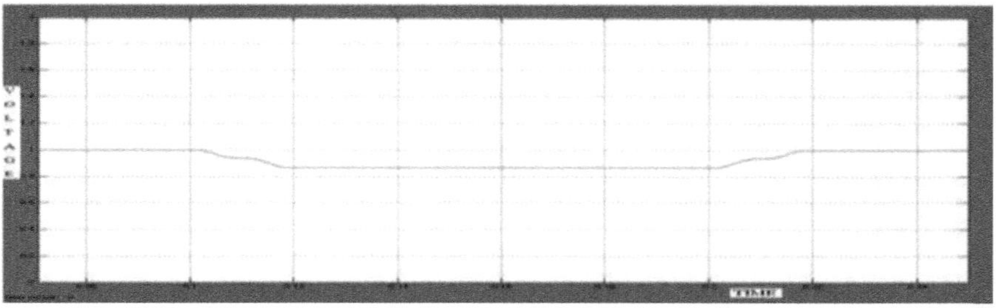

Fig. 10 Voltage at load point for single line to ground is 0.8257

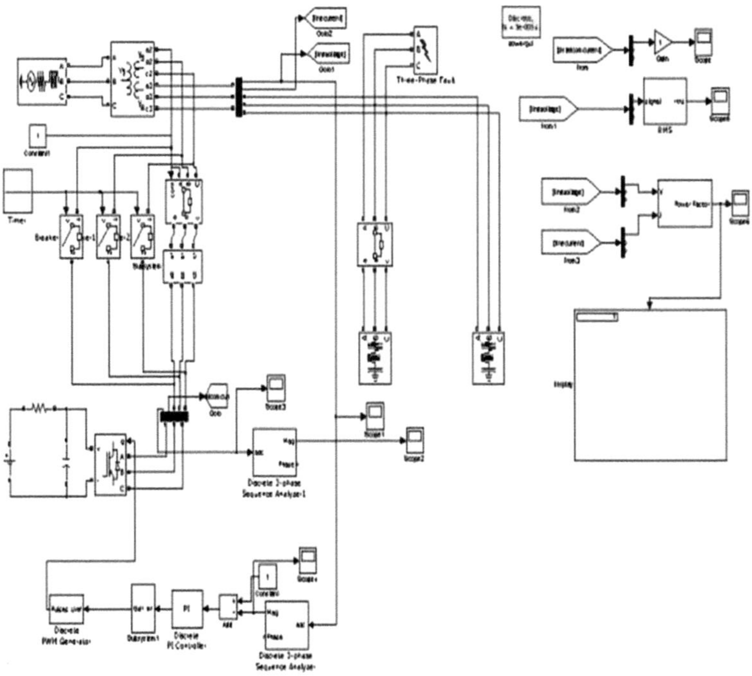

Fig. 11 Simulink model with insertion of DSTATCOM

Case 3: Insertion of DSTATCOM with Using Fuzzy Logic in Controller

See Fig. 20.

Results of Insertion of DSTATCOM with Using Fuzzy Logic in Controller

See Figs. 21, 22, 23, 24.

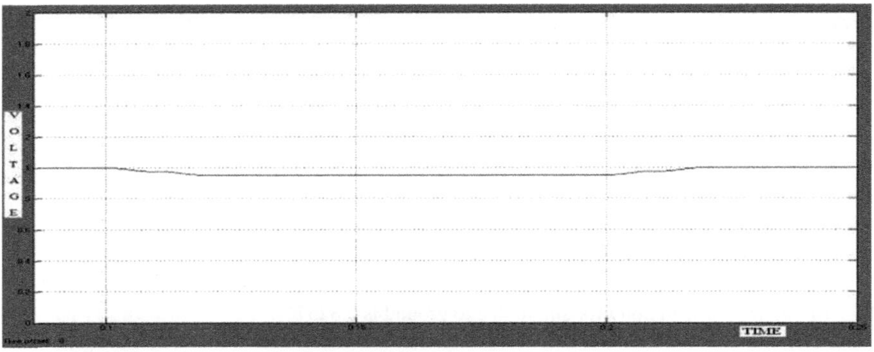

Fig. 12 Voltage at load point for three phase to ground is 0.9367 pu

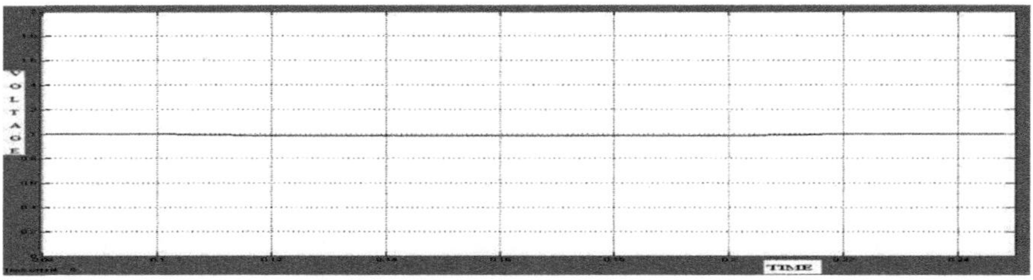

Fig. 13 Voltage at load point for double line to ground is 0.9800 pu

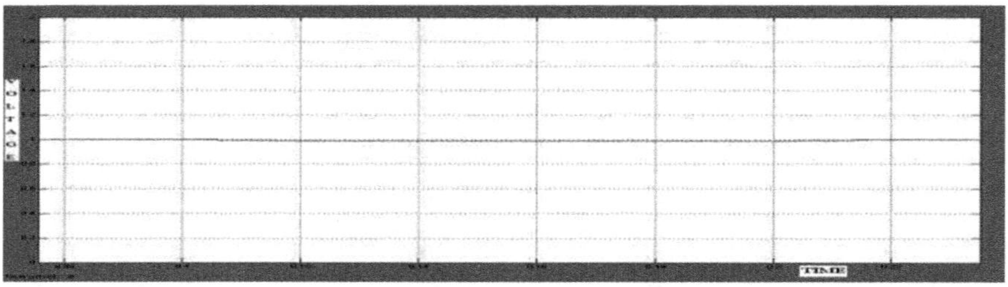

Fig. 14 Voltage at load point for line to line is 1.068

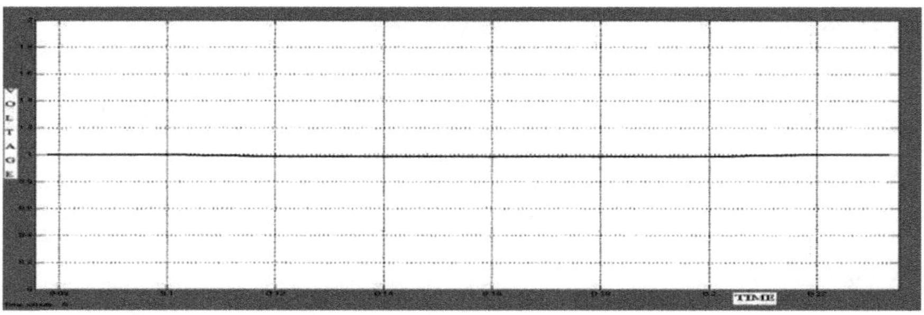

Fig. 15 Voltage at load point for single line to ground is 0.9837

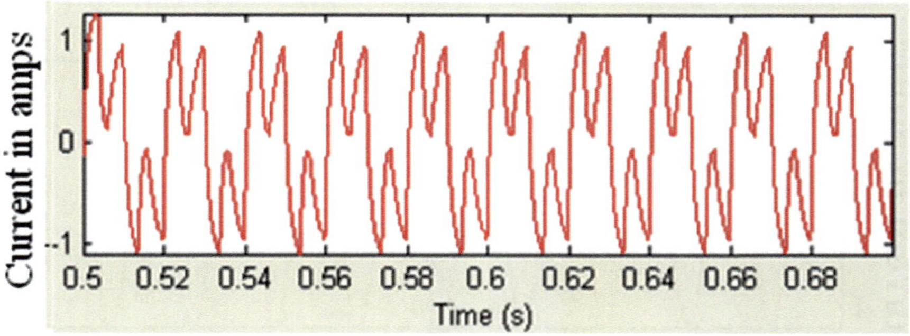

Fig. 16 Wave form of output current without LCL passive filter

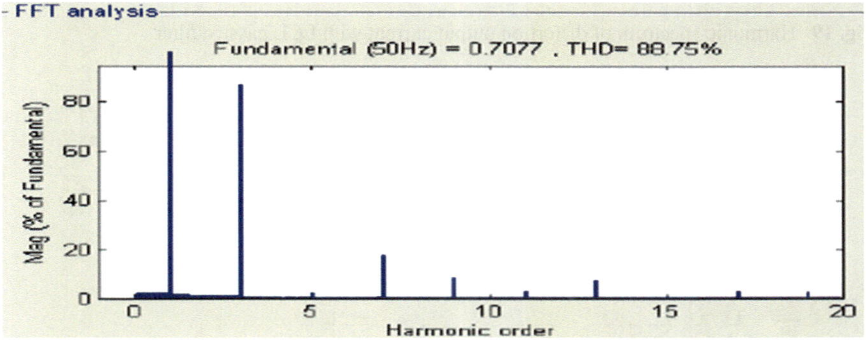

Fig. 17 Harmonic spectrum of distortion output current without LCL passive filter

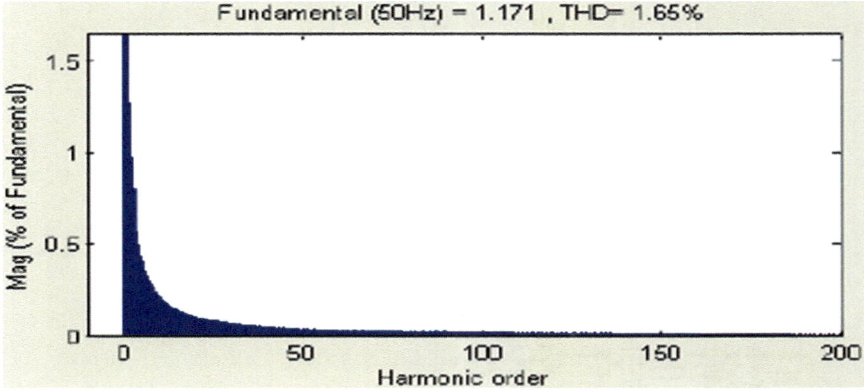

Fig. 18 Wave forms of output current with LCL passive filter

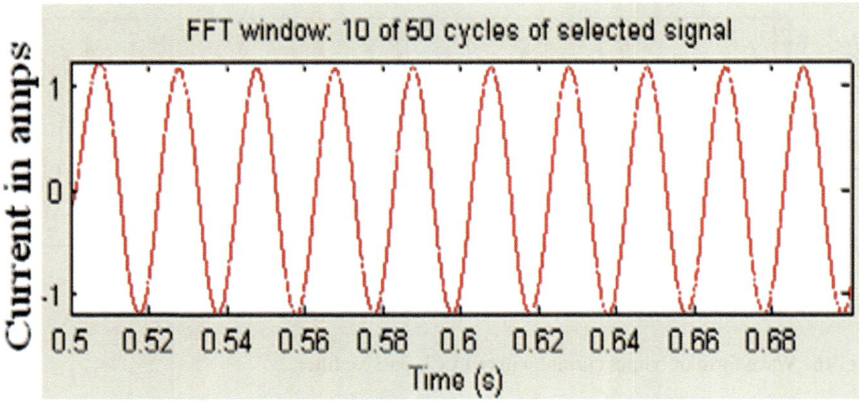

Fig. 19 Harmonic spectrum of distortion output current with LCL passive filter

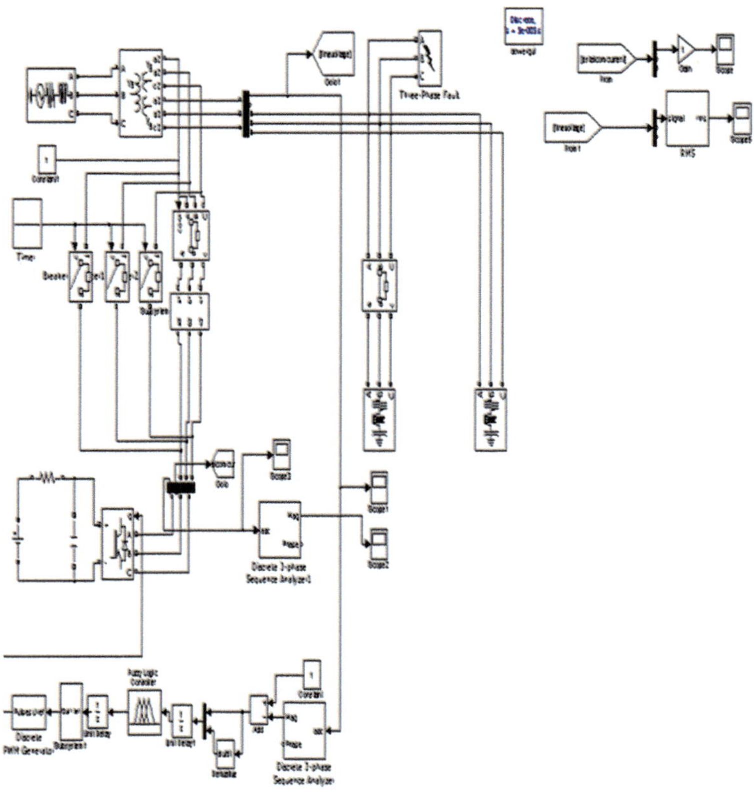

Fig. 20 Simulink design diagram of DSTATCOM with using fuzzy controller

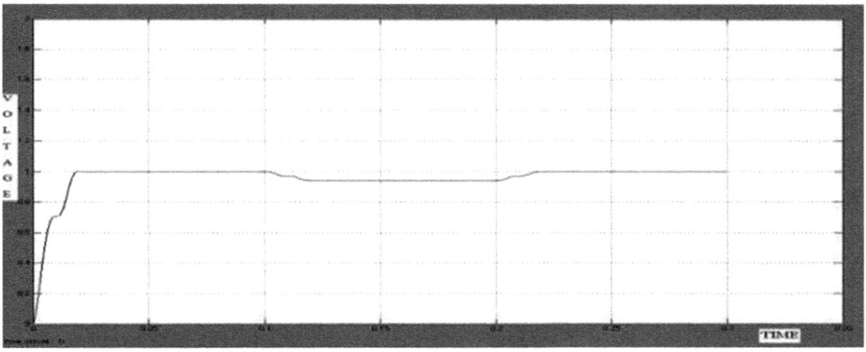

Fig. 21 Voltage at load point for three phase to ground is 0.9300 pu

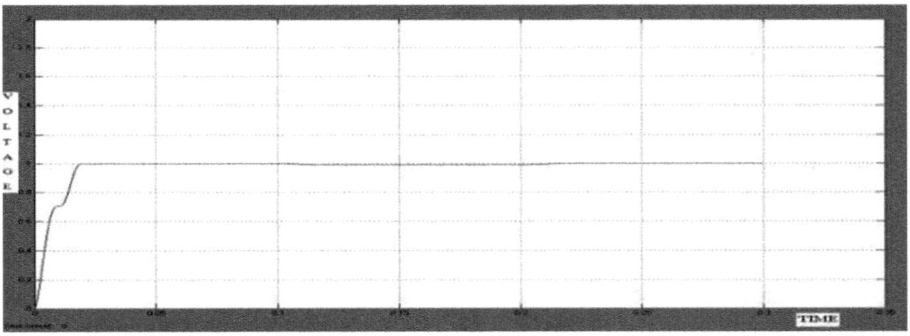

Fig. 22 Voltage at load point for double line to ground is 0.9908 pu

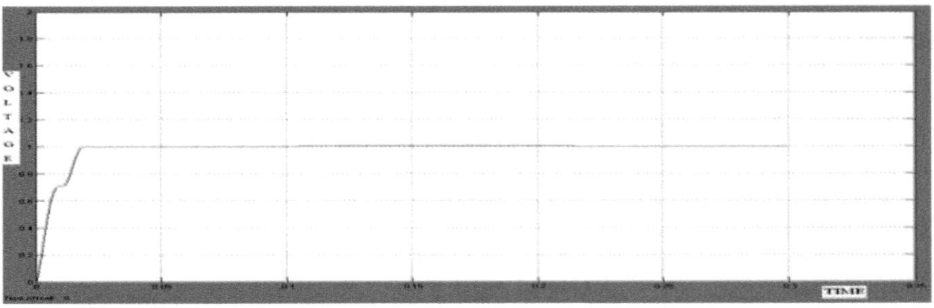

Fig. 23 Voltage at load point for line to line is 0.9987 pu

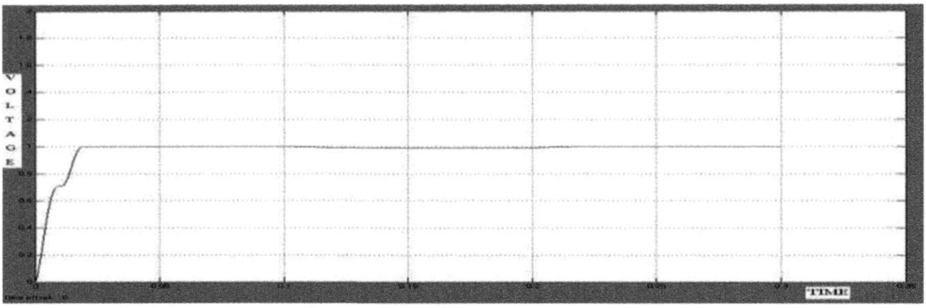

Fig. 24 Voltage at load point for single line to ground is 0.9914 pu

6 Conclusion

The proposed control strategy has been implemented and modeled by using MATLAB/SIMULINK. With the help of simulation results it can be concluded that in distribution system there will be a considerable improvement in voltage hangs and improvement in reduced power factor and also there is much improvement in reduction of harmonics with the help of DSTATCOM. The total harmonic distortion can be reduced further by inserting LCL filter in conjunction with DSTATCOM. It can be concluded that by insertion of DSTATCOM in conjunction with LCR filter there is considerable enhancement in quality of power and also enhancement in power factor. It also concluded that with the simulation results that instead of using PI controller in proposed control strategy, if we use fuzzy controller, the enhancement in quality of power is further improved.

References

1. Hammad AE (1995) Comparing the voltage source capability of present and future VAR compensation techniques in transmission system. IEEE Trans Power Deliv 1(1)
2. Yalienkaya G, Bollen MHJ, Crossley PA (1999) Characterization of voltage sags in industrial distribution system. IEEE Trans Ind Appl 34(4):682–688
3. Haque MH (2001) Compensation of distribution systems voltage sags by DVR and also D STATCOM. In: Power Tech Proceedings, 2001 IEEE Porto, vol 1, pp 10–13
4. Anaya-Lara O, Acha E (2002) Modeling and analysis of custom power systems by PSCAD/ EMTDC. IEEE Trans Power Deliv 17(2002):266–272
5. Bolle MHJ (2001) Voltage sags in three phase systems. IEEE Power Eng Rev 21(9):11–15
6. Madriga M, Acha E (2000) Modelling of custom power equipment using harmonics domain techniques. IEEE
7. Meinski R, Pawelek R, Wasiak I (2004) Shunt compensation for power quality improvement using a STATCOM controller modelling and simulation. In: IEEE Proce, vol 151

8. Nastran J, Cajhen R, Seliger M, Jereb P (2004) Active power filters for nonlinear AC loads. IEEE Trans Power Electron 9(1):92–96
9. Moran LA, Dixon JW, Wallace R (1995) A three phase active power filter with fixed switching frequency for reactive power and current harmonics compensation. IEEE Trans Ind Electron 42:402–408
10. Moran LT, Ziogas PD, Joos G. Analysis and design of three phase current source solid state VAR compensator. IEEE Trans Ind Appl

Islanding Detection in a Grid-Connected Photovoltaic System Using Zero-Sequence Impedance

Kumaresh Pal, Kasinath Jena, Aditya Prasad Padhy, Rahul Kumar, Nitish Ranjan, and Sachin Panda

Abstract Grid-connected photovoltaic (PV) power systems have the benefit of being rapid and dependable sources of electricity. The power industry has been obliged to transition over to more PV-penetrated distributed generation as a result of solar energy's favourable environmental effects in order to keep up with rising load demand. In grid-connected, PV systems, the problem of unintentional islanding in grid connectivity still presents a barrier. Therefore, quick islanding detection is required for effective and trustworthy operation of system. This paper proposes a passive islanding detection technique based on zero-sequence impedance computation at the Point of Common Coupling (PCC) for a PV interconnected grid system. The approach is simulated and tested under several Islanding test scenarios. By measuring the zero-sequence impedance at the PCC, which is calculated from the zero-sequence voltage and zero-sequence current, the islanding mode is identified. By identifying the Non-Detection Zone (NDZ), which fulfils the IEEE requirements of 2 s execution time, the results of this technique have been achieved and confirmed. A PV integrated grid system is used in the study, and many islanding situations were created using MATLAB/Simulink (version 2018b).

Keywords Photovoltaic (PV) system · Point of common coupling (PCC) · Unintentional islanding · Zero-sequence impedance · Non-detection zone (NDZ)

K. Jena · A. P. Padhy · R. Kumar · N. Ranjan · S. Panda
Department of EEE, Arka Jain University, Jamshedpur, India

K. Pal (✉)
Department of Electrical Engineering, NIT, Jamshedpur, India
e-mail: kumaresh.pal@rediffmail.com

J. Kumar et al. (eds.), *Recent Advances in Power Systems*, Lecture Notes in Electrical Engineering 1192, https://doi.org/10.1007/978-981-97-2788-9_4

1 Introduction

Renewable energy sources have exploded in popularity during the last decades. It produces less emissions and releases fewer greenhouse gases than fossil fuels and nuclear power [1]. Examples of distributed generation such as landfill gases, photovoltaic (PV) energy, wind energy, micro-hydro turbines, and biomass categorised as renewable energy [2]. Because of the world's fast expansion, distributed generation (DG) technology is advancing at a rapid pace, and as scientific, technological and economic progress advances, people are placing greater demands on the environment. Including these DGs in the distribution network provides several advantages, including improved reliability, increased energy efficiency, lower line losses and improved power quality. The adoption of emerging and advanced green energy technologies is critical to achieving a sustainable and resilient future grid. The combination of these technologies can enable the integration of renewable energy sources and energy storage systems into the grid, improving grid stability and reducing the need for fossil-fuel-based peaker plants [3]. Distributed generating technology is not only safe, clean, environmentally friendly, cost-effective and efficient, but it can also assess the whole power system's dependability and efficiency [4, 5].

A PV system becomes "islanded" when a portion of the utility system that includes load as well as distributed resources, such as a PV inverter, stays operational even though the utility system is disconnected [6]. Islanding mode can result in a number of challenges, including safety concerns, load damage, reclosing issues and more [7]. Grid voltage and frequency may fluctuate because a grid-connected inverter cannot regulate the grid's voltage and frequency when it is in islanding mode. Loads may sustain irreparable damage as a result of this [8, 9]. Grid synchronisation is also impossible in islanding mode due to the lack of a standard grid voltage. Additionally, it is challenging to synchronise an isolated grid with the utility grid during recloser operation due to the isolated grid's voltage being out of phase with the electrical utility grid, and the grid might trip again due to an overcurrent fault.

For grid-connected PV inverters, Anti-Islanding Detection (AID) is a necessary function since islanding might pose a hazard to the operation of the grid. When an island is detected, the PV inverter must stop energising the grid within the allotted period. A number of AID algorithms have been commercialised and have been developed to prevent islanding. These algorithms could be classified into passive techniques and active techniques [10]. Any significant system disturbance, a grid failure or intentional islanding for a planned or unexpected event might cause the islanding to happen. An islanding situation can be unintentional or intentional. Unintentional islanding situation is a threat to the electrical grid's protection, posing a risk to utilities, equipment and maintenance personnel. As a result, following island formation, the islanding system must be disconnected in 2 s in accordance with different islanding standards such as IEEE Std. 929-2000, IEEE Std. 1547-2003, IEC 62116 and distributed generating operations must be stopped [11, 12].

Islanding detection might be passive, active, hybrid or communication based, depending on the methodology. Monitoring of grid parameters is the foundation of passive approaches. Islanding can be easily detected using a relay of voltage and frequency, such as OUF (over- or under-frequency)/OUV (over- or under-voltage). Monitoring voltage and frequency that are not within the suitable range can avoid islanding. The detection of an islanding operation can also be done using a phase jump or a change in harmonics [13]. Active approaches disrupt the grid using a PV inverter and monitor how the grid's parameters respond. Two categories can be made for this. The first active approach uses positive feedback, which causes the grid to become unstable when islanding situation occurs. When an islanding condition arises, a positive feedback loop to power references or current phase angle starts and voltage or frequency instability would be tracked [14–16]. The frequency drift techniques (FDTs) are representative techniques, such as slip-mode frequency shift (SMS), active frequency drift (AFD), and Sandia-frequency shift (SFS) [8, 17]. The second active method type injects a signal into a grid and monitors how system parameters respond [18]. A signal used for the injection could be either the current or the voltage. After islanding, the voltage created by the injected components can be felt as the current is injected. Grid impedance can be determined and can be used as an indication for detection of islanding as high-frequency voltage is injected. Hybrid techniques combine both active and passive detection techniques. These methods have lower NDZ. Hybrid techniques for islanding detection combine two or more methods to improve the accuracy and reliability of the detection process. Two most accurate hybrid islanding detection schemes are voltage and frequency-based scheme and wavelet transform-based scheme [19].

Every day, a new technique emerges. Island detection technologies are being established in a variety of ways. Modern islanding detection techniques are classified into two categories: soft computing/intelligent-based techniques and modified passive/signal processing-based techniques [11].

This paper looks into a dilemma that arises at the point where a distributed generation facility meets the rest of the power system. The dilemma could be explained by the power systems' recognition of an islanding condition. In recent years, a great number of contemporary methods for determining islanding scenarios have been described.

2 Suggested Technique

Out of several detection methods, the essential requirement for the existence of any disturbances in the voltage signal and the current signal detected at the point of common coupling is the zero-sequence component approach [19]. Therefore, to be able to effectively identify islanding and notch perturbance caused by islanding occurrences, this research has analysed the zero-sequence component of voltage signals and the current signals discovered at PCC. The symmetrical component of analysis can be used to describe the zero-, negative-, and positive-sequence components of the voltage signal at the point of common coupling as [20]

$$V_n = \frac{1}{3}\left(V_a + \lambda^2 V_b + \lambda V_c\right) \quad (1)$$

$$V_p = \frac{1}{3}\left(V_a + \lambda V_b + \lambda^2 V_c\right) \quad (2)$$

$$V_z = \frac{1}{3}(V_a + V_b + V_c) \quad (3)$$

where V_a, V_b, and V_c are the three-phase voltages obtained at the PCC and V_n, V_p, and V_z are the negative-, positive-, and zero-sequence voltages, respectively, and $\lambda = 1\angle 120°$ is the intricate operator.

The symmetrical component of analysis can be used to indicate the zero-, negative-, and positive-sequence components of the current signal at the point of common coupling as [21]

$$I_n = \frac{1}{3}\left(I_a + \lambda^2 I_b + \lambda I_c\right) \quad (4)$$

$$I_p = \frac{1}{3}\left(I_a + \lambda I_b + \lambda^2 I_c\right) \quad (5)$$

$$I_z = \frac{1}{3}(I_a + I_b + I_c) \quad (6)$$

where I_a, I_b, and I_c are the three-phase currents obtained at the PCC and I_n, I_p, and I_z are the negative-, positive-, and zero-sequence currents, respectively, and $\lambda = 1\angle 120°$ is the intricate operator [22].

The zero-sequence impedance can be expressed by the symmetrical component of analysis given as

$$Z_o = \frac{V_z}{I_z} \quad (7)$$

The three-phase voltage signals and the current signals acquired at PCC are given as inputs to the three-phase sequence analyser in MATLAB block. The zero-sequence component is taken into consideration from among the three sequential components [23]. In this paper, zero-sequence impedance technique is used because it emulates the information following perturbance condition. Evaluation of the zero-sequence impedance at the point of common coupling provides excellent noise shielding for islanding detection, therefore it performs well [24]. Variations in ZSI magnitude can be used to identify the islanding mode. Zero-sequence voltage to zero-sequence current ratio measured at PCC can be used to compute the magnitude of ZSI (Fig. 1).

The technique is examined on a test system that includes a distributed generation unit powered by photovoltaic system and connected to a distribution network system through 100 kVA, 0.260/25 kV transformers. The total generation capacity of PV

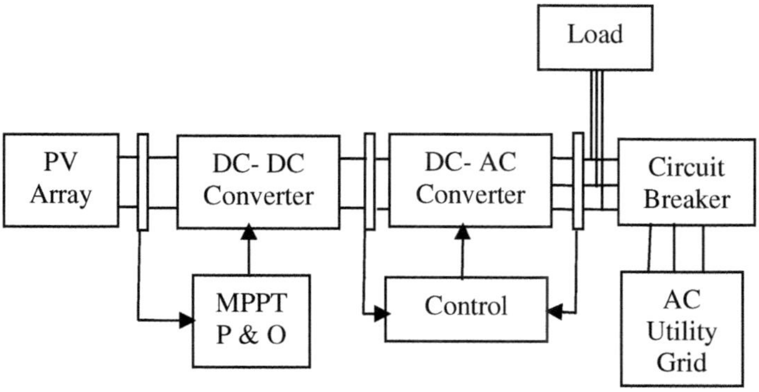

Fig. 1 Block diagram of islanding in the system

system is 100 kW. The distribution network system is operated at 25 kV. The PV system is generating 200 kW. Load is rated at 150 MW and 40 kVAR (Fig. 2).

3 Result and Discussion

The method is tested for various islanding scenarios like three conductors open, two conductors open and single conductor open. The system is simulated for 1 s. By opening the circuit breaker's contact for single conductor, two conductor and three conductors at 0.5 s with both 0 and 50% active power mismatch circumstances, the microgrid region is islanded.

3.1 Proposed Method Determined for Fifty Percent Active Power Mismatch (PV Capacity > Load Demand)

An islanding scenario occurring at 0.5 s with fifty percent power mismatch was simulated for the test system. Under these circumstances, it is discovered that the approach produces satisfactory results. In Fig. 3, the simulated waveforms are shown.

In Figs. 3, 4 and 5, the consequence of anti-islanding detection using the suggested methodology is displayed. At 0.5 s, the grid side's input current is blocked and then ZSI increased. The simulation results show that the suggested technique, using zero-sequence components, effectively estimates impedance.

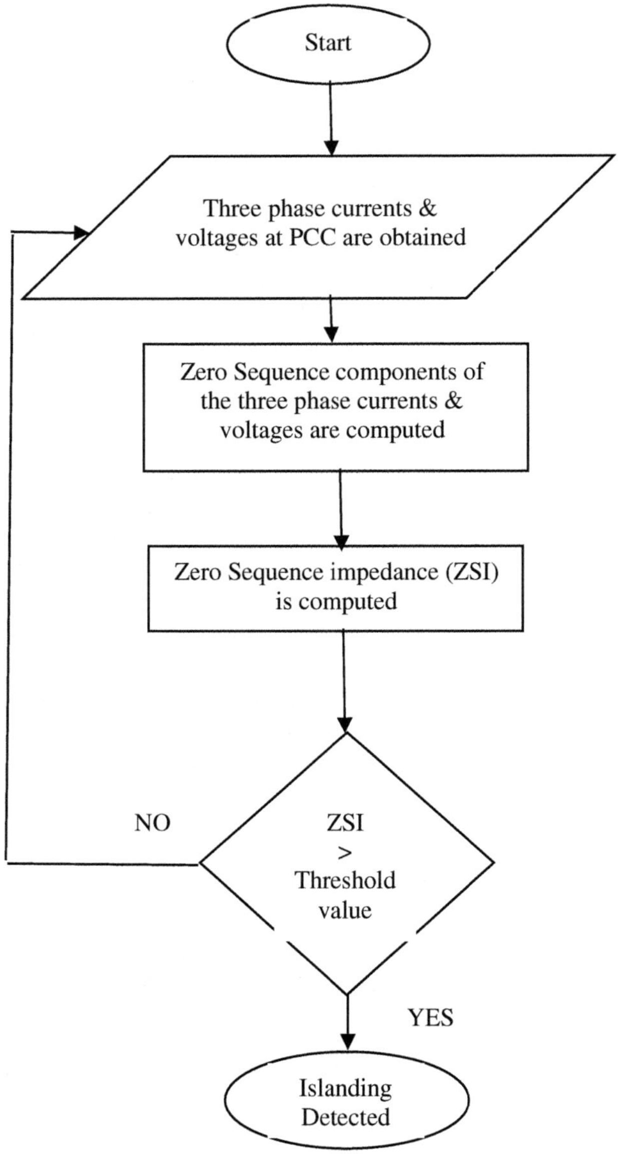

Fig. 2 Flowchart for the method described

3.2 *Proposed Method Determined for Fifty Percent Active Power Mismatch (PV Capacity < Load Demand)*

An islanding scenario occurring at 0.5 s with fifty percent power mismatch was simulated for the test system. It is discovered that, in these circumstances, the approach

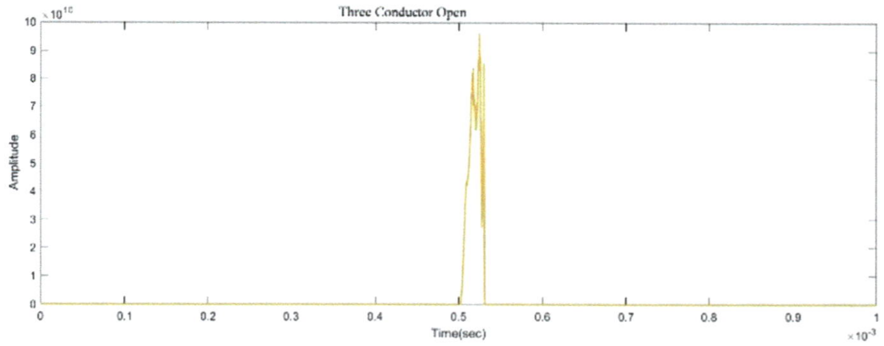

Fig. 3 Zero-sequence impedance simulation result for 50% mismatch in active power with islanding at 0.5 s for triple pole open

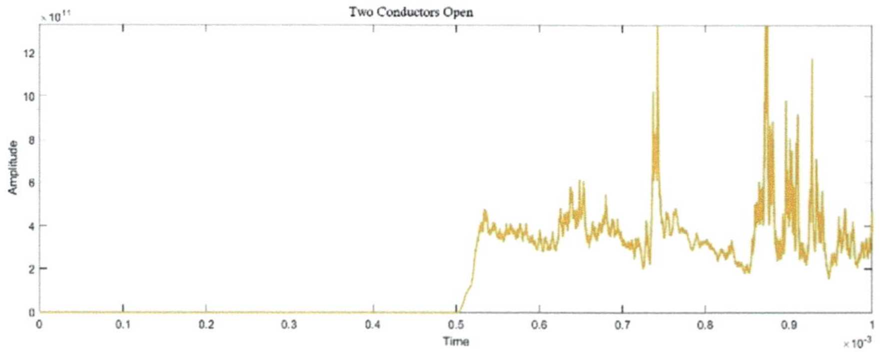

Fig. 4 Zero-sequence impedance simulation result for 50% mismatch in active power with islanding at 0.5 s for double pole open

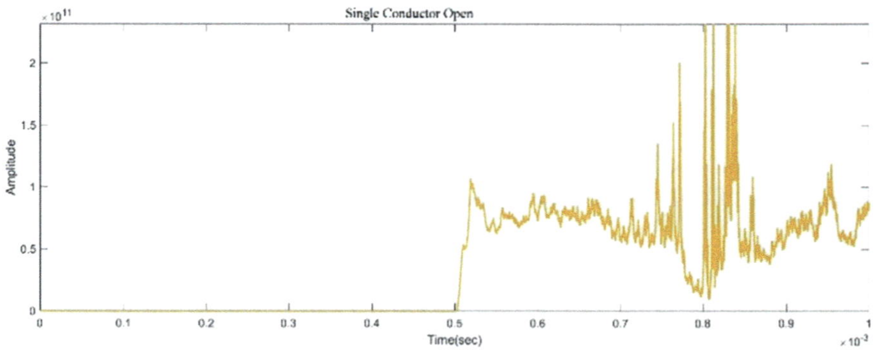

Fig. 5 Zero-sequence impedance simulation result for 50% mismatch in active power with islanding at 0.5 s for single pole open

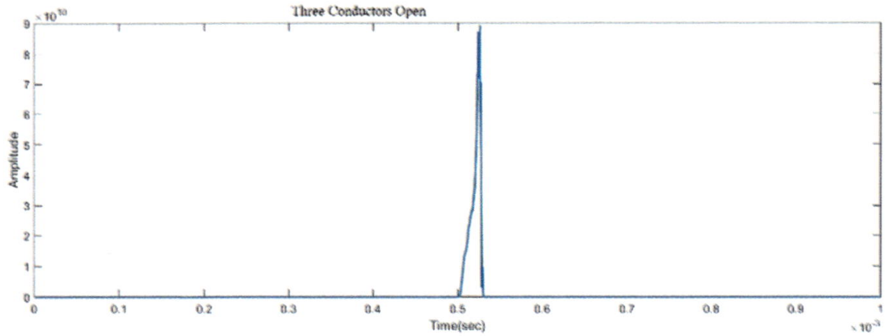

Fig. 6 Zero-sequence impedance simulation result for 50% mismatch in active power with islanding at 0.5 s for triple pole open

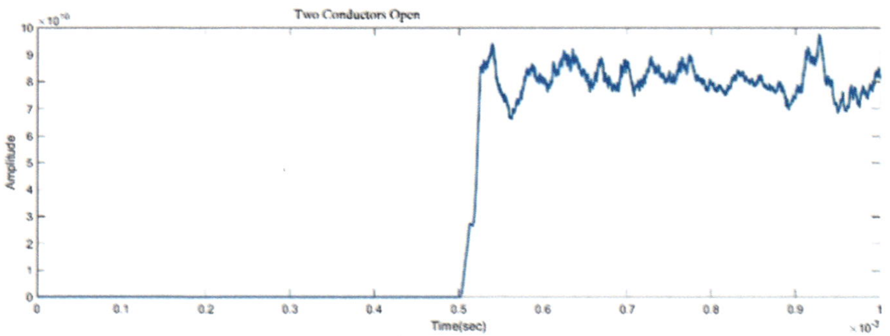

Fig. 7 Zero-sequence impedance simulation result for 50% mismatch in active power with islanding at 0.5 s for double pole open

delivers satisfactory results. In Fig. 4, the simulated waveforms are shown. At DG terminals, the method is investigated. It was found to work satisfactorily.

In Figs. 6, 7 and 8, the anti-islanding detection results using the suggested method are shown. At 0.5 s, the grid side's input current is blocked, the ZSI increased. The simulation results show that the suggested technique, using zero-sequence components, effectively estimates impedance.

3.3 Proposed Method Determined for Zero Percent Active Power Mismatch (PV Capacity = Load Demand)

An islanding scenario occurring at 0.5 s with no power mismatch was simulated for the test system. It is discovered that, in these circumstances, the approach delivers

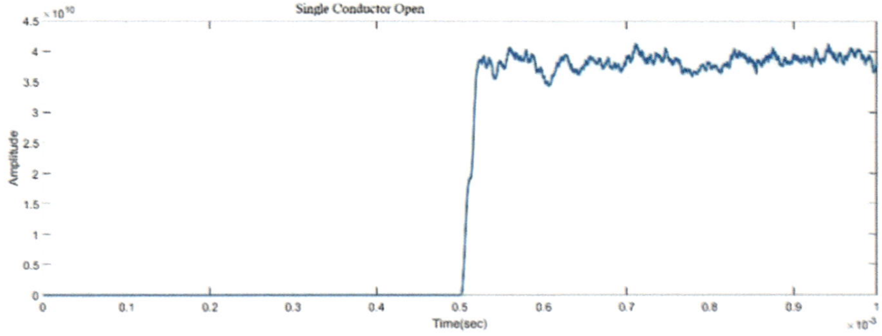

Fig. 8 Zero-sequence impedance simulation result for 50% mismatch in active power with islanding at 0.5 s for single pole open

satisfactory results. In Fig. 5, the simulated waveforms are shown. At DG terminals, the method is investigated. It was found to work satisfactorily.

In Figs. 9, 10 and 11, the anti-islanding detection results using the suggested method are shown. At 0.5 s, the grid side's input current is blocked, the ZSI increased. The simulation results show that the suggested technique, using zero-sequence components, effectively estimates impedance.

With the above results we can observe that during non-islanding condition the value of ZSI is zero, but the instant islanding condition arises in system, the ZSI value rises and crosses the threshold value of $3.5 \times 10^{10}\ \Omega$. For all the anti-islanding results, the ZSI is above the threshold value. Threshold value is chosen experimentally and it varies from system to system.

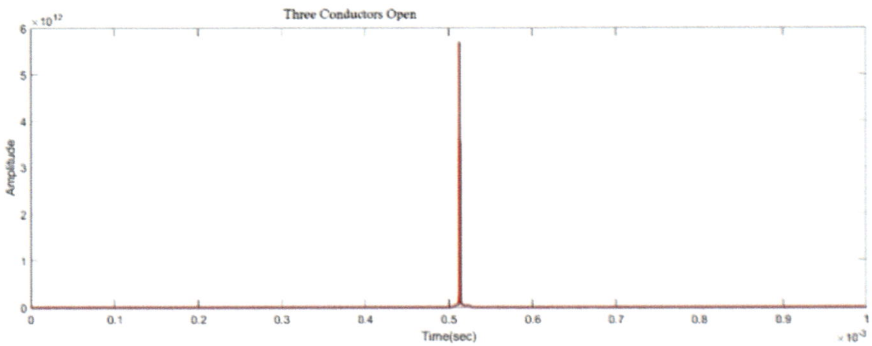

Fig. 9 Zero-sequence impedance simulation result for 0% mismatch in active power with islanding at 0.5 s for triple pole open

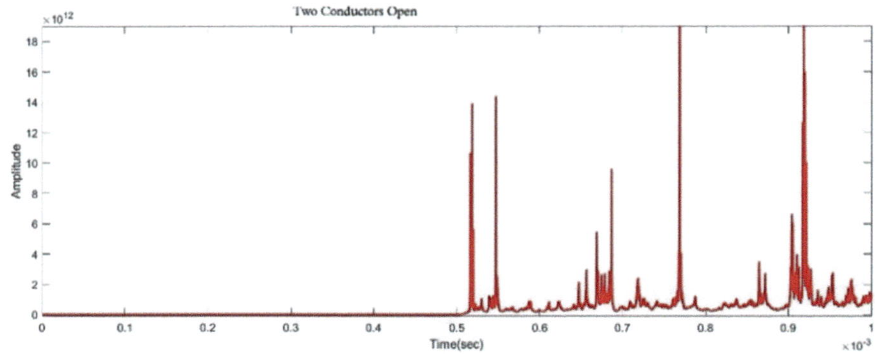

Fig. 10 Zero-sequence impedance simulation result for 0% mismatch in active power with islanding at 0.5 s for double pole open

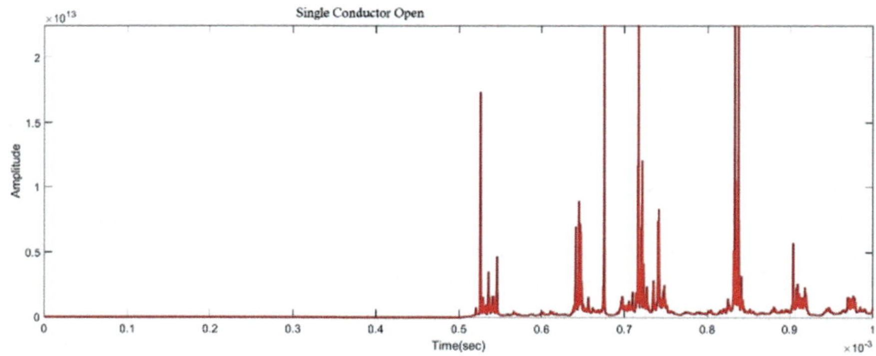

Fig. 11 Zero-sequence impedance simulation result for zero mismatch in active power with islanding at 0.5 s for single pole open

4 Conclusion

A method that is able to identify the islanding condition based on the computation of the magnitudes of zero-sequence impedance from zero-sequence components of current and voltage is discussed in this paper. This approach can identify islands when there is no mismatch in active power, lesser mismatch in active power and larger mismatch in active power. The method has proven to work satisfactorily in a variety of islanding circumstances. The ZSI is large enough to distinguish between islanding operation and regular operation. It is possible to identify islanding based on this variation. Variations in ZSI magnitude can be used to identify the islanding mode. The method also performs well for measurements taken at DG terminals and at point of common connection.

References

1. Swarnkar NK, Mahela OP, Khan B, Lalwani M (2021) Identification of islanding events in utility grid with renewable energy penetration using current based passive method. IEEE Access 9:93781–93794. https://doi.org/10.1109/ACCESS.2021.3092971
2. Panigrahy N, Ilamparithi T, Kashinath MV, Prakash R (2016) Comparison and review of islanding detection techniques for power distribution studies. Int J Adv Res Electr, Electron Instrum Eng 5(7):6485–6492
3. Fadzil NA, Hairi MH, Hanaffi F, Kamarudin MN, Isira AS, Packeer Mohamed MF (2019) A research of islanding detection method for distributed generation: mechanism, merits and demerits. Int J Innov Technol Explor Eng (IJITEE) 8(12S2). ISSN: 2278-3075
4. Dutta S, Sadhu PK, Jaya Bharata Reddy M, Mohanta DK (2018) Shifting of research trends in islanding detection method—a comprehensive survey. Prot Control Mod Power Syst
5. Dash PK, Padhee M, Panigrahi TK (2012) A hybrid time–frequency approach based fuzzy logic system for power island detection in grid connected distributed generation. Int J Electr Power Energy Syst 42(1):453–464
6. Ahmad KNEK, Selvaraj J, Rahim NA (2013) A review of the islanding detection methods in grid-connected PV inverters. Renew Sustain Energy Rev 21:756–766
7. Soumya AV, Edward JB (2016) Review of islanding detection techniques for distributed energy sources. Int Adv Res J Sci, Eng Technol; In: National conference on emerging trends in engineering and technology (NCETET'16), vol 3, No 3. Lourdes Matha College of Science and Technology
8. Pal K, kumar Panigrahi B, Mohapatra S, Mohapatra A (2017) Impact of STATCOM on voltage profile in a DG penetrated grid connected system. In: 2017 International conference on circuit, power and computing technologies (ICCPCT), Kollam, India, pp 1–5. https://doi.org/10.1109/ICCPCT.2017.8074213
9. Pal K, Akella AK, Namrata K, Pati S (2022) Face detection using artificial neural network and wavelet neural network. In: International conference on intelligent controller and computing for smart power (ICICCSP), Hyderabad, India, pp 1–6. https://doi.org/10.1109/ICICCSP53532.2022.9862349
10. Reigosa DD, Briz F, BlancoCharro C, Guerrero JM (2017) Passive islanding detection using inverter nonlinear effects. IEEE Trans Power Electron 32:8434–8445
11. Panigrahi BK, Bhuyan A, Shukla J, Ray PK, Pati S (2021) A comprehensive review on intelligent islanding detection techniques for renewable energy integrated power system. Int J Energy Res 45(10):14085–14116
12. Ahmadipour M, Hizam H, Othman ML, Radzi MAM, Chireh N (2019) A novel islanding detection technique using modified slantlet transform in multi-distributed generation. Int J Electr Power Energy Syst 112:460–475
13. Bhuyan A, Panigrahi BK, Pal K, Pati S (2022) Convolutional neural network based fault detection for transmission line. In: 2022 International conference on intelligent controller and computing for smart power (ICICCSP), Hyderabad, India, pp 1–4. https://doi.org/10.1109/ICICCSP53532.2022.9862446
14. Zhou Y, Haji MM, Xu W, Yong J (2018) A novel open-loop method to synchronize an islanded system with the main grid. IEEE Trans Smart Grid 9:1626–1635
15. Khamis A, Shareef H, Bizkevelci E, Khatib T (2013) A review of islanding detection techniques for renewable distributed generation systems. Renew Sustain Energy Rev 28:483–493
16. Pal K, Akella AK, Namrata K, Pati S (2022) Classification of fault using artificial neural network and power quality improvement using DVR in a PV integrated hybrid power system. In: 2022 International conference on intelligent controller and computing for smart power (ICICCSP), Hyderabad, India, pp 1–4. https://doi.org/10.1109/ICICCSP53532.2022.9862499
17. Abokhalil A, Awan A, Al-Qawasmi A-R (2018) Comparative study of passive and active islanding detection methods for PV grid-connected systems. Sustainability 10:1798

18. Prasanna SL, Pal K, Mandal D (2022) Predicting the monthly average incident shortwave solar energy for Hubli, India by using training functions in ANN. In: 2022 International conference on intelligent controller and computing for smart power (ICICCSP), Hyderabad, India, pp 1–4. https://doi.org/10.1109/ICICCSP53532.2022.9862510
19. Panigrahi BK, Ray PK, Rout PK, Mohanty A, Pal K (2017) Detection and classification of faults in a microgrid using wavelet neural network. J Inf Optim Sci. https://doi.org/10.1080/02522667.2017.1374738
20. Kumar Panigrahi B, Nandi R, Mahanta B, Pal K (2016) Islanding detection in distributed generation. In: 2016 International conference on circuit, power and computing technologies (ICCPCT), Nagercoil, India, pp 1–5. https://doi.org/10.1109/ICCPCT.2016.7530295
21. Kumar N, Namrata K, Samadhiya A (2023) Bi-level decision making in techno-economic planning and probabilistic analysis of community-based sector-coupled energy system. Appl Intell 53:6604–6628. https://doi.org/10.1007/s10489-022-03794-9
22. Gupta D, Kumari N, Samadhiya A (2020) Photovoltaic modeling using single diode model in MATLAB. In: 2020 IEEE international conference on computing, power and communication technologies (GUCON), Greater Noida, India, pp 734–739. https://doi.org/10.1109/GUCON48875.2020.9231165
23. Namrata K, Sharma SP, Seksena SBL (2014) Determining regression constants for calculating global solar radiation at Jharkhand (India) region. In: 2014 International conference on renewable energy research and application (ICRERA), Milwaukee, WI, USA, pp 795–797. https://doi.org/10.1109/ICRERA.2014.7016494
24. Hartmann NB, dos Santos RC, Grilo AP, Vieira JCM (2017) Hardware implementation and real-time evaluation of an ANN-based algorithm for anti-islanding protection of distributed generators. IEEE Trans Ind Electron 65(6):36, 5051–5059

A High-Gain Multi-input Multi-output Non-isolated DC–DC Converter for Microgrid Application

Stutee Patra, Aditi Chatterjee, and K. C. Patra

Abstract This paper presents a high-gain multi-input multi-output non-isolated DC–DC converter (HGMNDC) integrating hybrid resources into a microgrid. The converter operation is analyzed in continuous conduction mode with suitable waveforms and the governing equations are derived accordingly. A model predictive control (MPC)-based incremental conductance MPPT algorithm is implemented for faster response during varying input conditions. The proposed converter is simulated in MATLAB/Simulink to provide the necessary results. The comparison between traditional MPPT techniques and MPC-based MPPT techniques has been analyzed and discussed. Finally, the paper concludes the efficiency of the proposed technique in achieving the desired output with the speed in response.

Keywords Multi-input multi-output DC–DC converter · Maximum power point tracking (MPPT) · Model predictive control (MPC)

1 Introduction

The concept of hybrid renewable energy resources (HRER) has revolutionized the application of microgrids in the recent energy distribution scenario. The HRER concept integrates various intermittent renewable energy resources and energy storage systems (such as batteries, ultracapacitors, superconducting magnetic energy storage, pumped hydropower, and compressed air energy storage) into the power electronic interface and the indigenous grid that improves the reliability and efficiency of microgrids [1].

The incorporation of HRER systems into microgrids becomes more challenging and demanding for a versatile, unified, dynamic, and compact power electronic interface. Thus, the technology of multi-input multi-output (MIMO) converters has evolved and the DC–DC converters are ubiquitous among power electronic interfaces due to the exquisite range of applications. MIMO converters are broadly categorized

S. Patra (✉) · A. Chatterjee · K. C. Patra
C. V. Raman Global University, Bhubaneswar, Odisha 752054, India
e-mail: 2101070013@cgu-odisha.ac.in

J. Kumar et al. (eds.), *Recent Advances in Power Systems*, Lecture Notes in Electrical Engineering 1192, https://doi.org/10.1007/978-981-97-2788-9_5

into three types, i.e., isolated MIMO converters, partially isolated MIMO converters, and non-isolated MIMO converters [2]. Solar PVs, the highly resourced renewable energy, have the limitation of providing low unregulated voltage to the DC distribution system. This has urged the requirement of high-gain converters to raise the voltage levels from 36 V (single PV cell) to 400–800 V (DC distribution grid).

Various configurations of high-gain DC–DC converters are already available in the research literature, leading to the development of advanced high-gain DC–DC converter technology. The advancement in high-gain non-isolated DC–DC converter technology includes quadratic converters [3, 4], Dickson charge pump-based voltage multipliers (VMs) [5], Cockcroft–Walton-based VMs [6, 7], switched capacitor (SC)-based, switched capacitor and coupled inductor (SC and CI)-based [8, 9], transformer with switched capacitor-based [10, 11] high-gain converters, super-lift Luo [12] or ultra-lift Luo converters [13] and new age converters (X–Y family configuration [14], hybrid configuration [15, 16], and multi-modular configuration [17]). The MIMO converters are advantageous over the single-input single-output converters so highly preferred in microgrids [18].

The prototype of a high-gain non-isolated DC–DC converter (based on switched inductor modification) was developed and the MPC-based MPPT algorithm was implemented using the Dspace controller [19]. Another non-isolated high-gain three-port DC–DC buck–boost converter (based on the integration of buck–boost and bidirectional boost converter) was fabricated, the control logic was implemented with Xilinx FPGA board, and experimental results demonstrated the applicability of converter in an HRER system [20]. A new type of high-gain three-port DC–DC converter (based on coupled inductors) was developed for boosting the voltage level up to 10–20 times with 93% operational efficiency [21]. The digital control for different modes of converter operation was devised using an algorithm that included PV voltage, battery voltage, and output voltage regulation. A high-gain multilevel boost single-input single-output DC–DC converter (based on VMs) was controlled with MPC-based INC maximum power point tracking for dynamic operating conditions [22]. This technique presented excellent tracking results and higher power conversion efficiency.

This paper proposes a high-gain multi-input multi-output non-isolated DC–DC converter (HGMNDC) suitable for the HRER system as an application in microgrids. Section 2 explains the operating principle and the voltage stress across the components of the proposed converter. In Sect. 3, the proposed converter is realized in a microgrid system with the open-loop control using Model Predictive Control (MPC)-based incremental conductance MPPT algorithm. The simulated results are validated in Sect. 4 and the conclusion is presented in Sect. 5.

2 High-Gain Multi-input Multi-output Non-isolated DC–DC Converter (HGMNDC)

2.1 Operating Principle

The configuration of the proposed HGMNDC is presented in Fig. 1. The converter is a boost-derived modified version of a MIMO non-isolated DC–DC converter with Dickson charge pump-based VMs to elevate the voltage gain ratio. The converter consists of four input voltage sources ($V_1/V_2/V_3/V_4$), each connected to an inductor ($L_1/L_2/L_3/L_4$) and a switch ($S_1/S_2/S_3/S_4$); now the combination of these forms four input ports. There are two output ports with resistive load (R_1/R_2). This circuit is a combination of diode-capacitor VMs based on the Dickson charge pump configuration at various power conversion stages.

The converter configuration is analyzed for a continuous conduction mode of operation under the assumption that the power semiconductor devices are ideal, and the resonant components are of large value to make the currents and voltages across them constant. The HGMNDC has three modes of operation as discussed below.

Mode I In this mode of operation, the switches S_1 and S_3 are gated on with pulses G_1 and G_3 as shown in Fig. 5a and the switches S_2 and S_4 are off. The input voltage sources V_1 and V_3 charge inductors L_1 and L_3, thus making the inductor currents i_{L1} and i_{L3} rise linearly as shown in the waveform in Fig. 5c. The inductor currents i_{L1} and i_{L3} circulate in the input port-1 and port-3 but the inductor current i_{L2} charges the capacitors C_1, C_2, and C_3 and the inductor current i_{L4} charges the capacitors C_4, C_5, C_{o1}, and C_{o2}. The voltage across switches S_2 and S_4 (V_{DS2} and V_{DS4} as shown in Fig. 5a) reverse biases the diodes D_1, D_3, and D_6. Thus, diodes D_2, D_4, D_5, and D_7 conduct (as indicated in Fig. 5b with diode voltages V_{D2}, V_{D4}, V_{D5}, and V_{D7} becoming zero and diode currents I_{D2}, I_{D4}, I_{D5}, and I_{D7} rising to a finite value) the inductor currents i_{L2} and i_{L4} to charge the respective set of capacitors as shown in Fig. 2. The output currents I_{o1} and I_{o2} circulate in the output port-1 and port-2 that supplies the load at output ports. The equations for voltage across inductors L_1, L_2, L_3, and L_4 are given below:

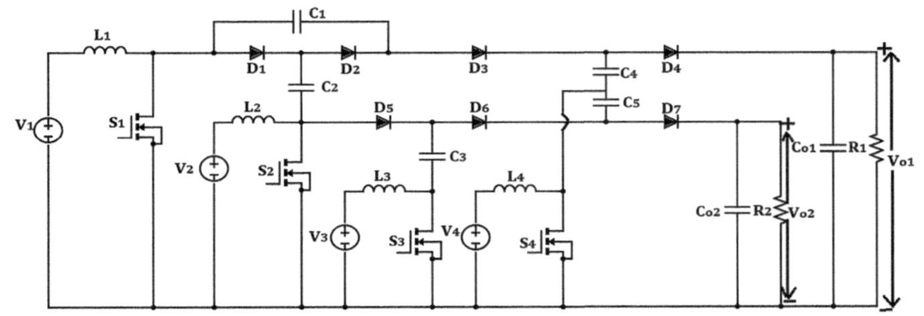

Fig. 1 Circuit configuration of the high-gain multiport non-isolated DC–DC converter (HGMNDC)

$$V_{L1} = L_1 \frac{di_{L1}}{dt} = V_1 \tag{1}$$

$$V_{L2} = L_2 \frac{di_{L_2}}{dt} = V_2 - \frac{1}{C_3} \int i_{C3} dt = V_2 - V_{C3} \tag{2}$$

$$V_{L2} = V_2 + \frac{1}{C_2} \int i_{C2} dt - \frac{1}{C_1} \int i_{C3} dt = V_2 + V_{C2} - V_{C1} \tag{3}$$

$$V_{L3} = L_3 \frac{di_{L3}}{dt} = V_3 \tag{4}$$

$$V_{L4} = L_4 \frac{di_{L4}}{dt} = V_4 + V_{C4} - V_{o1} = V_4 + V_{C4} - V_{C_{o1}} \tag{5}$$

$$V_{L4} = V_4 + V_{C5} - V_{o2} = V_4 + V_{C5} - V_{C_{o2}} \tag{6}$$

Mode II In this mode of operation, the switches S_1 and S_3 are already conducting and switches S_2 and S_4 are gated ON with pulses G_2 and G_4 as shown in Fig. 5a. Hence, all four switches are conducting simultaneously. The input voltage sources V_1 and V_3 are already supplying the inductor currents i_{L1} and i_{L3} and now voltage sources V_2 and V_4 are supplying the inductor currents i_{L2} and i_{L4}. The inductor currents i_{L1}, i_{L2}, i_{L3}, and i_{L4} circulate through their individual input ports port-1, port-2, port-3, and port-4 as shown in Fig. 3. As a result, the diodes get reverse biased and the capacitors C_1–C_5 get disconnected from the configuration. But the remaining output port capacitors C_{o1} and C_{o2} discharge their currents through the loads connected to the output ports.

$$V_{L1} = L_1 \frac{di_{L1}}{dt} = V_1 \tag{7}$$

$$V_{L2} = L_2 \frac{di_{L2}}{dt} = V_2 \tag{8}$$

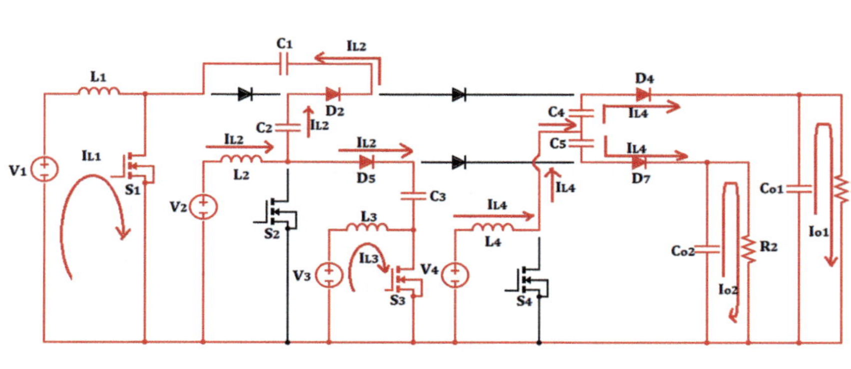

Fig. 2 Operation of HGMNDC in mode I

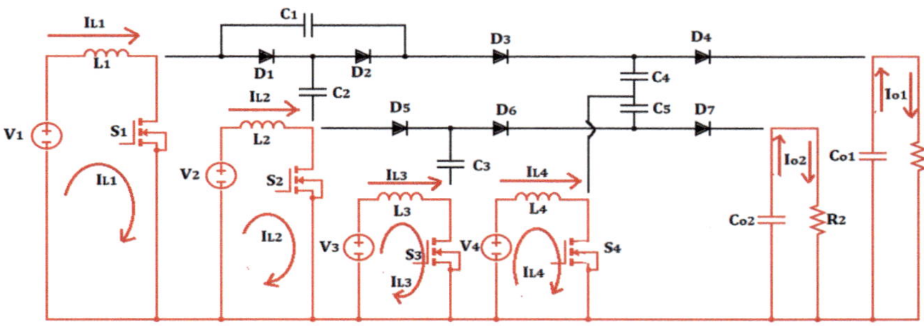

Fig. 3 Operation of HGMNDC in mode II

$$V_{L3} = L_3 \frac{di_{L3}}{dt} = V_3 \tag{9}$$

$$V_{L4} = L_4 \frac{di_{L4}}{dt} = V_4 \tag{10}$$

$$i_{C_{o1}} = C_{o1} \frac{dV_{o1}}{dt}, \ I_{o1} = \frac{V_{o1}}{R} \tag{11}$$

$$i_{C_{o2}} = C_{o2} \frac{dV_{o2}}{dt}, \ I_{o2} = \frac{V_{o2}}{R} \tag{12}$$

Mode III In this mode of operation, the switches S_2 and S_4 are already conducting and the switches S_1 and S_3 are off. The inductor currents i_{L2} and i_{L4} continue getting charged through input voltage sources V_2 and V_4 thus circulating in the input ports 2 and 4. Whereas the inductor currents i_{L1} and i_{L3} get discharged through the path as indicated in Fig. 4. The inductor current i_{L1} charges capacitors C_1 and C_2 and inductor current i_{L3} charges capacitors C_3, C_4, and C_5 due to the conducting diodes D_1, D_3, and D_6 (as hinted in Fig. 5b with diode voltages V_{D1}, V_{D3}, and V_{D6} becoming zero and diode currents I_{D1}, I_{D3}, and I_{D6} rising to a finite value). The voltage across the switches S_1 and S_3 (V_{DS1} and V_{DS3} as shown in Fig. 5a) reverse biases the diodes D_2, D_4, D_5, and D_7. The equations for inductor voltages V_{L1}, V_{L2}, V_{L3}, and V_{L4}; output capacitor currents i_{Co1} and i_{Co2}; and output currents are presented below:

$$V_{L1} = L_1 \frac{di_{L1}}{dt} = V_1 - V_{C2} = V_1 + V_{C1} - V_{C4} \tag{13}$$

$$V_{L2} = L_2 \frac{di_{L2}}{dt} = V_2 \tag{14}$$

$$V_{L3} = L_3 \frac{di_{L3}}{dt} = V_3 + V_{C3} - V_{C5} \tag{15}$$

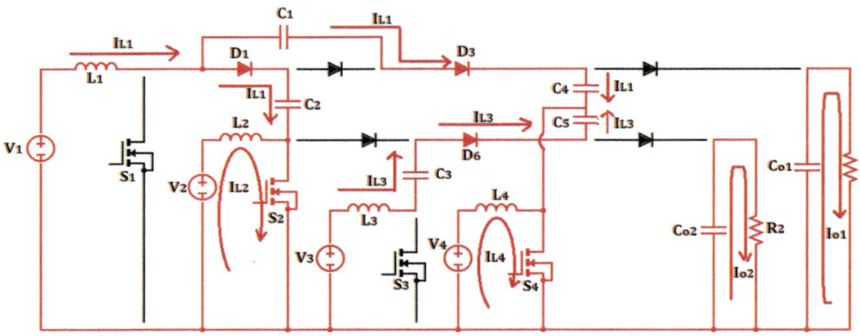

Fig. 4 Operation of HGMNDC in mode III

$$V_{L4} = L_4 \frac{di_{L4}}{dt} = V_4 \tag{16}$$

$$i_{Co1} = C_{o1} \frac{dV_{o1}}{dt}, \; I_{o1} = \frac{V_{o1}}{R} \tag{17}$$

$$i_{Co2} = C_{o2} \frac{dV_{o2}}{dt}, \; I_{o2} = \frac{V_{o2}}{R} \tag{18}$$

On applying the principle of volts-seconds balance on (1)–(6), (7)–(10), and (13)–(16), we can calculate the output voltages at the two output ports. We also consider that the switches S_1, S_2, S_3, and S_4 are having duty ratios D_1, D_2, D_3, and D_4.

$$V_{o1} = 2\frac{(1 + D_1)}{(1 - D_1)} V_1 + \frac{(1 + D_2)}{(1 - D_2)} V_2 + \frac{(1 + D_4)}{(1 - D_4)} V_4 \tag{19}$$

$$V_{o2} = \frac{(1 + D_2)}{(1 - D_2)} V_2 + \frac{(1 + D_3)}{(1 - D_3)} V_3 + \frac{(1 + D_4)}{(1 - D_4)} V_4 \tag{20}$$

The voltages across the capacitors C_1, C_2, C_3, C_4, and C_5 are presented below:

$$V_{C1} = \frac{(1 + D_1)}{(1 - D_1)} V_1 + \frac{(1 + D_2)}{(1 - D_2)} V_2 \tag{21}$$

$$V_{C2} = \frac{(1 + D_1)}{(1 - D_1)} V_1 \tag{22}$$

$$V_{C3} = \frac{(1 + D_2)}{(1 - D_2)} V_2 \tag{23}$$

$$V_{C4} = 2\frac{(1 + D_1)}{(1 - D_1)} V_1 + \frac{(1 + D_2)}{(1 - D_2)} V_2 \tag{24}$$

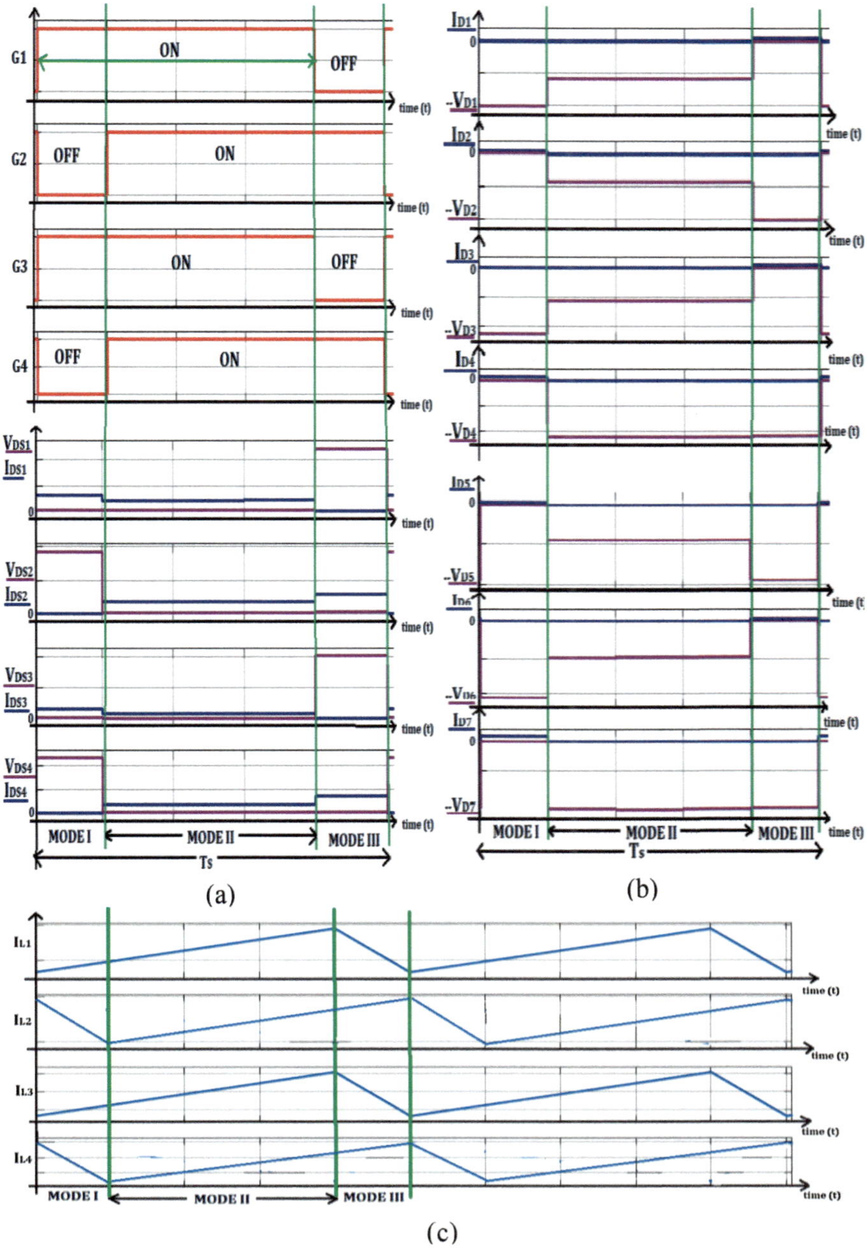

Fig. 5 Waveforms for a switching cycle of HGMNDC under continuous conduction **a** gating pulses with MOSFET voltages and currents, **b** diode voltages and currents, **c** inductor current

Table 1 Voltage stress across different switching components

Components	Voltage stress	Parts count
Switches	$\frac{V_n}{1-D_n}$	4
Diodes D_1, D_2	$\frac{V_1}{1-D_1}+\frac{V_2}{1-D_2}$	2
D_4, D_7	$\frac{V_4}{1-D_4}$	2
D_3	$\frac{V_1}{1-D_1}+\frac{V_4}{1-D_4}$	1
D_5	$\frac{V_2}{1-D_2}+\frac{V_3}{1-D_3}$	1
D_6	$\frac{V_3}{1-D_3}+\frac{V_4}{1-D_4}$	1

where n = number of input ports to which the respective switch is connected

$$V_{C5}=\frac{(1+D_2)}{(1-D_2)}V_2+\frac{(1+D_3)}{(1-D_3)}V_3 \tag{25}$$

The voltages across the output capacitors $V_{C_{o1}}$ and $V_{C_{o2}}$ are the same as that of voltages V_{o1} and V_{o2} as given in (19) and (20).

The waveforms of voltages and currents across various switching components and gating pulses are shown below.

2.2 *Voltage Stress Across Semiconductor Devices of HGMNDC*

It is important to evaluate the voltage stress across the MOSFETs and the diodes of HGMNDC because a converter design depends on the accurate selection of components. The voltage stress across switches and diodes is derived and tabulated in Table 1.

3 HGMNDC in a Microgrid System

The proposed HGMNDC has been applied to a microgrid system with input ports interfacing a solar PV and a proton exchange membrane fuel cell (PEMFC)—1.26 kW–24 Vdc as the resources and a supercapacitor and Li–ion battery as the storage devices that is portrayed in the block diagram below. The open-loop control of HGMNDC is performed by implementing a model predictive control (MPC)-based incremental conductance MPPT technique (Figs. 6 and 7).

Due to faster computation advancements, the model predictive control (MPC) strategy is mostly adopted for power electronics control [23]. The MPC-based MPPT is employed for DC–DC converters throughout the research literature [19, 22–24]. The steps involved in an MPC-based controller are mentioned below:

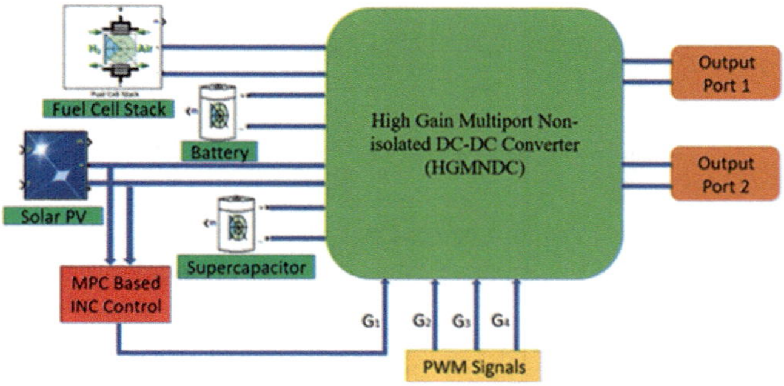

Fig. 6 HGMNDC in a microgrid system

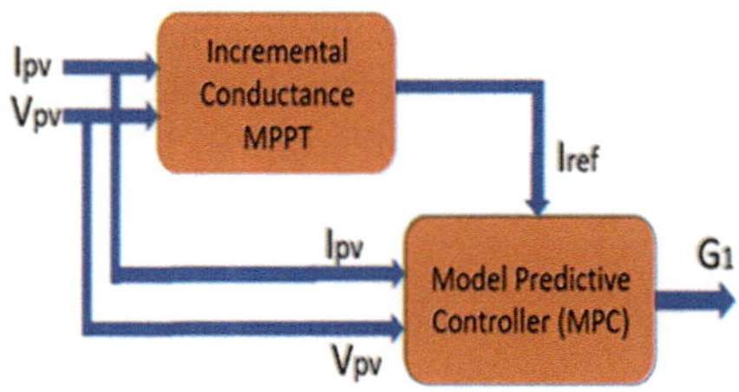

Fig. 7 Proposed open-loop control of PV module

- Discretization of the converter system model on the switching action of power MOSFETs.
- Prediction of the future states of the PV current value to be injected into the proposed converter.
- Formulation of the cost function to err between the reference current generated from the MPPT algorithm and the predicted PV current value.
- Optimization of the cost function to obtain exact switching states for the power MOSFET interfacing the PV module and the proposed converter circuit. The optimization of the cost function is done by a receding horizon strategy also known as receding horizon optimization [24].

The discretized system model of the proposed HGMNDC is derived from the different modes of operation as presented below. When switch S_1 is gated on, (1) and (7) represent the inductor voltage equation of port-1 interfacing solar PV with the HGMNDC. Hence, the predicted state for PV current when S_1 is on:

$$I_{\_pv}^{G_1=1} = I_{\_pv}(k) + \frac{T_s}{L_1} V_{\_pv}(k) \tag{26}$$

When S_1 is turned off, (13) represents the inductor voltage equation of port-1 and the predicted state for PV current is written as (27) or (28):

$$I_{\_pv}^{G_1=0} = I_{\_pv} + \frac{T_S}{L_1}\left(V_{\_pv}(k) - V_{C2}\right) \quad (27)$$

$$I_{\_pv}^{G_1=0} = I_{\_pv} + \frac{T_S}{L_1}\left(V_{\_pv}(k) + V_{C1} - V_{C4}\right) \quad (28)$$

From (22) the value of V_{C2} can be replaced in (27) or from (21) the value of V_{C1} and from (24) the value of V_{C4} can be replaced in (28) to obtain predicted PV current when the switch S_1 is off.

$$I_{\_pv}^{G_1=0} = I_{\_pv}(k) - \frac{T_S}{L_1}\left(\frac{2D_1}{1 - D_1}\right)V_{\_pv}(k) \quad (29)$$

The reference current value is generated from the incremental conductance algorithm as presented in the flowchart as shown in Fig. 8 and the model predictive control logic is implemented to generate the gating pulses for the switch S_1 as depicted in the flowchart shown in Fig. 9.

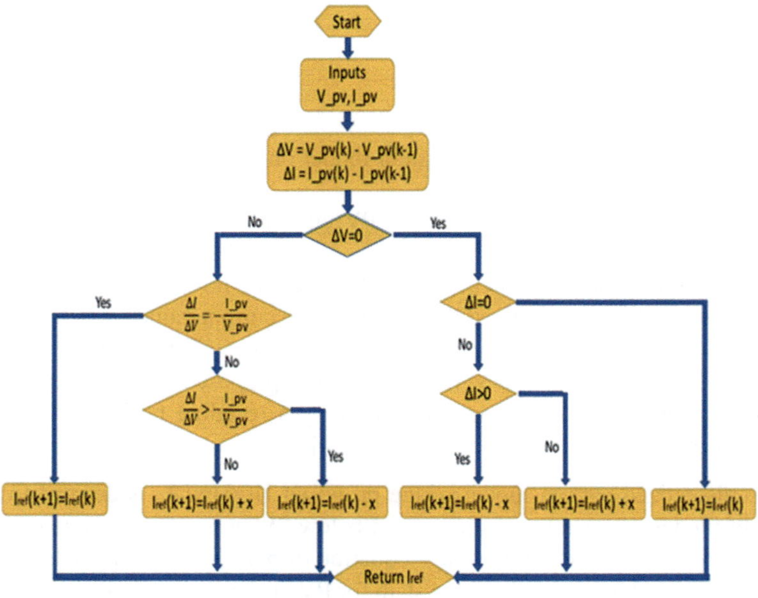

Fig. 8 INC algorithm for generating I_{ref}

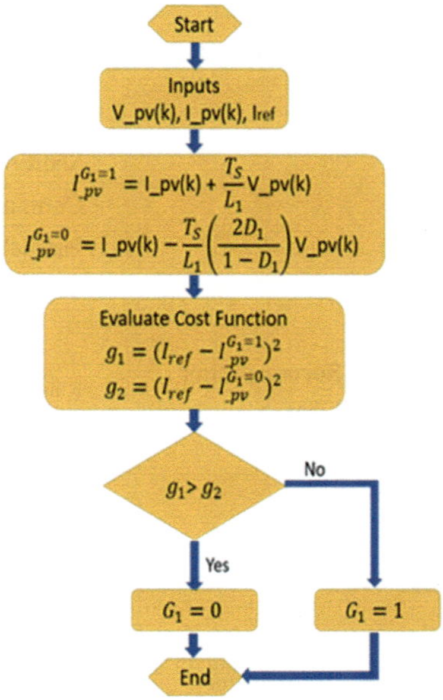

Fig. 9 MPC logic for generating gate pulse

4 Simulation Results and Discussion

The open-loop control of the proposed HGMNDC was performed in MATLAB/ Simulink environment. The system specifications chosen for performing the simulation are presented in Table 2. Tables 3, 4 and 5 present the solar PV, supercapacitor, and battery specifications.

The HGMNDC consists of four input ports interfacing a solar PV controlled with MPC-based incremental conductance MPPT technique and a supercapacitor, a fuel cell, and a battery controlled with PWM technique. HGMNDC consists of two output ports connected to a resistive load, providing different output voltage values. The output voltages are nearly 10 times the supplied input voltages as depicted in the results obtained in Figs. 10 and 11.

The fuel cell output variables, battery output variables, and supercapacitor output variables are presented as the supporting results for the multi-input multi-output operation of the proposed HGMNDC converter as depicted in Figs. 12, 13 and 14.

Table 2 Details of system components

Output port voltages (V)	$V_{o1} = 738$, $V_{o1} = 193$
Switching frequency (kHz)	40
Output port resistors (R_1/R_2) (Ω)	$R_1 = 100$, $R_2 = 75$
Inductors ($L_1/L_2/L_3/L_4$) (mH)	0.4
Capacitors ($C_1/C_2/C_3/C_4/C_5$) (μF)	44
Filter capacitors (C_{o1}/C_{o2}) (μF)	440

Table 3 PV 1Soltech 1STH-215-P specification

Max. power (W)	213.15
Open-circuit voltage (V)	36.3
Voltage at max. power point (V)	29
Short-circuit current (A)	7.84
Current at max. power point (A)	7.35
Parallel strings	10
Series-connected modules per string	3

Table 4 Supercapacitor specifications

Rated capacitance (F)	99.5
Equivalent DC series resistance (mΩ)	8.9
Rated voltage (V)	48
No. of series capacitors	18
No. of parallel capacitors	1
Initial voltage (V)	0

Table 5 Li-ion battery specifications

Nominal voltage (V)	7.2
Rated capacity (Ah)	5.4
Initial SoC (%)	100
Battery response time (s)	30

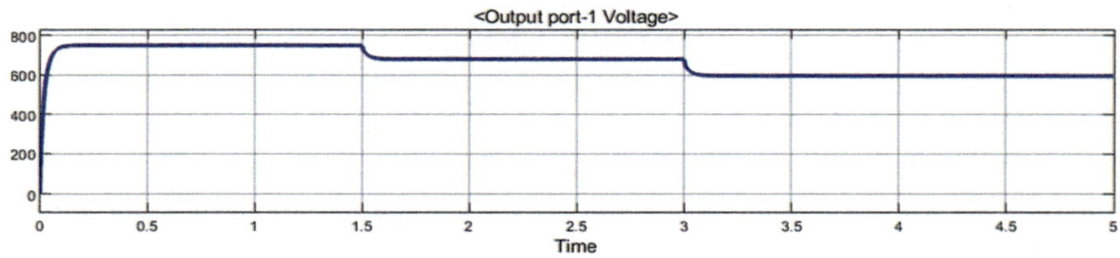

Fig. 10 Output port-1 voltage (Vo_1) of HGMNDC converter in volts

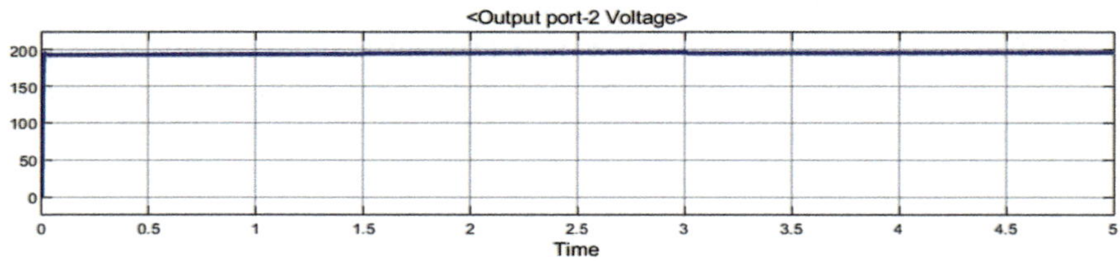

Fig. 11 Output port-2 voltage (Vo_2) of HGMNDC converter in volts

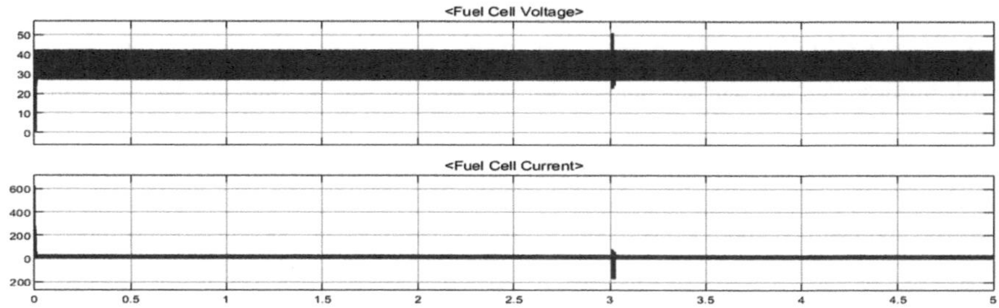

Fig. 12 Fuel cell output voltage and current

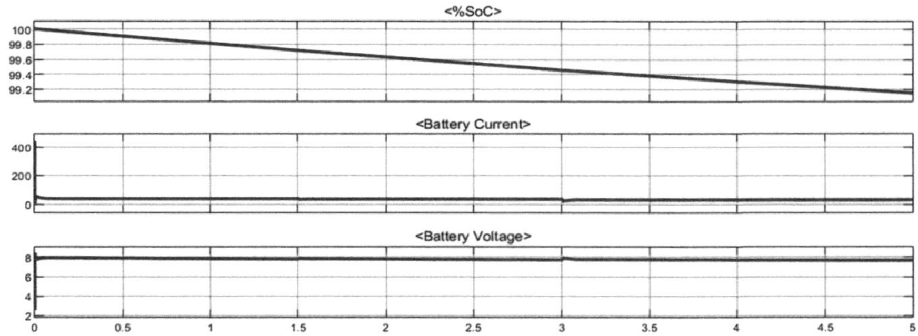

Fig. 13 Battery output in % SoC, current and voltage

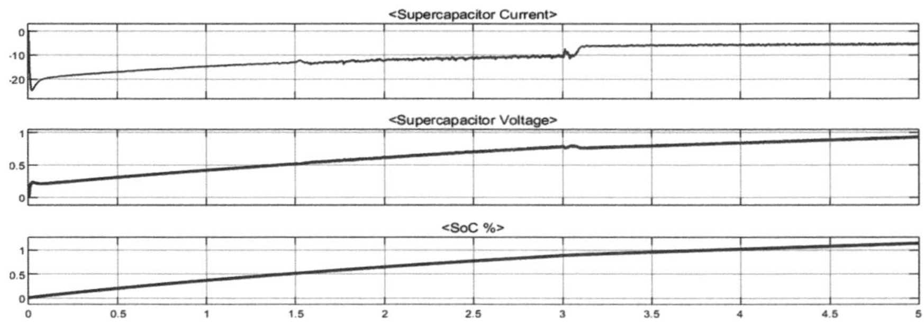

Fig. 14 Supercapacitor output current, voltage, and SoC%

The performance of MPC-based MPPT technique is tested for changing values of solar irradiance at different intervals of time. The solar irradiance value is fixed at 1000 W/m^2 initially, at t = 1.5 s irradiance value is changed to 800 W/m^2, and at t = 3 s again the irradiance value is changed to 600 W/m^2. The output voltage V_{o1} does vary according to the output power available at the end of the solar PV panel during the instants t = 1.5 s and t = 3 s, as shown in Fig. 10. The output port-1 voltage has greater variation because of input port-1 voltage whereas output port-2 voltage almost remains constant because of constant input voltage sources (supercapacitor, battery, and fuel cell) on output port-2. In this scenario, the battery is discharging (with a current range of 37–30 A), the supercapacitor is charging (with a current range of −25 to −5 A), and the fuel cell is providing an average voltage of 24 V.

The effectiveness of the proposed MPC-based incremental conductance MPPT method compared to the traditional MPPT methods could be visualized from the results of output voltage (Fig. 15c), output current (Fig. 16c), and output power (Fig. 17c) of the PV panel. There are fewer fluctuations observed in tracking output parameters of the PV panel in the MPC-based INC method as shown in Figs. 15, 16 and 17. The comparison of the P&O method, INC method, and MPC-based INC method is done based on settling time, power oscillations, and efficiency exhibited during maximum power point (MPP) tracking, as presented in Table 6.

The output voltage of the PV panel remains constant during t = 1.5 s as the irradiance value changes from 1000 to 800 W/m^2 as well as during t = 3 s. It could be noticed clearly that the MPC-based INC method provides results with very less fluctuations compared to the other two traditional methods.

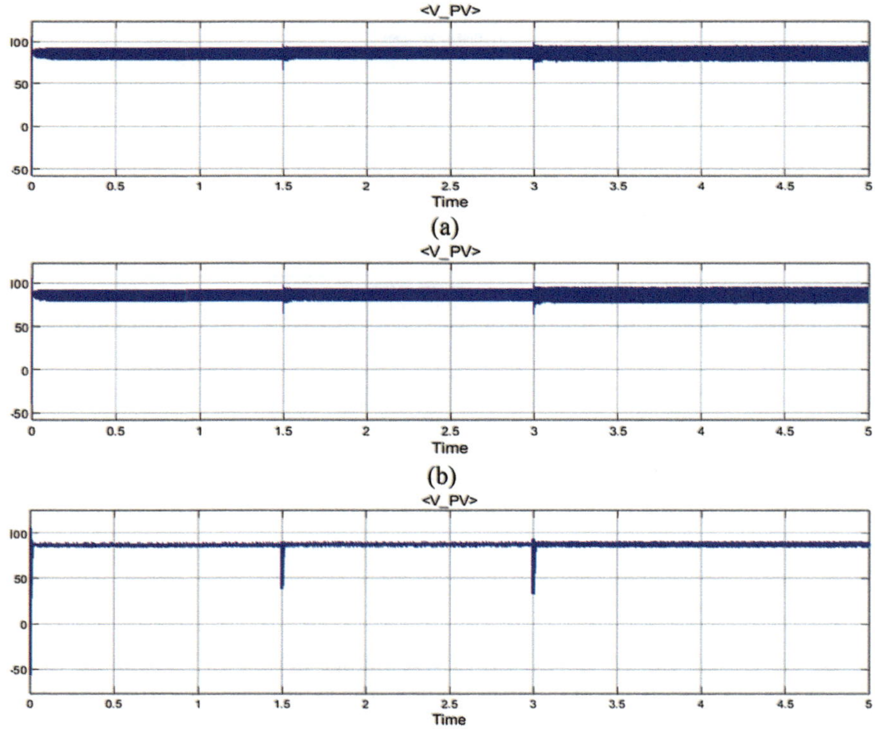

Fig. 15 Output voltage of PV panel with **a** P&O method, **b** INC method, **c** MPC-based INC

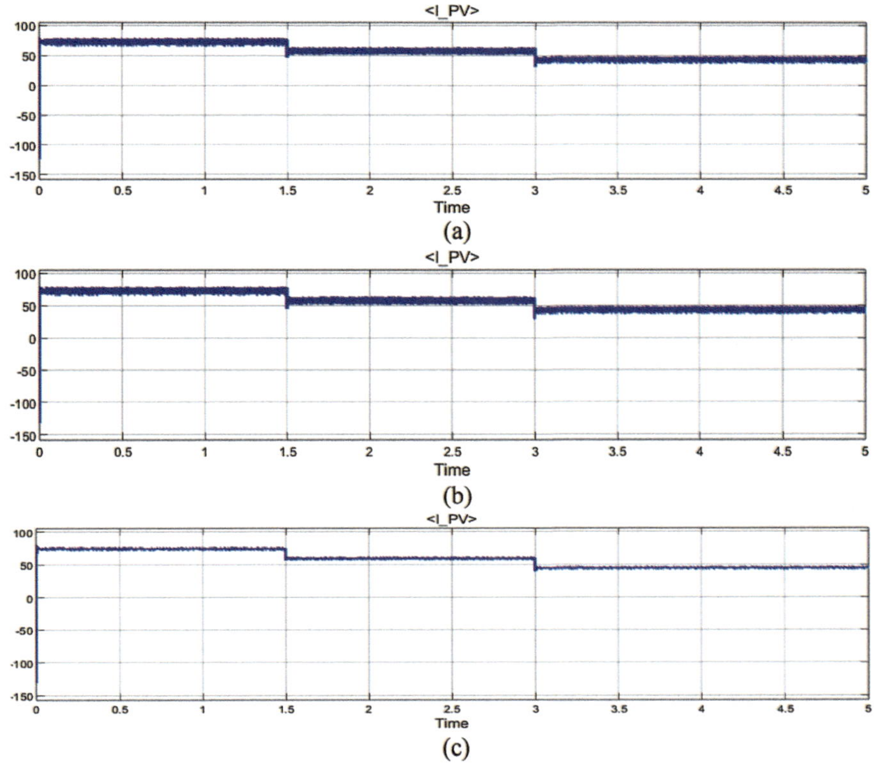

Fig. 16 Output current of PV panel with **a** P&O method, **b** INC method, **c** MPC-based INC

The tabular representation presents a comparative analysis of the P&O method, INC method, and MPC-based INC method during the condition irradiance changes from 1000 to 800 W/m^2 in the PV array.

As per the analysis presented, the MPC-based INC method has a faster settling time with transients dying out within 8 ms whereas the P&O and the INC method have a longer settling time with receding transient undershoot (this criticality can be noticed in Figs. 18a, 19a and 20a). This analysis was done for changing irradiance condition from 1000 to 800 W/m^2 during 1.5 s time instant. MPC-based INC method exhibited a lesser range of power oscillation as compared to the P&O method and INC method. This could be observed in Figs. 18b, 19b and 20b. Finally, the efficiency calculation by evaluating power samples from the maximum power point value P_{max} proved the MPC-based INC method efficient over P&O and INC methods.

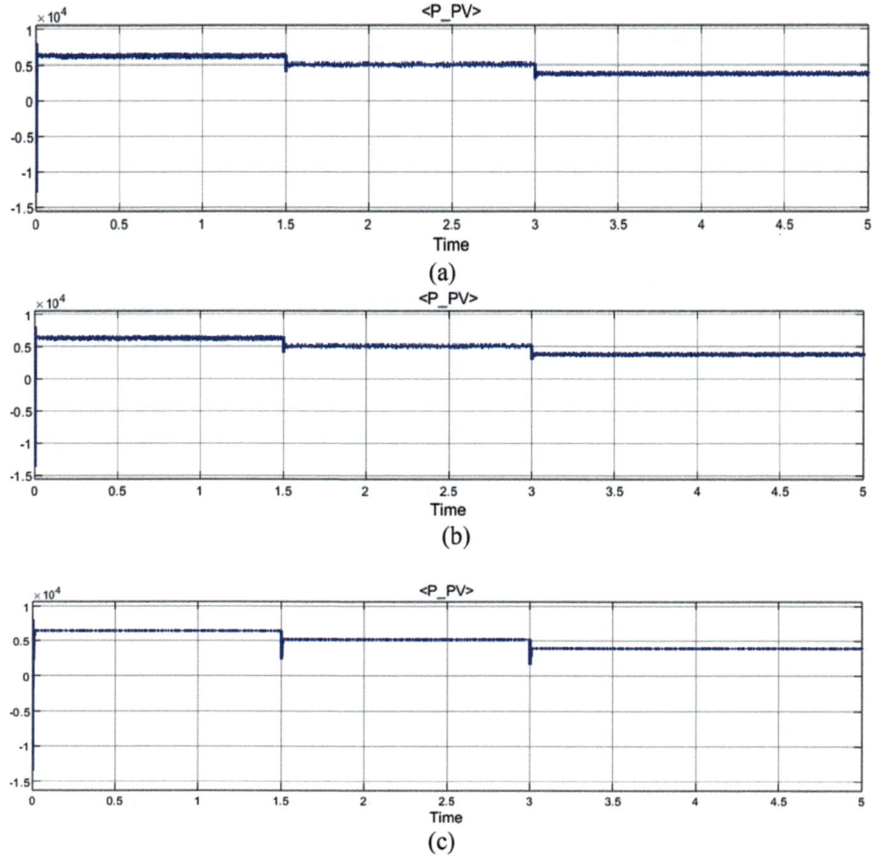

Fig. 17 Power output of PV panel with **a** P&O method, **b** INC method, **c** MPC-based INC

Table 6 Comparative analysis of P&O-, INC-, and MPC-based INC method

MPPT techniques	Settling time (ms)	Power oscillations in a PV module (at 1000 W/m^2) (W)	Efficiency $\eta = \frac{\sum_{i=1,2}^{n} P_i / n}{P_{max}}$ n = no. of samples (%)
P&O	48	11.2	94.77
INC	55	10	96.26
MPC-based INC	8	0.767	99.09

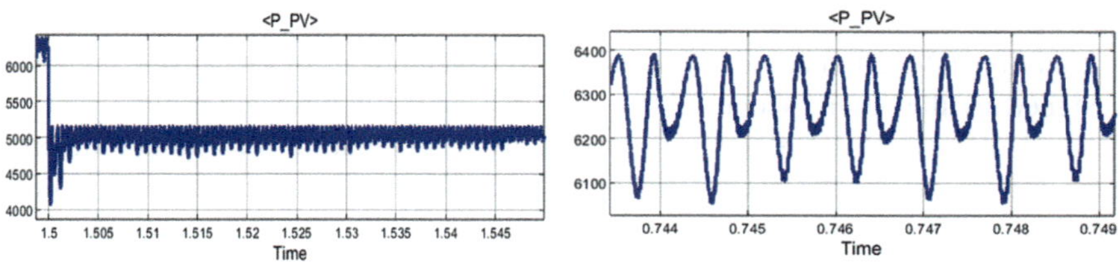

Fig. 18 Settling time and power oscillations of P&O method

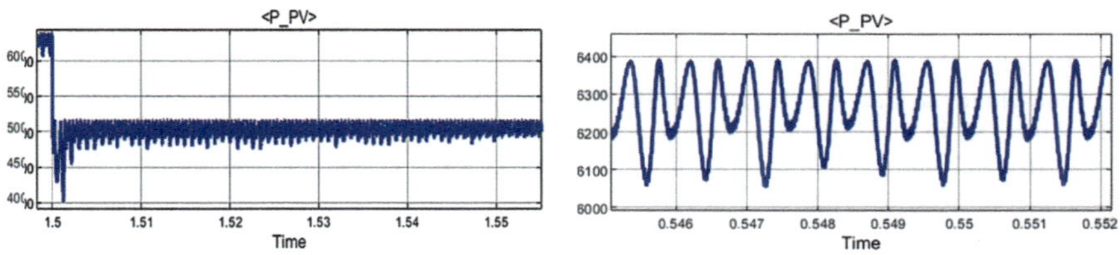

Fig. 19 Settling time and power oscillations of INC method

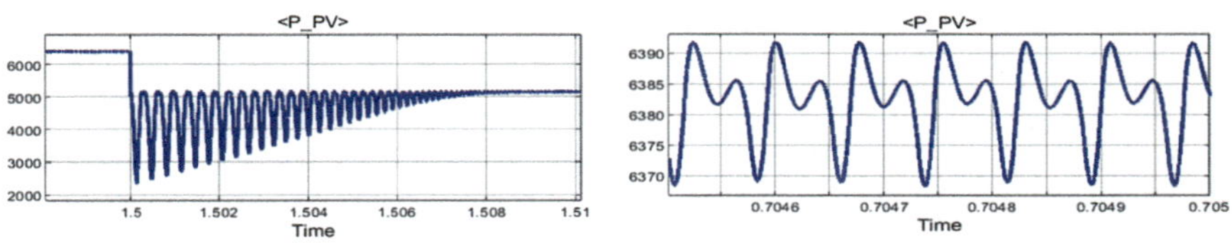

Fig. 20 Settling time and power oscillations of MPC-based INC method

5 Conclusion

A high-gain multi-input multi-output non-isolated DC–DC converter (HGMNDC) configuration was developed to integrate hybrid renewable energy resources with DC loads. The converter's continuous conduction mode of operation was studied to prove the inherent features, i.e., high-gain conversion ratio, minimum voltage stress, and multiport operability. The HGMNDC application in the microgrid system was enhanced by implementing a model predictive control (MPC)-based incremental conductance MPPT technique. The MPC control of HGMNDC was performed in MATLAB/Simulink environment. The smooth control of output voltage was validated with the presented results despite changing irradiance conditions. A comparative analysis of traditional MPPT methods (i.e., P&O method and INC method) and MPC-based INC method was carried out. The MPC-based INC method exhibits almost zero power oscillations such that the proposed algorithm accurately tracks the maximum power point at $P_{max} = 6393$ W. Finally, the critical analysis of MPPT techniques reveals the efficacy of MPC-based MPPT methods with an accuracy of

99.09%. Hence it is proved that MPC-based MPPT methods are better compared to traditional MPPT methods. The hardware development of HGMNDC could be designed as the future scope of this research and a control algorithm could be developed to establish different modes of converter operation (i.e., during the presence or absence of PV power, variation in fuel cell's input parameters, and battery/supercapacitor's charging or discharging modes).

References

1. Bartolucci L, Cordiner S, Mulone V, Luis Rossi J (2019) Hybrid renewable energy systems for household ancillary services. Electr Power Energy Syst (Elsevier) 107:282–297
2. Wu H, Xu P, Hu H, Zhou Z, Xing Y (2014) Multiport converters based on integration of full-bridge and bidirectional DC–DC topologies for renewable generation systems. IEEE Trans Ind Electron 61(2):856–869
3. Gopinathan S, Rao VS, Sundaramoorthy K (2022) Family of non-isolated quadratic high gain DC–DC converters based on extended capacitor-diode network for renewable energy source integration. IEEE J Emerg Sel Top Power Electron 10(5):6218–6230
4. Pires VF, Cordeiro A, Foito D, Silva JF (2019) High step-up DC–DC converter for fuel cell vehicles based on merged quadratic boost–Cuk. IEEE Trans Veh Technol 68(8):7521–7530
5. Baddipadiga BP, Ferdowsi M (2017) A high-voltage-gain DC–DC converter based on modified Dickson charge pump voltage multiplier. IEEE Trans Power Electron 32(10):7707–7715
6. Park S, Yang J, Rivas-Davila J (2020) A hybrid Cockcroft–Walton/Dickson multiplier for high voltage generation. IEEE Trans Power Electron 35(3):2714–2723
7. Muller L, Kimball JW (2015) High gain DC–DC converter based on the Cockcroft–Walton multiplier. IEEE Trans Power Electron
8. Zhang M, et al (2022) A high step-up DC–DC converter with switched-capacitor and coupled-inductor techniques. IEEE J Emerg Sel Top Ind Electron 3(4):1067–1076
9. Ajami A, Ardi H, Farakhor A (2015) A novel high step-up DC/DC converter based on integrating coupled inductor and switched-capacitor techniques for renewable energy applications. IEEE Trans Power Electron 30(8):4255–4263
10. Rahimi R, et al (2022) An interleaved high step-up DC–DC converter based on integration of coupled inductor and built-in-transformer with switched-capacitor cells for renewable energy applications. IEEE Access 10:32–45
11. Padmanaban S, Bhaskar MS, Maroti PK, Blaabjerg F, Fedák V (2018) An original transformer and switched-capacitor (T&SC)-based extension for DC–DC boost converter for high-voltage/low-current renewable energy applications: hardware implementation of a new T&SC boost converter. Energies 11(783):1–23
12. Sarasvathi K, Divya K (2018) Analysis and design of super-lift Luo boost converter. In: Lund H, et al (eds) 4th international conference on electrical energy systems (ICEES), pp 591–595
13. Pearl Nightingale, Deshpande M (2019) Design of ultra lift Luo converter using incremental conductance technique for PV application. In: 2nd international conference on power and embedded drive control (ICPEDC), pp 201–207
14. Sadaf S, Al-Emadi NA, Iqbal A, Bhaskar MS, Meraj M (2019) New high gain 2LC-Y multilevel-boost-converter (2LC-Y MBC) topologies for renewable energy conversion: members of X–Y converter family. In: Al-Haddad K, et al (eds) 28th international symposium on industrial electronics (ISIE), pp 2647–2652
15. Zhang Y, Sun J-T, Wang Y-F (2013) Hybrid boost three-level DC–DC converter with high voltage gain for photovoltaic generation systems. IEEE Trans Power Electron 28(8):3659–3664
16. Gupta N, et al (2020) Novel hybrid high gain converter: combination of Cuk and buck–boost structures with switched inductor for DC microgrid. In: Aktekin S, et al (eds) 2nd global power, energy and communication conference (IEEE GPECOM2020), pp 47–52

17. Alfares A, Lehman B, Amirabadi M (2022) A Cúk-based modular DC–DC converter for medium voltage direct current (MVDC) applications. IEEE Open J Power Electron 3:560–573
18. Patra S, Chatterjee A, Patra KC (2022) A comparative analysis of multiport DC–DC converters in microgrid using small-signal modelling. In: IEEE 2nd international symposium on sustainable energy, signal processing and cyber security (iSSSC), pp 1–6
19. Abdel-Rahim O, Wang H (2020) A new high gain DC–DC converter with model-predictive-control based MPPT technique for photovoltaic systems. CPSS Trans Power Electron Appl 5(2):191–200
20. Chandrasekar B, et al (2020) Non-isolated high-gain triple port DC–DC buck–boost converter with positive output voltage for photovoltaic applications. IEEE Access 8:113649–113666
21. Chien L-J, Chen C-C, Chen J-F, Hsieh Y-P (2014) Novel three-port converter with high-voltage gain. IEEE Trans Power Electron 29(9):4693–4703
22. Mosa M, Shadmand MB, Balog RS, Rub HA (2017) Efficient maximum power point tracking using model predictive control for photovoltaic systems under dynamic weather condition. IET Renew Power Gener 11(11):1401–1409
23. Chatterjee A, Thakre K (2020) Investigation analysis of dual loop controller for grid integrated solar photovoltaic generation systems. In: Singh AK, Tripathy M (eds) Control applications in modern power systems. LNEE 710, pp 153–164
24. Chatterjee A, Mohanty K, Kommukuri VS, Thakre K (2016) Model predictive current controller for performance enhancement of grid-integrated single-phase photovoltaic distributed generation plants. Trans Inst Meas Control (SAGE Publication) 40(3):762–775

Severity of Delamination on Single-Phase Gas-Insulated Bus Duct

Vempalle Rafi, G. V. Nagesh Kumar, P. Meenakshi, S. Hussain Vali, A. Nagaraju, and P. Janaki

Abstract Currently, the breakdowns of Gas-Insulated Bus duct (GIB) are caused by a variety of issues resulting from various faults on the solid spacers. The influence of delamination on the disc style spacer for a single-phase GIB is examined in this work. The different permittivity values using functionally graded materials (F G M) are spatially scattered throughout the spacer profile to achieve consistent electrical current, which enhances the dielectric property of the insulators. The simulation is carried out on the spacer at various voltages and F G M gradings. The effect of delamination is later prevented by inserting the Metal Insert (MI) into the F G M insulator.

Keywords Gas-insulated bus duct · Functionally graded materials · Electrical field stress · Solid spacers · Metal insert

1 Introduction

Rising energy needs are placing a strain on both traditional and unconventional energy sources. For critical substations in areas vulnerable to hurricanes, environmental conditions, earthquakes, floods, wildlife, and other natural hazards, as well as physical attacks and criminal activity, conventional substations, which are based on predetermined air-insulated designs, are not the best option. Substation owners have been forced to consider design strategies to lessen the possibility of damage to their

V. Rafi (✉) · G. V. Nagesh Kumar · P. Meenakshi · S. H. Vali
Department of Electrical Engineering, JNTUA College of Engineering, Pulivendula, Andhra Pradesh 516390, India
e-mail: vempallerafi@gmail.com

A. Nagaraju
GITAM School of Technology, Bengaluru, India

P. Janaki
LENDI Institute of Engineering and Technology, Vizianagaram, India

J. Kumar et al. (eds.), *Recent Advances in Power Systems*, Lecture Notes in Electrical Engineering 1192, https://doi.org/10.1007/978-981-97-2788-9_6

facilities as a result of recent weather events like superstorms, widespread tornado outbreaks, coastal and inland river flooding, and intense thunderstorms followed by strong straight-line winds.

The popularity of the Gas-Insulated Bus duct (GIB) is increasing in this setting as a result of the fast rise in energy consumption and urbanization. In order to address the energy needs of modern society, numerous authors [1, 2] claim that gas-insulated transmission lines are a workable substitute for overhead transmission lines. These lines can be used in conjunction with pipeline technology in GIB. A limited group of researchers [3, 4] found that the density of the interior is mostly responsible for the high field stress in the GIB. Despite the GIB's history of excellent work, many issues remain unresolved. According to Matinyan A. Metal [5], one of the many causes of gas-insulated line circuit breakers failing is that using GIS equipment reduces the amount of labour that can be found on high electric field locations to switch, maintain, and repair high-voltage equipment. The failure of insulating insulators is the most prevalent GIB issue, according to a small group of authors [6–8], as it is the insulating connection with the least strength due to a variety of flaws, including protrusion, gap, depression, delamination, and so forth. So, it is necessary to employ a method for lessening the strain these breakdowns put on the electrical field. According to Misrikhanov examination M. Sh et al.'s of numerous SF6 and conventional isolation reliability metrics [9], the usage of gas-insulated switchgears is favourable for power system conditions.

This paper investigates the impact of stress under an F G M insulator in a disc-shaped insulator for a single-phase gas-insulated bus duct. To reduce stress at TJs and preserve a constant field across its surface, metal inserts have been placed to the enclosing end of the proposed F G M insulator. The triple connection was successfully insulated by the metal inlay, which also decreased the electric tension at TJ. This method of replacing the insulator made it simpler and less expensive thanks to MI. To duplicate a single-phase GIB, the conductors and housing must be distributed in accordance with the appropriate charge conservation [10–15]. The suggested model was then used to investigate the stress features at the enclosure end of the F G M insulator near the TJ. It is possible to map out each person's stress on the delaminated F G M insulator surface. Electric strain is examined at the enclosure end, which includes the disc-insulator surface, the conductor, and the triple junction. The three F G M grading methods (GL, GH, and GU) are uniformly but independently examined at the enclosing end of a disc insulator. Hence Conductor B is inserted into the F G M insulator with a metal insert at its enclosing end in order to maintain a consistent field stress at the insulator's critical/weak junction.

2 Governing Equation

The electric field intensity is determined by solving Laplace's and Poisson's equations for the boundary conditions. Other methods for calculating electric fields can be found in the literature. Because the analytical method is difficult, the computational method is used to determine the electric field intensity from these equations. Despite the availability of other numerical techniques, the finite element method (FEM) is used to address challenging issues.

The electric flux density $\vec{D}$ with Q_i as electrical charge can be expressed as Eq. (1) in a closed surface

$$\oint_S \vec{D} \cdot \overrightarrow{dS} = Q_i \quad (1)$$

where the electric field intensity is given by

$$\vec{E} = -\nabla V \quad (2)$$

According to divergence theorem,

$$\nabla \cdot \vec{D} = \rho_v \quad (3)$$

$$\nabla \cdot \varepsilon \vec{E} = \rho_v \quad (4)$$

$$\nabla \cdot \varepsilon(-\nabla V) = \rho_v \quad (5)$$

$$\nabla^2 V = \frac{-\rho_v}{\varepsilon} \quad (6)$$

Volume charge density is reduced when there is uniform charge distribution, and Eq. (6) is simplified to become Eq. (7)

$$\nabla^2 V = 0 \quad (7)$$

3 Results and Discussion

In this section, we use COMSOL to simulate the impact of delamination on a disc-type spacer. Single-phase GIB with delamination defect for MI-free and MI-containing disc insulator is shown in Figs. 1 and 2, respectively. Metal Inserts (MIs) of varying shapes and sizes were developed to mitigate this effect, and one MI that was found to be optimal had rectangle dimensions of 12 mm width and 1 mm height, and an ellipse with a-axis of 9.4 mm and b-axis of 1 mm, respectively. It was for this reason that this MI was developed to serve both the enclosure and conductor ends. Putting a metal insert on either end of the F G M disc insulator can also reduce the stress on the insulator. The proposed metal insert is shown in Fig. 2 with the aforementioned dimensions.

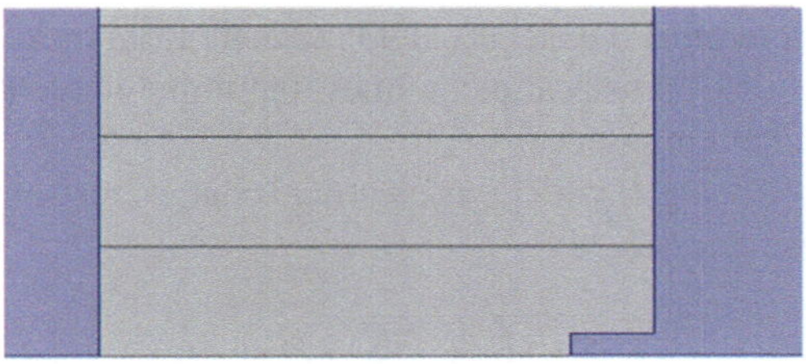

Fig. 1 Disc insulator incorporating delamination and excluding MI

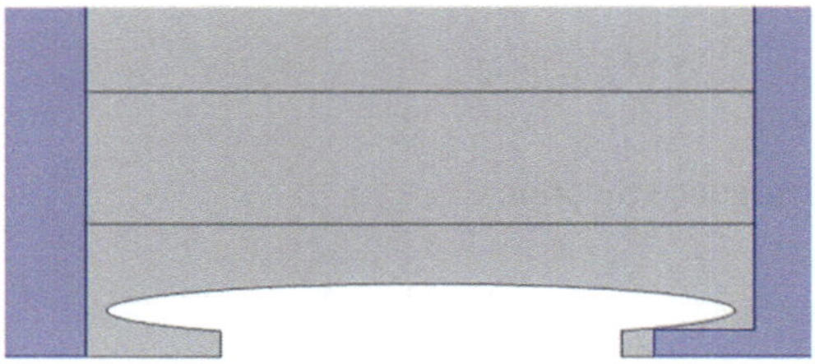

Fig. 2 Disc insulator incorporating delamination and including MI

The value of field stress at the enclosure, or at zero arc length, is higher than that at the conductor end, as illustrated in Fig. 3. The value of the stress is plainly visible in the enlarged version, as illustrated in Fig. 4. Using MI at the TJs reduces the stress at the enclosure caused by delamination, as demonstrated in Fig. 5. When compared to Fig. 4, the stress's amplitude is similarly reduced. The zoomed plot of Fig. 6 makes this obvious. Table 1 indicates the delamination effect for F G M disc insulator.

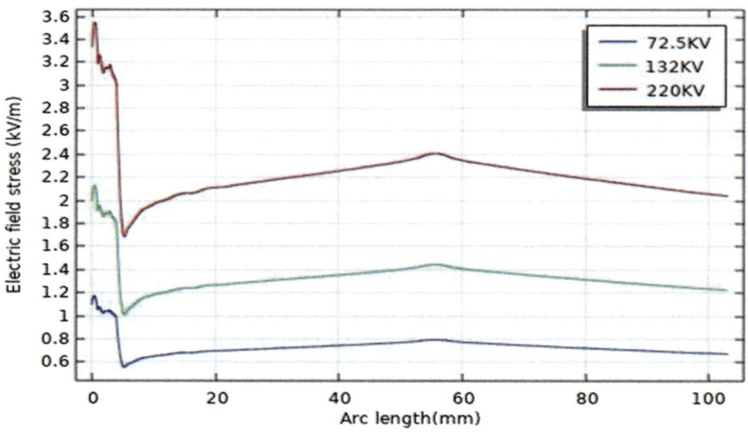

Fig. 3 Disc insulator stress diagram without MI

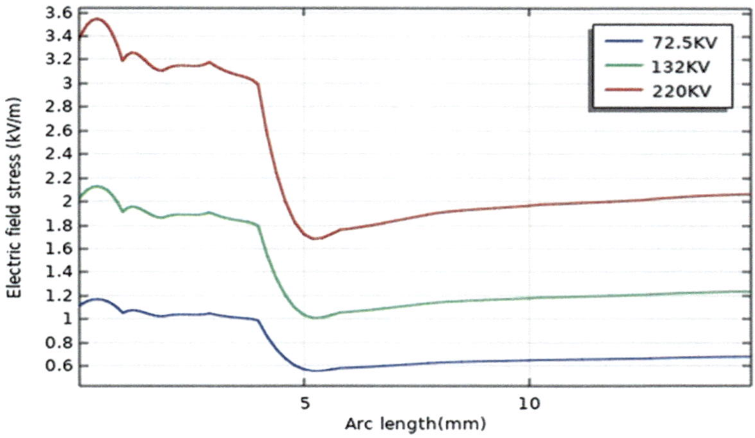

Fig. 4 Zoom plot of the disc insulator's stress map without MI

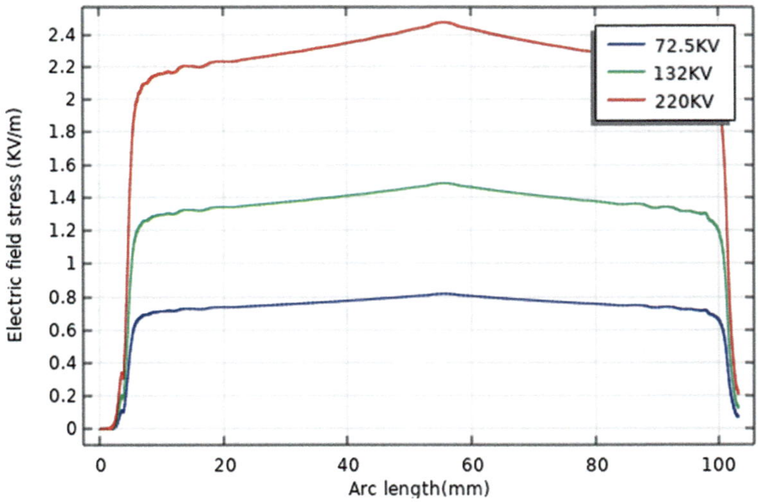

Fig. 5 Disc insulator stress plot with MI

The surface plots of the intended F G M disc insulator are shown in Fig. 7. Due to its high potential, the colour red indicates strong stress at the conductor end, whereas the colour dark blue indicates significant stress at the ends of the enclosure. When comparing Figs. 6 and 7, it can be seen that MI reduces the electric tension at the conductor end. Figures 8 and 10 show the corresponding zoomed surface plots of Figs. 7 and 9. Table 2 shows field stress reduction for F G M disc insulator with MI.

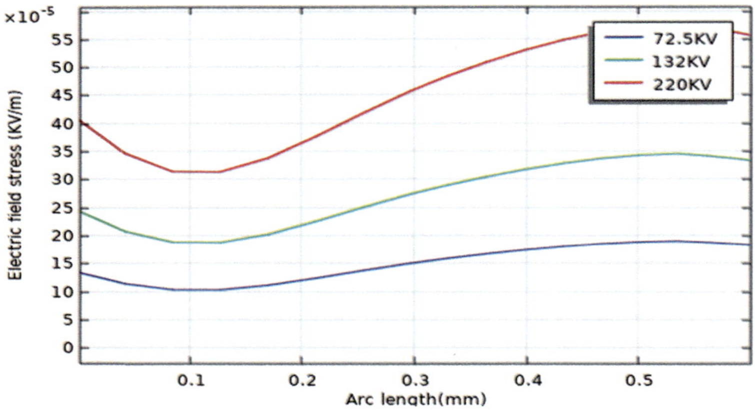

Fig. 6 Zoom plot of stress plot of disc insulator with MI

Table 1 Delamination effect for F G M disc insulator

Type	Applied voltage	Stress for 72.5 kV		Stress for 132 kV		Stress for 220 kV	
		Inner electrode end	Outer enclosure	Inner electrode end	Outer enclosure	Inner electrode end	Outer enclosure
GH-F G M	GH Case 1	1.0826	0.78	1.971	1.419	3.285	2.366
	GH Case 2	1.0896	0.7787	1.984	1.418	3.3065	2.363
	GH Case 3	1.095	0.7778	1.994	1.415 5	3.324	2.359
GL- F G M	GL Case 1	1.178	0.6785	2.1446	1.235 2	3.5742	2.058
	GL Case 2	1.1852	0.679	2.158	1.236 2	3.596	2.06
	GL Case 3	1.1922	0.6796	2.171	1.237 3	3.6177	2.0622
GU-F G M	GU Case 1	1.0842	0.6712	1.974	1.222	3.29	2.0368
	GU Case 2	1.0914	0.672	1.987	1.223 7	3.3117	2.0395
	GU Case 3	1.0983	0.673	1.9995	1.225	3.3328	2.042

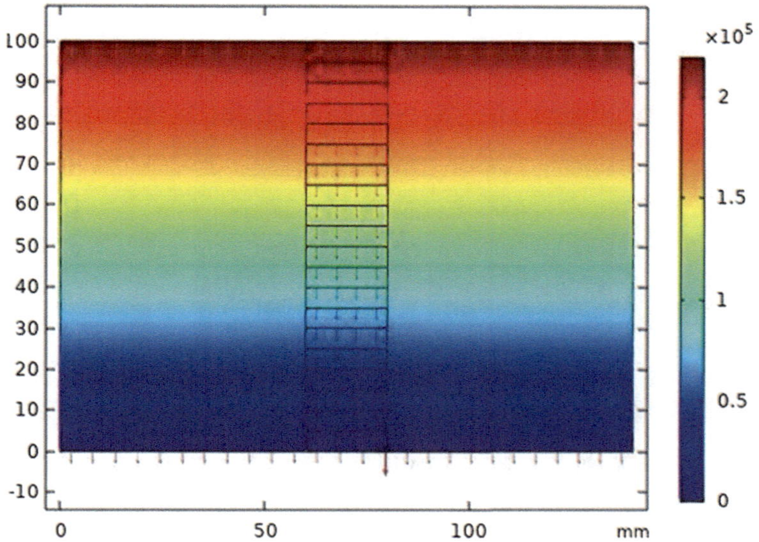

Fig. 7 Surface plot of a MI-free disc insulator

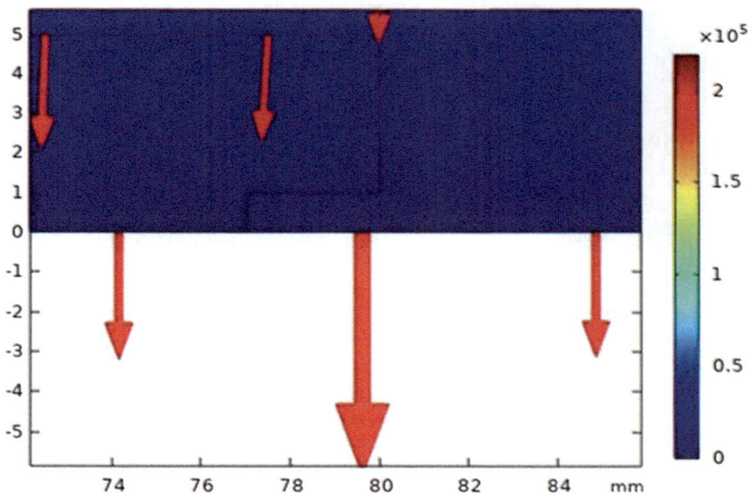

Fig. 8 Surface plot of a disc insulator without MI zoomed in

4 Conclusions

In this study, we explore the use of an MI at the TJ of the enclosure and conductor ends of the insulator to protect a single-phase GIB against delamination. The disc insulator is subjected to more stress at the enclosure end than the conductor end when

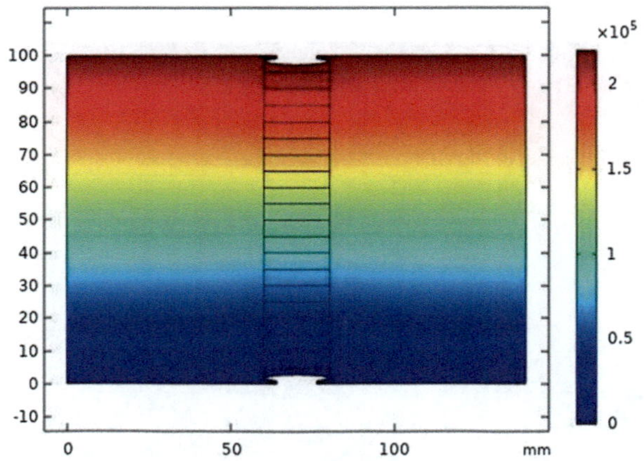

Fig. 9 Surface map of a disc insulator with MI

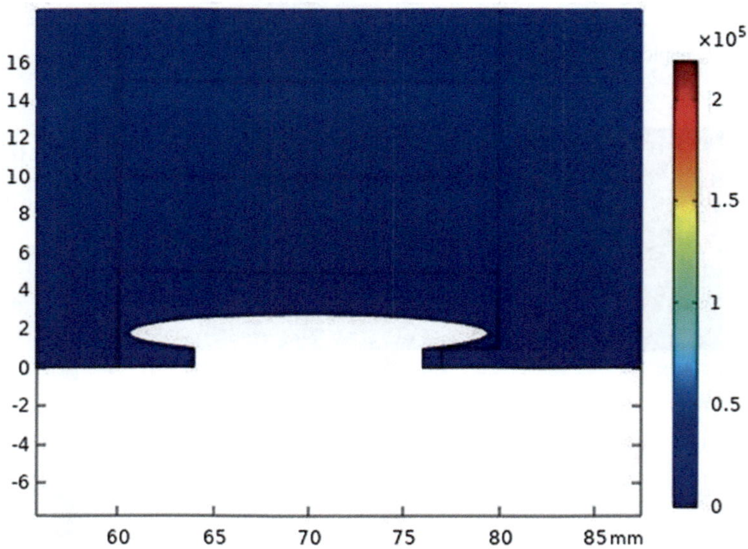

Fig. 10 A disc insulator surface plot with MI

operating voltages at TJs for the fault delamination are varied. According to GL-F G M grading, Case 3 exhibits higher stress at the TJs of the enclosure and conductor ends across all applied voltages than other cases with the delamination problem. The TJ of the enclosure end experiences the biggest reduction in stress across all F G M classes.

Table 2 Field stress reduction for F G M disc insulator with MI

Type	Applied voltage	Inner electrode end		Outer enclosure		Inner electrode end	
		Inner electrode end	Outer enclosure	Inner electrode end	Outer enclosure	Inner electrode end	Outer enclosure
GH-F G M	GH Case1	0.00014	0.0935	0.00025	0.17	0.000423	0.2835
	GH Case 2	0.000137	0.091	0.00025	0.167	0.000416	0.277
	GH Case 3	0.000135	0.0894	0.000246	0.163	0.00041	0.271
GL-F G M	GL Case1	0.000196	0.069	0.000357	0.1252	0.000595	0.2087
	GL Case 2	0.000191	0.0677	0.0003478	0.1232	0.00058	0.2056
	GL Case 3	0.000186	0.067	0.000339	0.1216	0.000565	0.2025
GU-F G M	GU Case1	0.00014	0.0689	0.0002545	0.125	0.000424	0.2089
	GU Case 2	0.000137	0.0678	0.00025	0.1236	0.000417	0.206
	GU Case 3	0.000135	0.0669	0.000246	0.1217	0.00041	0.203

References

1. Koch H, Schuette A (1998) Gas insulated transmission lines for high power transmission over long distances. Electr Power Syst Res 44(1):69–74
2. Vikharev AP (2019) Transmission capacity of gas-insulated power lines laid in earthen trenches. Power Technol Eng 53(1):101–105
3. Li S, Zhong L, Li J, Chen G (2010) A brief history and research progress on solid engineering dielectrics in china. IEEE Electr Insul Mag 26(6):14–21
4. Farag AS, Cheng TC, Hu L, Penn D, Thompson J (1997) Study for optimal design of gas insulated substations (GIS) to reduce magnetic fields. Int J Electr Power Energy Syst 19(3):185–194
5. Matinyan AM, Peshkov MV, Karpov VN, Alekseev NA, Pugachenko ZE (2017) TCSR features facilitating safe testing of 500 kV power transmission lines. Power Technol Eng 51(2):226–229
6. Adari J, Krishna Rao Gadi VS, Gundavarapu VNK (2018) Mitigation of field stress with metal inserts for cone type insulator in a gas insulated busduct under delamination. Eng Sci Technol Int J 21:850-861
7. Tanaka T, Imai T (2013) Advances in nano dielectric materials over the past 50 years. IEEE Electr Insul Mag 1(29):10–23
8. Tanaka T (2005) Dielectric nano composites with insulating properties. IEEE Trans Dielectr Electr Insul 12(5):914–928
9. Misrikhanov MSh, Mozgalev KV, Shuntov AV (2003) Reliability of gas insulated substations and switchgears with traditional insulation. Power Technol Eng 37(6):377–383
10. Nagarsheth R, Singh S (2014) Study of gas insulated substation and its comparison with air insulated substation. Int J Electr Power Energy Syst 55:481–485

11. Rafi V, Dhal PK (2022) Optimal analysis of time varying load radial distribution system with photovoltaic and wind generating system using novel hybrid optimization technique. Renew Ener Focus 41:246–257
12. Rafi V, Dhal PK (2020 Aug) Optimal placement of distributed generators in optimized reconfigure radial distribution network using PSO-DA optimization algorithm. In: 2020 International Conference on Advances in Computing, Communication & Materials (ICACCM), IEEE, pp. 239–246
13. Rafi V, Dhal PK, Rajesh M, Srinivasan DR, Chandrashekhar M, Madhava Reddy N (2023) Optimal placement of time-varying distributed generators by using crow search and black widow–Hybrid optimization. Meas: Sens 30. https://doi.org/10.1016/j.measen.2023.100900
14. Rafi V, Dhal PK (2020) Maximization savings in distribution networks with optimal location of type-I distributed generator along with reconfiguration using PSO-DA optimization techniques. Mater Today: Proceedings 33:4094–4100
15. Rafi V, Dhal PK (2020) Loss minimization by reconfiguration along with distributed generator placement at radial distribution system with hybrid optimization techniques. Technol Econ Smart Grids Sustain Ener 5. https://doi.org/10.1007/s40866-020-00088-2

Load Forecasting with LSTM and Loss of Life Estimation of Distribution Transformer Considering the Impacts of Electric Vehicle

Lekshmi R. Chandran, Nachiket V. Padwal, Manjula G. Nair, and K. Ilango

Abstract Electric vehicle is becoming an extra load on the existing power grid. Integration of electric vehicle in electrical network increases power quality issues and deterioration of electrical assets like transformer. The increase in load on the electrical system and its impact on electrical assets need to be analyzed properly. Hence, this work focuses on LSTM—a deep learning ANN model-based load forecasting and loss of life (LoL) estimation considering the impact of electric vehicles (EV). The load forecast RMSE for LSTM model is obtained as 0.72411.

Keywords Electric vehicles · Distribution transformers · Load forecasting · LSTM · LoL

Abbreviations

LSTM	Long Short-Term Memory
LoL	Loss of Life
ANN	Artificial Neural Networks
EV	Electric Vehicle
BEV	Battery Electric Vehicle

L. R. Chandran (✉) · N. V. Padwal · M. G. Nair
Department of Electrical and Electronics Engineering, Amrita Vishwa Vidyapeetham, Amritapuri, India
e-mail: lekshmichandran@am.amrita.edu

N. V. Padwal
e-mail: amenp2pre21009@am.students.amrita.edu

M. G. Nair
e-mail: manjulagnair@am.amrita.edu

K. Ilango
Department of Electrical and Electronics Engineering, Amrita School of Engineering, Amrita Vishwa Vidyapeetham, Coimbatore, India
e-mail: k_ilango@cb.amrita.edu

J. Kumar et al. (eds.), *Recent Advances in Power Systems*, Lecture Notes in Electrical Engineering 1192, https://doi.org/10.1007/978-981-97-2788-9_7

HEV	Hybrid Electric Vehicle
ICE	Internal Combustion Engine
RES	Renewable Energy Sources
DT	Distribution Transformer
GHG	Green House Gases
PV	Photovoltaic

1 Introduction

Electric vehicles are growing in popularity as well as in number in recent years because of rising concerns about the adverse effects of climate change on ecology. Electric vehicles are known for their low emissions than their ICE counterparts even with non-renewable power sources. When combined with RES charging solutions, they theoretically produce zero emissions. There have been many research studies in the field of electric vehicles. They highlight the demand boost for this technology in the near future. Simultaneously, the EV market and industry are also growing with the help of government policies and more concern for the environment. Another aspect is the continual rise in the costs of oil prices and innovation leading to a decline in the production cost of EV batteries. All these factors make EV adoption more beneficial and economical. It is expected that around 30% of vehicles on the roads will be EVs by 2030. Even the International Energy Association has proposed 36% EV penetration by 2030 as a Sustainable Development Scenario goal [1].

In the field of EVs, the majority of research takes place in the fields of EV improvements like energy storage and safety and reliability improvements. Other research scenarios are about EV penetration from a consumer perspective. Yet, it is also equally important to study the case from a grid perspective. One bigger problem with EV adoption is its adverse effects on the grid infrastructure. The inevitable integration of EVs in the current transportation system is going to have its impact on the grid and these problems need to be addressed. Studies show that simultaneous use of charging infrastructure for EVs during peak load hours overloads the transformers. Distribution Transformers (DTs) are the devices used for stepping down high-voltage power from the grid to low voltage for residential applications. As a solution to this problem, the charging can be shifted to off-peak hours. But the problem with this solution is that it does not allow time for transformers to cool down before the next cycle [2]. DTs are capital-intensive investments. Also, while replacing them, they have to be disconnected which has technical as well as financial difficulties. Thus, to cause less harm to distribution transformer health, it is imperative that necessary preparations are considered to accommodate the charging load of EVs.

As there is growth in EV adoption, time calls for study in this field. DT life deterioration due to EV charging and additional load demand estimation caused by EV charging and grids' ability to accommodate it are a few of the areas of concern.

Also, for safe operation of DTs, EV load demand management and EV charging efficiency should be investigated.

An uninterrupted and reliable power supply is an important aspect of power distribution from the perspective of both the utility and the end user. Unreliable power can have harmful effects on the grid equipment, and it is undesired. To avoid this, load forecasting is an essential method. It is important to consider planning and operation from the utility side. Apart from this, it is also required for maintenance scheduling, outage planning, prospective energy contracts, unit commitment, and economic load dispatch. Load forecasting is one of the main areas of study in power systems. Since we cannot store electrical energy, the generation of energy should match the demand from the energy market. Not matching the generation and demand can have devastating effects on energy-generating units. Thus, we need to have an idea about the amount of energy to be generated.

The structure of the remaining article is as follows. Section 2 is comprised of work done previously, describing the impact of the effects of EVs on distribution transformers. Section 3 describes the LSTM model based on load forecasting. Section 4 explains the mathematical modeling of the transformer for loss of life estimation. Results and findings are discussed in Sect. 5. Section 6 discusses the conclusion. Section 7 is the reference materials studied for this work.

2 Literature Review

2.1 Climate Change and Electric Vehicles

Due to the rising concerns about climate change, reduction in GHG emissions has never been important as it is today. Earth's temperatures are rising by 0.14 °C since the 1880s and by 0.32 °C since the last 40 years. While 2020 was recorded as the second warmest year [3]. Looking at significant share of mobility sector in global emissions, it seems logical that more sustainable and efficient mobility solutions are needed of the hour. Transportation sector accounts for 24% of total global emissions worldwide. Whereas road vehicles cause 2/3rds of the transportation emissions [4]. Figure 1 shows comparative data for the year 2016. In such scenario, electric vehicles seem a promising solution for road transport, with its lower carbon footprint than its ICE counterparts. As a result, EV sales in 2019 were at 2.1 million globally and market shows 40% year on year growth. It is expected that by 2030 global share of EVs will be around 36%, which is also sustainable development scenario goal [1]. A detailed survey to estimate trends in EV mobility sector is carried out using database from German Aerospace Centre called Vehicle Concept Database that studies HEV and BEVs [5].

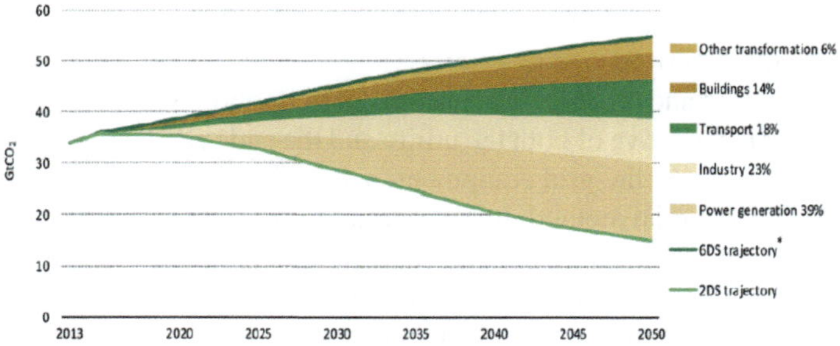

Fig. 1 GHG emissions by sector. *Source* International Energy Agency (2016)

2.2 *Problem with EV Charging*

EVs are promising in terms of reduction in emissions, showing emissions about 11–50% less than their ICV counterparts. Despite this advantage, there are certain challenges that hinder the path to mass adoption of EVs [6]. Yet, high penetration of electric vehicles (EVs) has its own challenges. EVs contain battery packs that have higher energy density, and they need to be charged frequently. This is done at low voltage distribution grid. These pose severe problems to grid assets such as transformers which were designed and implemented in pre-EV era [7]. High power demands posed by EVs put power grids under pressure. Critical grid infrastructure such as power distribution transformers (DTs) are more vulnerable in such cases. High power demand consisting of high-capacity batteries as well as fast charging speeds, hysteresis losses, ever increasing consumer and industrial power demands and rising temperatures are causing damage to the DTs. The main problems are in terms of higher temperatures and hence deterioration of transformer cooling oil. Studies show that LoL is affected when transformers overload at certain level of EV penetration. After certain level, accelerated deterioration of oil is expected [8].

2.3 *State of the Art: Loss of Life Improvement of Distribution Transformer*

To reduce stresses on the Power Distribution Transformers, there are few methods that are mainly discussed and studied. One very common method is implementing more advanced equipment that is capable of handling and processing more data, which is necessary for status monitoring of transformers. This adds augmentation costs as these equipment are relatively expensive. Another method is integration of charging system with the Renewable Energy Sources (RES). This will form a distributed generation system that will allow EVs to charge on the energy produced by RES locally. This will disconnect them from the grid and reduce their impact on the transformers. The third method is about scheduling of the EV load. It proposes the idea of scheduling loads throughout the day such that they reduce overheating of transformers in turn reducing their LoL. There are several ways of doing this at utility level, either by strict scheduling or developing customer behavior through incentives.

A study in [9] discusses the thermal modeling of distribution transformer. They propose is simpler thermal model than more complicated IEEE Annex G model which tracks heating in distribution transformers. For the service transformer fleet, this model allows accurate as well as computationally inexpensive description, says conclusion. Another thermal modeling is shown in [10], which is solved using Differential and Exponential methods. It is concluded that exponential methods perform better than differential methods. The idea of using Smart Monitoring Systems, Smart Controlling Systems, Big Data algorithms is proposed in [11]. For better condition monitoring and integration with RES, the more data we have, the better it is for predictions and better RES integration. But this will add augmentation costs for highly accurate sensors and data storage devices.

A Health Index-based review and case study is conducted in [12]. Two different health indices are considered and for a fleet of transformers, and reliability of two indices is evaluated. A probabilistic health index-based method is proposed in similar work that estimated apparent transformer age. A Bayesian belief network is constructed that calculates probabilistic health index. It uses data from condition monitoring of transformer. This method has three models, viz., an end-of-life model based on statistics, physical winding degradation model and condition monitoring data-based health indexing model [13]. Similar work is shown in [14]. This article presents a new method that determines health index, breakdown probability and remaining useful life of a transformer. This method has three models, viz., winding degradation model, health index model and end of life model. Health index model is based on condition monitoring and end of life model is based on statistics. End of life model has used data of scrapped transformers in Norway. Combining all three models, probability of breakdown is condition dependent as well as individual. Furthermore, maintenance and replacement optimization are solved using stochasticity of this method [14]. Yet, when there is less data available for prediction or available data in intermittent, then the results gained might not be accurate. To overcome this and deal with the uncertainty of data for assessment of health index [15] investigates a transformer insulation health index model. Random Forest method is used for prediction of data using oil interfacial tension (IFT). The certainty model is designed in order to accommodate the predicted values while also considering prediction accuracy.

Several studies show the use of Machine Learning algorithms and Optimization methods to reduce the damage to the transformer. Various ML algorithms are used to classify the faults occurring in transformers in [16]. A 2KVA transformer is considered for study for which best suitable ML method is found out. One study particularly focuses on health management of dry-type transformers. This study proposed a method that evaluates health score of a transformer by determining earlier unhealthy behaviors. The study uses dynamic-weighted-feedback broad learning system (DW-FB-BLS) method [16]. Jose Ignacio Aizpurua et al. [17] propose an approach for condition assessment of transformer, which integrates model-based experimental models, forecasting models that are data driven and uncertainty modelling. This increases accuracy of prediction and better handling of uncertainty. This approach is using XGB (Extreme Gradient Boosting) algorithm. Advantage of this is that XGB improves accuracy of prediction and captures thermal model nonlinearities best among the various forecasting models. Co-optimization algorithm is used for reduction of the effects of PEV charging on power transformer, which also considers transformer operating costs [18]. Another study that uses multi-optimization is shown in [19]. Here EV charging costs and emission costs are considered as objective function. A model of EV charging was developed that considers various parameters such as Vehicle-to-Grid, degradation of EV battery and transformer capacity. This study is implemented at Utrecht, the Netherlands. The study observes that in majority of the cases, EV charging benefits in terms of emissions or extra costs under transformer of high capacity don't outweigh emissions as well as costs of transformer upgrades [19].

Integration of RES is also one way by which we can reduce the EV charging impact on distribution transformer. In [20], PEV charging is considered in the presence of photovoltaic solar roof. To address the solar PV uncertainties caused from solar irradiance and temperature and probabilistic aspect emulation of PEV charging, a Monte Carlo simulation method is proposed. This work considers about 20 different scenarios of PEV and solar PV penetration of different levels [20]. Another study mentioned in [21] focuses on determining mains charging points for EV charging in RES charging in low voltage networks. Multi-objective optimization is used. It is also considered that when EVs are not charging, charging station can generate energy which can be fed to external grid in to smooth out the load curve.

Another way is maintenance scheduling and demand response management. A study focusing on this approach is shown in [22]. Particle Swarm Optimization algorithm is used to solve an optimization problem and determine the best strategy for maintenance. The method establishes maintenance decision-making model considering economy and reliability for oil-filled transformers. Study conducted in [2] shows that accelerated aging of distribution transformers can be avoided without the need for grid augmentation. Here, demand response-based thermal modeling is done. This model is solved as an optimization problem. It proposes a strategy that uses Time of Use (ToU) tariff, which is an incentive approach that shapes consumer behavior for EV charging considering thermal loading of power transformer.

Based on the study of these various research studies and experiments, we have analyzed that grid augmentation is a difficult option as it has high investment costs. With the integration of RES and proper load forecasting and scheduling, the adverse effects of EV charging on power transformers can be reduced up to a certain level of EV penetration.

3 LSTM

Load forecasting is usually divided into three types based on duration as shown in Fig. 2. These are long-term, medium-term and short-term forecasting [23]. Since this work relates to load scheduling and maintenance scheduling of distribution transformers, we will focus on short-term scheduling of the load.

LSTM stands for Long Short-Term Memory. The LSTM is part of Recurrent Neural Networks (RNNs). LSTM contains memory blocks, which are special units in hidden layers of the neural network. These hidden blocks have hidden cells that store the temporal state of a network. It also consists of gates which control the flow of information. They are special multiplicative units. Gates are of three types, viz., Input gate, Output gate and the Forget gate. The job of input gate is to control the flow of input activation into the memory cell. The output gate has to control the output flow of cell activations into the rest part of a network. The job of a forget gate is to scale the internal state of the cell and then add it as an input to the cell through the self-recurrent connection of the cell. This allows a cell to adaptively forget and reset the memory of a cell. In addition, the modern LSTM architectures contain peephole connections. They stretch from architecture's internal cells to the gates within the same cell to learn about the precise timing of any output [24].

Authors in [24] similar approach of using LSTM for load forecasting. According to [24], the load forecasting models used are based on point forecasting. They can be subcategorized as statistical, AI-based and hybrid models. But the focus of these models, i.e., point forecasting is not enough while considering the fluctuations and uncertainties in the load demand. In such cases, probabilistic models should be used. This is where LSTM comes into the picture.

A comparison work is done on three neural networks in [25] for load forecasting. Authors have compared Recurrent Neural Networks (RNN), Gated Recurrent Unit (GRU) and LSTM. All these algorithms are fed with the hourly energy consumption data to train the models. Authors conclude that out of these three methods, LSTM is the most accurate one. LSTM is found to have the least Mean Absolute Error

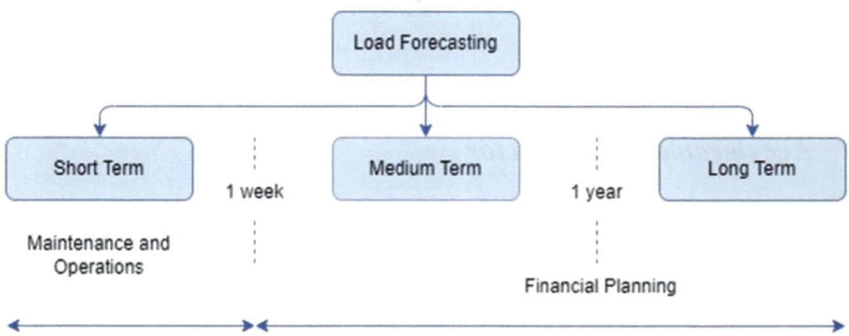

Fig. 2 Load forecasting classification

(MAE), Mean Absolute Percentage Error (MAPE) and Root Mean Square Error (RMSE) [25].

4 Mathematical Modeling of Distribution Transformer

Equations for thermal aging of transformer from the IEEE Std. C57.91 are used for dynamic modeling of Distribution Transformer.

4.1 Loss of Life

The Loss of Life (LoL) of a transformer is calculated using Eq. (1).

$$Percentage\ LoL = \frac{F_{EQA}.T.100}{L_{\text{normal}}} \quad (1)$$

where,

T is the number of time slots.

L_{normal} is the nameplate insulation life of a transformer at rated load.

4.2 Average Aging Factor

The average aging factor FEQA mentioned in Eq. (1) is given as in Eq. (2).

$$F_{EQA} = \frac{\sum_{n=1}^{N} F_{AA} \cdot n \cdot \Delta t_n}{\sum_{n=1}^{N} \Delta t_n} \quad (2)$$

4.3 Accelerated Aging Factor

For each time step, the accelerated aging factor, FAA is given by the Eq. (3).

$$F_{AA} = e^{\frac{15000}{383} - \frac{15000}{\theta_H + 273}} \quad (3)$$

The use of FAA is for the determination of lifetime degradation of a transformer. This is an empirical relation, which is dependent on the hot-spot temperature of a transformer. FAA is greater than 1 when hot-spot temperature is found to be higher than the reference value of hot-spot temperature, and FAA is less than 1 when hot-spot temperature is found to be less than the reference value of hot-spot temperature.

4.4 Hot-Spot Temperature

Transformer accelerated aging factor is dependent on hot-spot temperature as mentioned in Eq. (3) and is given by Eq. (4).

$$\theta_H = \theta_A + \Delta\theta_{TO} + \Delta\theta_g \tag{4}$$

Last two terms, viz., $\Delta\theta_{TO}$ and $\Delta\theta_g$ in above relation depend upon the machine rated data and various losses inside the transformer (ohmic losses PDC, load losses PLL, eddy current losses PEC, other stray losses POSL and no load PNL losses).

4.5 Losses

Equation (5) and (6) give the values of top oil temperature rise with respect to the ambient temperature and conductor temperature rise relative to top oil temperature respectively.

$$\Delta\theta_{TO} = \Delta\theta_{TO-R}\left(\frac{P_{LL} + P_{NL}}{P_{LL-R} + P_{NL}}\right)^{0.8} \tag{5}$$

And

$$\Delta\theta_g = \Delta\theta_{g-R}\left(\frac{P_{DC} + P_{EC}}{P_{DC-R} + P_{EC-R}}\right)^{0.8} \tag{6}$$

where, $\Delta\theta_{TO-R}$ is the rated oil temperature rise relative to the ambient temperature, $\Delta\theta_{g-R}$ is the rated conductor temperature rise with respect to top oil temperature and P_{DC-R}, P_{LL-R}, P_{EC-R} are the rated ohmic losses, load losses and eddy current losses respectively.

The load factor β which can be computed as RMS current I with respect to transformer sinusoidal current, I_{R2} at low voltage winding. Harmonic Loss Factor (F_{HL}) and Harmonic loss factor for other stray losses ($F_{HL\text{-}STR}$) is used to calculate the load losses or operating losses of transformer. They are mentioned in Eqs. (7), (8), (9) and (10).

$$P_{LL} = P_{DC} + P_{EC} + P_{OSL} \tag{7}$$

where,

$$P_{DC} = \beta^2 P_{DC-R} \tag{8}$$

$$P_{EC} = \beta^2 P_{EC-R} F_{HL} \tag{9}$$

$$P_{OSL} = \beta^2 P_{OSL-R} F_{HL-STR} \tag{10}$$

With,

$$F_{HL} = \frac{1}{I^2} \sum_{k=0}^{N} k^2 I_k^2 \tag{11}$$

$$F_{HL-STR} = \frac{1}{I^2} \sum_{k=0}^{N} k^{0.8} I_k^2 \tag{12}$$

$$\beta = \frac{S}{S_R} \cong \frac{I}{I_{R2}} \tag{13}$$

$$I^2 = \sum_{k=0}^{N} I_k^2 \tag{14}$$

where, I_k is the effective value of the kth current harmonic, N is the highest harmonic order, S is the apparent power and SR represents the transformer rated power under normal (sinusoidal) condition.

The rated value of winding eddy current losses $P_{EC\text{-}R}$ and rated value of other stray losses $P_{OSL\text{-}R}$ are estimated in accordance with the international standard recommendations, explained as in Eqs. (15) and (16).

$$P_{EC-R} = 0.33(P_{LL-R} - P_{DC-R}) \tag{15}$$

$$P_{OSL-R} = 0.67(P_{LL-R} - P_{DC-R}) \tag{16}$$

Moreover, the rated ohmic losses of transformer ($P_{DC\text{-}R}$) are computed from the values of primary and secondary rated currents (I_{R1}, I_{R2}) and the DC ohmic resistances of the windings indicated by the manufacturer (R1, R2), given by Eq. (17).

$$P_{DC-R} = 3\left(R_1 I_{R1}^2 + R_2 I_{R2}^2\right) \tag{17}$$

The no-load losses (P_{NL}) of transformer are invariable with respect to the current harmonic spectrum as in case of the actual industrial power distribution system, the supply voltage waveforms are balanced and sinusoidal.

5 Results and Discussions

Figure 3 shows the methodology followed in the study. The methodology consists of three stages.

Stage 1: Data collection regarding the ambient temperature, EV load and Residential load.

Stage 2: Load forecasting using LSTM.

Stage 3: Analysis of Loss of life of transformer for present and forecasted load considering the absence and presence of harmonics.

Load forecasting is essential for predicting the load on transformers. Loss of life estimation of the distribution transformer is mandatory for ensuring the reliability of the asset. Load forecasting is done using the LSTM method due to its ability to handle uncertainty factor during prediction and LoL estimation is carried out as per the modeling mentioned in section IV. The distribution transformer with parameters shown in Table 1 has been considered for mathematical modeling.

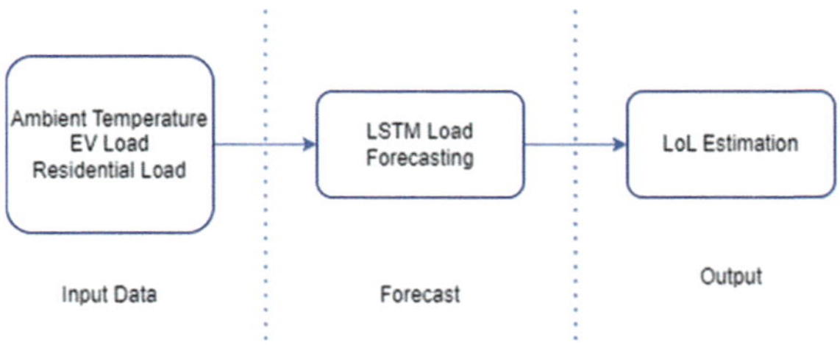

Fig. 3 LSTM-based load forecasting and LoL estimation

Table 1 Transformer parameters (AS PER IEC60076-7)

Parameter	Value
Rated power	500 KVA
Frequency	50 Hz
Type	ONAN
Primary voltage	11 kV
Secondary voltage	0.4 kV
No load losses	0.68 KW
On load losses	5.15 KW
Winding material	Copper
TOT rise over ambient temperature	55 °C
HST rise over TOT	23 °C
Oil time constant	180
Winding time constant	4
Oil exponent	0.8
Winding exponent	1.6
Nominal life	40 years

One-year averaged temperature and load power data per month per hour data points of a distribution transformer are considered for this study. For EV load data, two EVs are considered with different battery capacities. Ambient temperature and both load profiles are shown in Figs. 4, 5 and 6.

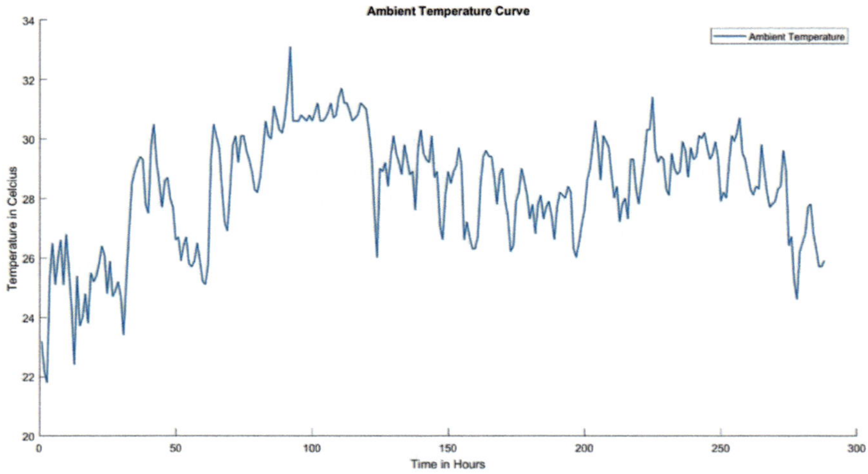

Fig. 4 Ambient temperature curve

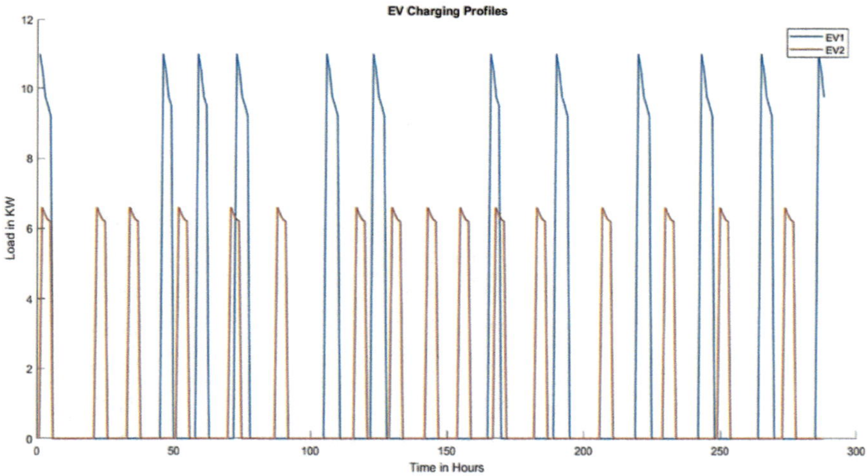

Fig. 5 EV charging profiles

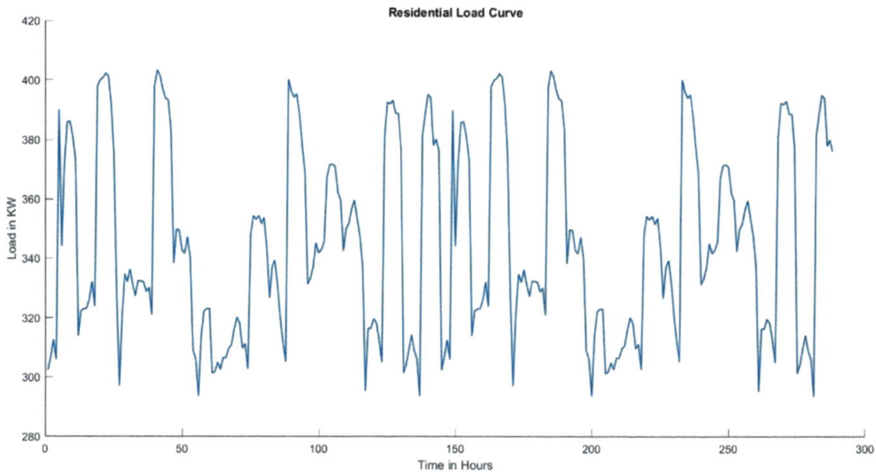

Fig. 6 Residential load curve

5.1 Load Forecasting

As explained in chapter III, LSTM is used to forecast the load on a transformer for a specific time frame. Training data sets are normalized to ensure reliable forecasting outcomes of the model. 400 iterations are done to train the model. The normalized data is fed to LSTM model with 1000 hidden neurons. The final evaluation indicator is root mean square error (RMSE) as shown in Eq. (19).

$$MSE = \frac{1}{T^2} \sum_{t=T_1+1}^{T_1+T_2} e_t^2 \tag{18}$$

$$RMSE = \sqrt{MSE} \tag{19}$$

Figure 7 shows the LSTM training RMSE and loss plot. The training model is converging with loss factor of 0.35 after 400 iterations. Based on this training, the model predicts the load for remaining data points with RMSE of 0.72411. Figure 8 shows the training and predicted load of transformer.

5.2 Loss of Life Estimation of Distribution Transformer

Figure 9 shows the total transformer load for various loads including residential, EV considering harmonics.

Figures 10 and 11 plot the critical top oil temperature and the winding hot spot temperature with respect to time for load without harmonics and with harmonics,

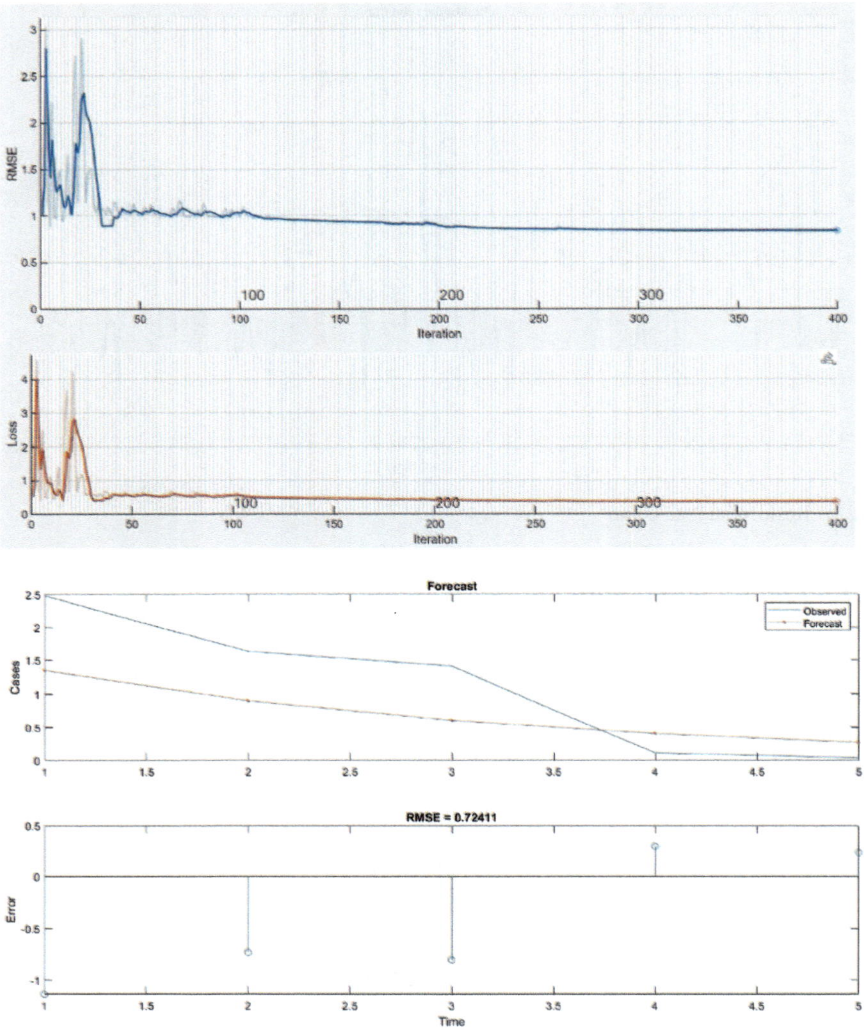

Fig. 7 LSTM training RMSE and loss

respectively, estimated using modeling equations as discussed in Sect. 4. For case without harmonics, maximum top oil temperature is found to be 54.58 °C and maximum hot spot temperature is 66 °C. In case of harmonics, maximum top oil temperature is 72.5 °C and maximum hot spot temperature is 95.9 °C.

Furthermore, we are also calculating the accelerated aging factor and average aging factor of the transformer considering the harmonic losses. Figures 12 and 13 show how accelerated aging factor varies with time for two cases of harmonics. It is seen that with the rise in top oil temperature and winding hot spot temperature, the accelerated aging factor of distribution transformer also increases.

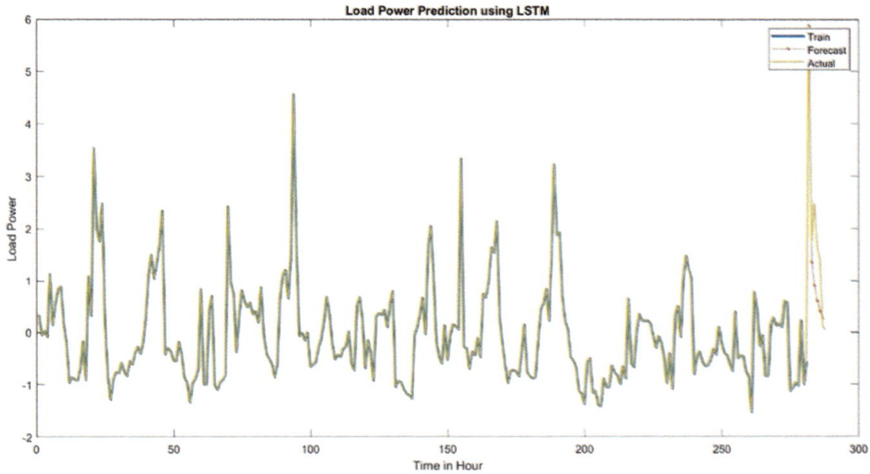

Fig. 8 LSTM training forecasting of transformer load

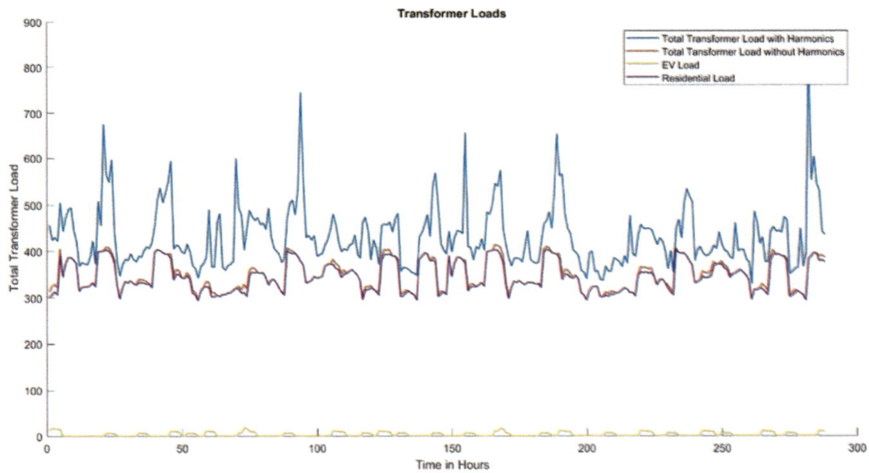

Fig. 9 Total transformer load

For without harmonics case, Average Aging Factor is found to be 0.0058, whereas, for with harmonics case, it is found to be 0.1415. Based on the values of the accelerated aging factor and average aging factor, the loss of life (LoL) of the transformer for load without harmonics is 0.00048/year and with harmonics is 0.0118/year.

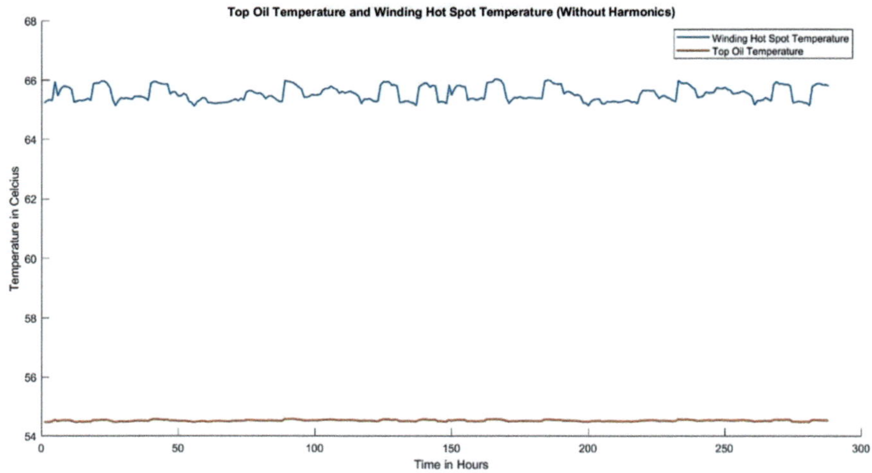

Fig. 10 Top oil temperature and winding hot spot temperature curve (without harmonics)

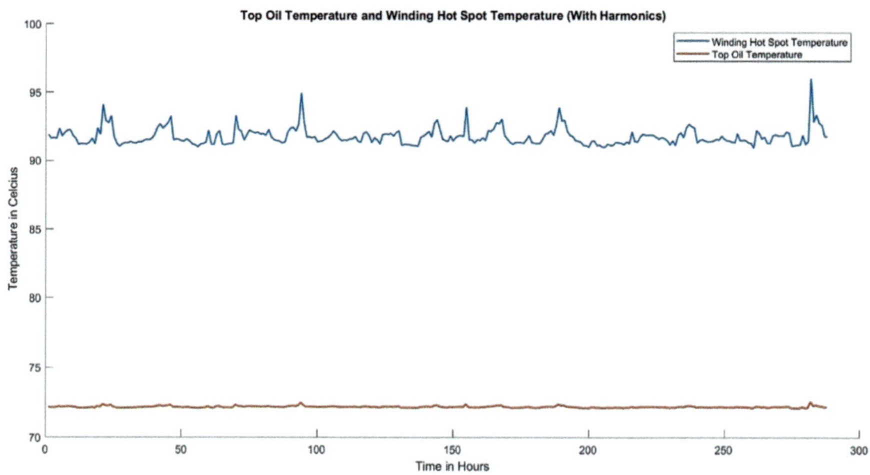

Fig. 11 Top oil temperature and winding hot spot temperature curve (with harmonics)

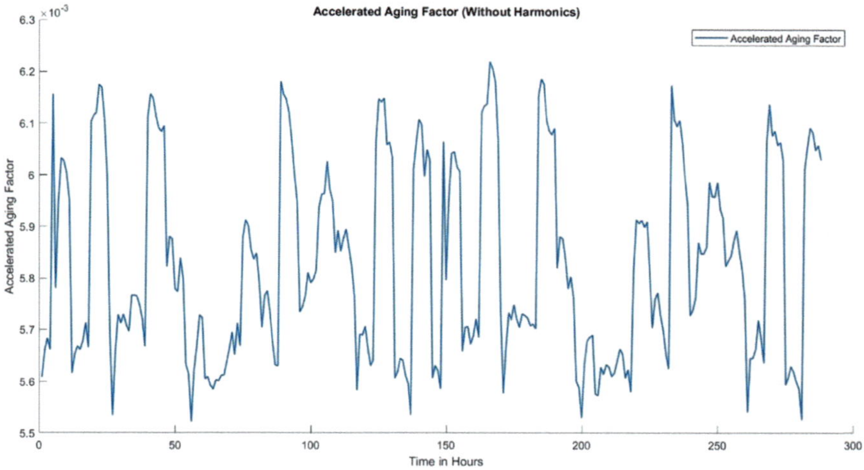

Fig. 12 Accelerated aging factor (without harmonics)

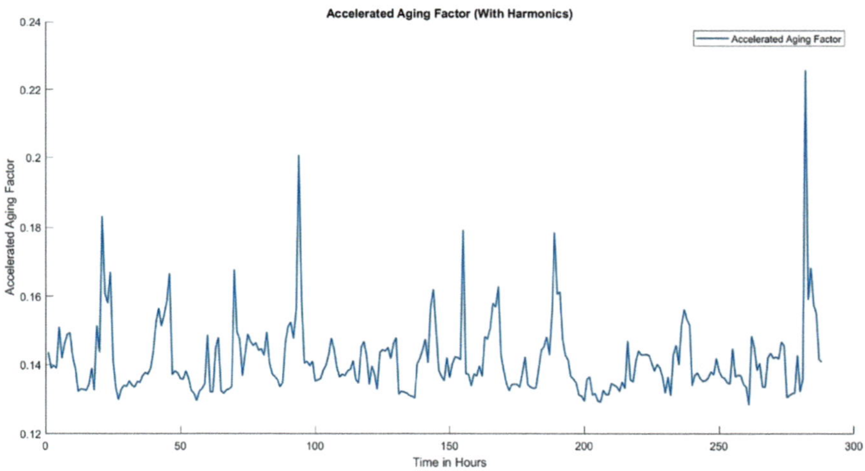

Fig. 13 Accelerated aging factor (with harmonics)

6 Conclusion

This study emphasizes the importance of load forecasting and loss of life estimation of distribution transformer. It is found that the ambient temperature and power demand affect the percentage loss of life of distribution transformer as it causes top oil temperature and winding hot spot temperature to rise. This results in an increase in accelerated aging factor and average aging factor. The higher temperatures and load result in an increase in the percentage loss of life of distribution transformers.

The loss of life of distribution transformer is found to be 0.00048/year and with harmonics is 0.0118/year for this study. Moreover, the LSTM model is used to forecast the load on the transformer with RMSE 0.7241. This enables us to predict the future loads and schedule the maintenance as well as the loads accordingly.

References

1. IEA (2020) Tracking electric vehicles report
2. Pradhan P, Ahmad I, Habibi D, Kothapalli G, Masoum MAS (2020) Reducing the impacts of electric vehicle charging on power distribution transformers. IEEE Access 8
3. NCEI (2022) Global climate report, May
4. IEA (2020) Tracking transport report
5. Frieske B, Kloetzke M, Mauser F. Trends in vehicle concept and key technology development for hybrid and battery electric vehicles. In: EVS27 international battery, hybrid and fuel cell electric vehicle symposium
6. Franzò S, Nasca A (2020) The environmental impact of electric vehicles: a comparative LCA-based evaluation framework and its application to the Italian context. In: 2020 fifteenth international conference on ecological vehicles and renewable energies (EVER)
7. van den Berg MA, Lampropoulos I, AlSkaif TA (2021) Impact of electric vehicles charging demand on distribution transformers in an office area and determination of flexibility potential. Sustain Energy, Grids Netw 26:100452
8. Godina R, Rodrigues EMG, Paterakis NG, Erdinc O, Catalão JPS (2016) Innovative impact assessment of electric vehicles charging loads on distribution transformers using real data. Energy Convers Manag 120:206–216
9. Seier A, Hines PDH, Frolik J (2015) Data-driven thermal modeling of residential service transformers. IEEE Trans Smart Grid 6(2)
10. Chandran LR, Ajith Babu GS, Nair MG, Ilango K (2021) A review on status monitoring techniques of transformer and a case study on loss of life calculation of distribution transformers. Mater Today: Proc 46:4659–4666
11. Suwanapingkarl P, Prakobkit S, Srivallop K, Boonthienthong M (2020) Reviews: the impacts of Electric Vehicles (EVs) and Renewable Energy Resources (REs) on the distribution power network. In: 2020 international conference on power, energy and innovations (ICPEI 2020) October 14–16, Chiangmai, Thailand
12. Ortiz F, Fernández I, Ortiz A, Renedo CJ, Delgado F, Fernández C. Health indexes for power transformers: a case study
13. Li S, Wu G, Dong H, Yang L, Zhen X. Probabilistic health index-based apparent age estimation for power transformers. In: IEEE access special section on artificial intelligence technologies for electric power systems.
14. Foros J, Istad M (2020) Health index, risk and remaining lifetime estimation of power transformers. IEEE Trans Power Deliv 35(6)
15. Prasojo RA, Suwarno, Abu-siada A (2021) Dealing with data uncertainty for transformer insulation system health index. IEEE Access 9
16. Ge Q, Jiang H, Lu Z, Yao G, Sun C (2022) Health management of dry-type transformer based on broad learning system. IEEE Trans Ind Electron 69(3)
17. Aizpurua JI, McArthur SDJ, Stewart BG, Lambert B, Cross JG, Catterson VM (2019) Adaptive power transformer lifetime predictions through machine learning and uncertainty modeling in nuclear power plants. IEEE Trans Ind Electron 66(6)
18. Karimi Madahi SS, Nafisi H, Askarian Abyaneh H, Marzband M. Co-Optimization of energy losses and transformer operating costs based on smart charging algorithm for plug-in electric vehicle parking lots. IEEE Trans Transp Electrif

19. Brinkela NBG, Schrama WL, AlSkaifa TA, Lampropoulosa I, van Sarka WGJHM (2020) Should we reinforce the grid? Cost and emission optimization of electric vehicle charging under different transformer limits. J Appl Energy 276:115285
20. Abdelsamad SF, Morsi WG, Sidhu TS (2015) Probabilistic impact of transportation electrification on the loss-of-life of distribution transformers in the presence of rooftop solar photovoltaic. IEEE Trans Sustain Energy 6(4)
21. Sidorova VT (2021) Determining mains connection points charging stations on RES for electric vehicles. In: 2021 international conference on industrial engineering, applications and manufacturing (ICIEAM)
22. Dong M, Zheng H, Zhang Y, Shi K, Yao S, Kou X, Ding G, Guo L (2019) A novel maintenance decision making model of power transformers based on reliability and economy assessment. IEEE Trans Ind Electron 66(6)
23. Chandran LR, Jayagopal N, Shyma Lal L, Narayanan C, Deepak S, Harikrishnan V. Residential load time series forecasting using ANN and classical methods. In: Proceedings of the 6th international conference on communication and electronics systems
24. Sak H, Senior A, Beaufays F. Long short-term memory recurrent neural network architectures for large scale acoustic modeling. Google, USA
25. Angel TS, Praveen A, Hashim Mohideen S, Lakshmi Narasimhan M. Comparison of deep learning-based methods for electrical load forecasting. In: 2022 third international conference on intelligent computing, instrumentation and control technologies
26. Daminov I, Prokhorov A, Caire R, Alvarez-Herault M-C´ (2021) Assessment of dynamic transformer rating, considering current and temperature limitations. Electr Power Energy Syst 129:106886
27. Sun B (2021) A multi-objective optimization model for fast electric vehicle charging stations with wind, PV power and energy storage. J Clean Prod 288:125564
28. Eriksson ELV, Gray EMA (2022) Optimization of renewable hybrid energy systems: a multi-objective approach. IEEE Trans Ind Electron 69(3)
29. Longo L. Optimal design of an EV fast charging station coupled with storage in Stockholm. Master of Science Thesis KTH School of Industrial Engineering and Management Energy Technology EGI-2017-MSC Division of Energy and Climate Studies
30. Sun Z, Li K, Yang Z, Niu Q, Foley A (2015) Impact of electric vehicles on a carbon constrained power system—A post 2020 case study. J Power Energy Eng 3:114–122
31. Ebenezer M, Rajkumar MR, Chandramohanan Nair PS (2013) Determination of hot spot temperature of a single-phase transformer subjected to harmonic pollution. The Institution of Engineers (India)
32. Leone C, Longo M, Fernández-Ramírez LM, García-Triviño P. Multi-objective optimization of PV and energy storage systems for ultra-fast charging stations. IEEE Access 3147672
33. Ravikrishnan G, Kiran P, Kanakasabapathy P. Internal fault discrimination in power transformers using curvelet. In: 2013 4th international youth conference on energy (IYCE)

Microgrid Frequency Stabilization Study with Approaching SCA Tuned Fuzzy Controller

Kiran Bhuyan, Prakash Chandra Sahu, Ramesh Chandra Prusty, and Sidhartha Panda

Abstract The grid with fractional MW rated is termed as microgrid. The microgrid is normally located at the distribution level and able to generate electrical energy with penetrating different renewable source-based generating units. The renewable energies such as wind power, solar energy, tidal power, geothermal power are now most convenient source to produce electricity. The limitations of renewable power generating plants are uncertainty in wind velocity and variation in the solar radiation power. These uncertainties produce abnormal in the microgrid frequency and also in voltage. This article has employed a noble fuzzy PID approach to manipulate frequency oscillation issues under such uncertainties. The controller is also implemented to maintain standard frequency environment under frequency load dynamic issues. Further, the controllability of the proposed fuzzy controller over standard PID and PI controllers is studied by different results and responses. The research has also applied a sine cosine algorithm (SCA) to get the most fit parameters of the controller. Finally, the effective action of the suggested SCA technique over PSO and GA has been synthesized to validate superiority of the proposed algorithm.

Keywords Frequency stability · Robust fuzzy controller · AC Microgrid · Sine cosine algorithm · Particle swarm optimizer · Solar PV system

---

K. Bhuyan
School of Electrical Science, VSSUT, Burla 768018, India

P. C. Sahu (✉)
Electrical Department, Aryan Institute of Engineering and Technology, Bhubaneswar 752050, India
e-mail: prakashsahu.iter@gmail.com

R. C. Prusty · S. Panda
School of Electrical Engineering, VSSUT, Burla 768018, India

J. Kumar et al. (eds.), *Recent Advances in Power Systems*, Lecture Notes in Electrical Engineering 1192, https://doi.org/10.1007/978-981-97-2788-9_8

1 Introduction

In the present electrical power generating scenario, power engineers are getting huge challenges for producing reliable electrical power. The traditional way for electrical power generation seems to be more questionaries' for getting power in near future. The coal and hydro are main conventional sources for producing electricity [1]. In recent environment, coals are not available plentily and also create huge pollution and transport cost as compared to other sources. In other hand, the collection of water to build hydro power plant is a tough task to produce bulk electrical power. So conventional way power generation is not so economic and reliable in present electrical market [2]. Now researchers are searching for various resources to produce electricity effectively [3]. The researchers are now planning to produce electricity as reliable source with employing fuel cell, solar energy and wind energy [4]. Besides these various small power-generating stations like micro-turbine, diesel generator and fuel cells have been implemented for the generation of green electrical energy [5]. Individually, these power generating units could able to electrify only low rated load and also able to supply small area [6]. The power engineers planned how to electrify certain moderate populated area by these renewable energy source-based power plants judiciously [7]. Then power engineers integrated individual power-generating units together through a small grid which in turn referred as microgrid [8]. This microgrid voltage level is normally is at distribution level that's why this grid is also known as distribution grid. As per the interconnection concern, microgrids undergo two different modes of connection [9]. These modes area utility grid connected mode and grid isolated mode [10]. Microgrids are also getting huge uncertainty issues due to the penetration of renewable energy-based power units [11]. The large unpredictable wind fluctuation puts adverse effect on wind plant thereby causing serious frequency dynamics in the system [12]. The performance of solar PV cell also gets affected due to the variation in solar power [13]. In electrical environment, loads are completely dynamic and always changes with respect to the time [14]. Since microgrid is a completely inertia less system it can't able to sustain stable performance immediately with such load dynamics especially in the off-grid mode of operation [15]. So various control techniques are essential to obtain steady and stable performance of the microgrid system under different uncertainties [16]. The concept of basic PI, PID and fuzzy PID approaches is well addressed in different articles for controlling frequency of the microgrid in different abnormal conditions and uncertainties [17–20]. Further, the control parameters are tuned optimally with employing suitable soft computing techniques. Sahu et al. developed a novel sine cosine algorithm and grey wolf algorithm to tune the parameters of fuzzy controller. The novelty of various improved and hybridized techniques is also presented through different articles for frequency monitoring of electrical power system [21–24]. The system performance basically depends upon the controller behaviors and consequently controller action depends upon the selection of proper optimization technique. This article has employed a noble type-1 fuzzy control technique for suitable control in the microgrid system

[25–28]. In response to the improved access time and good convergence ability, the study has adopted an advanced sine cosine algorithm (SCA) for optimal parameter design of fuzz PID controller.

2 System Investigated

Microgrid is a small distribution grid and is structured with connecting various small power-generating sources together. As the power rating is very small, this grid is termed as microgrid. In this proposed study of microgrid frequency control, a single area isolated system is modeled in software-based platform. The system is incorporated with various small capacity-based plants like diesel generator, tidal power plant, micro-turbine, fuel cell, wind power plant and solar power-based PV plant [29, 30]. A saturation band of limit 500–500 is put before the governor for accessing of signal smoothly. The tidal plant carries a permanent magnet linear synchronous generator (PMLSG) for the generation of electricity. In order to use electrical power judiciously, some energy storing components such as battery storage, flywheel energy storage units are integrated with the microgrid system. The power generating units always inject power to the microgrid, however, energy soring devices inject as well as absorb power as per the requirement. The net rating of the microgrid system is 10 MW with an effective load of 8 MW (Fig. 1).

Wind Generator (WGe) System

The empirical expression for wind power generation by the wind turbine with the help of wind velocity, blade area and wind speed is demonstrated in Eq. (1), however, the transfer function model of the wind plant is given in Eq. (2)

$$P_W = \frac{1}{2}.s.A.C.S^3 \tag{1}$$

Here, s = Blade sweep area in m^2, A = density of the wind, C = Air co-efficient, S = wind velocity

$$G_{WGe}(s) = \frac{\Delta P_{WGeO}}{\Delta P_{WGeI}} = \frac{K_{WGe}}{1 + s.T_{WGe}} \tag{2}$$

K_{WGe} = Gain of wind plant, T_{WGe} = Time constant of wind plant.
Photo voltaic system is structured as

$$G_{PV}(s) = \frac{\Delta P_{PVO}}{\Delta P_{PVI}} = \frac{K_{PV}}{1 + s.T_{PV}} \tag{3}$$

Diesel energy power is structured as

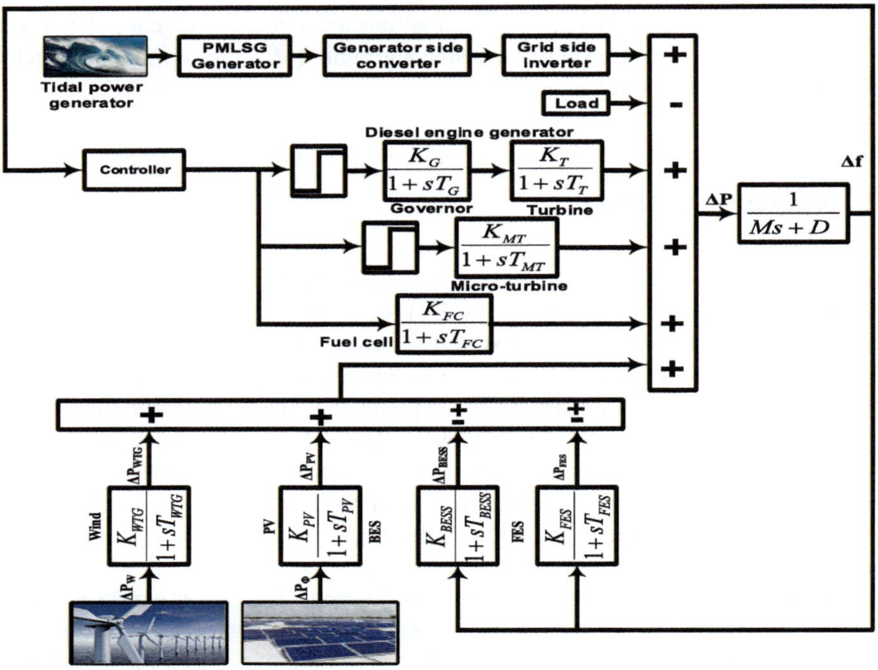

Fig. 1 Microgrid model

$$G_T(s) = \frac{\Delta P_{DEG}(s)}{\Delta P_g(s)} = \frac{1}{1 + sT_d} \tag{4}$$

Battery is structured as

$$G_B(s) = \frac{\Delta P_{BOUT}(s)}{\Delta P_{BIN}(s)} = \frac{K_{BESS}}{1 + sT_{BESS}} \tag{5}$$

The PMLSG is designed as

$$G_{PMLSG} = \frac{\Delta P_{ELECT}}{\Delta P_{MECH}} = \frac{K_{TPG}}{1 + sT_{TPG}} \tag{6}$$

3 Fuzzy-PID Controller (Fuzzy PID)

In the beginning stage of the control scenario basic proportional, integral and derivative controllers are much more popular for controlling various parameters in their respective plant area. Later on, the plants become more complex and exhibit most non-linearity in nature. Under such non-linearity environment, these conventional PI and PID controllers get poor performance in order to improve controllability of the system behavior. Then control engineers adopted fuzzy-ruled PI and PID control approaches to control necessary parameters of the non-linear power system. Such type of controller passes through different processes for its execution. Particularly, at the initial stage, the numerical data inputs are converted to equivalent fuzzy sets and the process incurred is called fuzzification. Then these fuzzy sets pass through different fuzzy rules to get the desired output in the system, and the concerned process is called fuzzy inference system. The associated fuzzy rules are gathered in Table 1.

Finally, through the process of defuzzification, again the fuzzy variables are converted to useful output data. These above processes are involved in the action of the proposed fuzzy PID controller, which has been implemented for this study. The suggested fuzzy controller gives outstanding results especially under working non-linear system. Since the designed model of microgrid is a non-linear nature, this fuzzy controller fulfills the objective through desired results (Fig. 2).

The work of this research study also has employed an improved objective function named integral of time multiplied absolute error (ITAE) to produce the best fit parameters of this fuzzy controller.

Table 1 Rules for the study

d.E *E*	MN	FN	N	FP	MP
MN	MN	MN	N	FP	N
FN	MN	FN	FN	N	FP
N	FN	FN	N	FP	FP
FP	FN	N	FP	FP	MP
MP	M	FP	FP	MP	MP

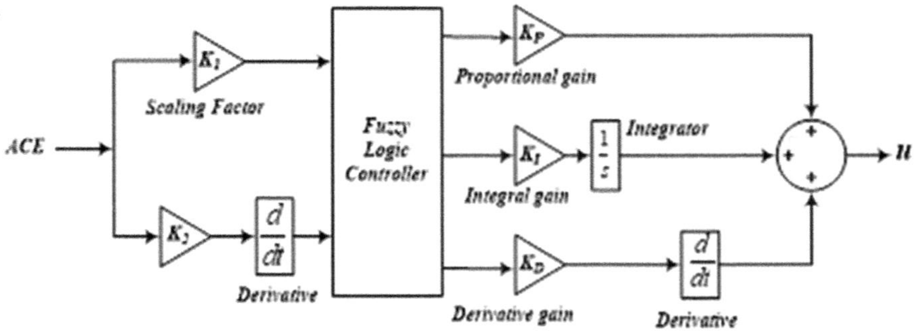

Fig. 2 Model of fuzzy PID

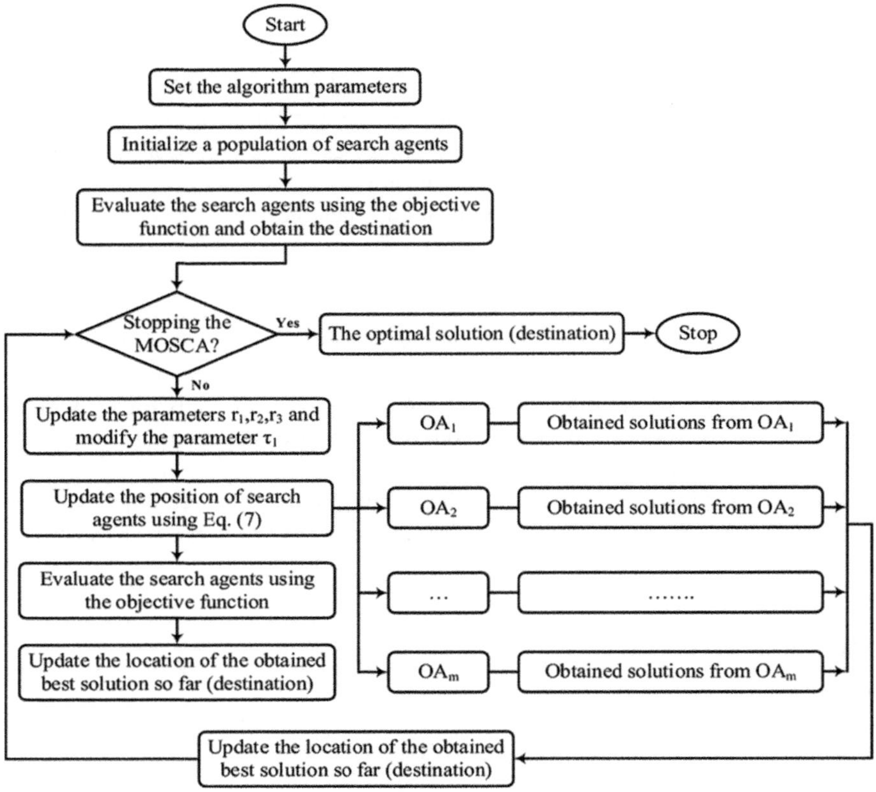

Fig. 3 SCA algorithm

4 Sine Cosine Algorithm (SCA)

The trigonometric function sine and cosine is the main inspiration to design the sine cosine algorithm. It is one of the best population-based algorithms. At the beginning level, the SCA technique produces few random solutions and tries to keep it towards most best observation or solution. The flowchart of this suggested algorithm is shown in Fig. 3.

5 Results and Analysis

This work of research study is conducted in an application software called MATLAB/ SIMULINK software. At first, the mathematical model of the suggested microgrid dynamic model is designed in the MATLAB forum. The controller structure is also modeled in this software to progress this research activity in an effective manner.

Since the micrigrid is involved with different renewable energy-based power plants, the proposed frequency regulation study has been advanced in concern to various disturbances and uncertainties. As the controller has a great role to manage the system frequency, it is needed to justify the activeness and importance of the proposed fuzzy controller. The controller performances are also judged by the nature of the optimization technique. So, this section also presented the supremacy of the suggested SCA technique in different relative studies and analyses.

Study 1: Controller activity study

It is found that the system behavior is highly dependent on the activity of the suggested controller. In order to show the most potential behavior of the suggested fuzzy controller, it is required to make relative study of this proposed controller with few standard controllers like PID and PID controllers. In this section activity, the leadership of the proposed fuzzy PID controller has been tested over basic PI and PID approaches in different abnormal conditions like dynamics in step load, stochastic load disturbance, wind oscillation and fluctuation in solar intensity power.

5.1 Section 1: Under Step Noise

This validation section of research study justifies the excellency of the employed fuzzy PID approach over PI and PID controllers under the injection of a step load noise of 1%. At first, all the three employed controller parameters are obtained in MATLAB software with applying an artificial technique-oriented soft computing technique called SCA algorithm. Then the controller parameters are adopted in the whole system model to extract dynamic frequency response of the system. With this in the same optimal scenario, the response of fluctuation in microgrid frequency response with all three employed controllers is given in Fig. 4a. In continuation with simulation result, the dynamics in diesel generator output power is shown in Fig. 4b. The response which is resulted due to fuzzy PID approach settles faster with little oscillation. This clearly suggests novel action of the fuzzy PID approach for frequency monitor of the AC microgrid.

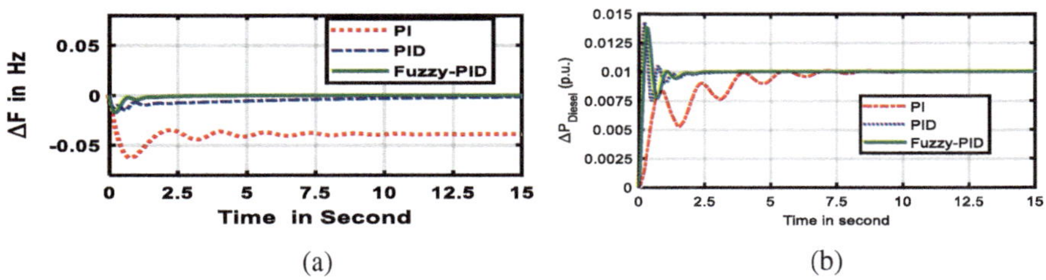

Fig. 4 **a** Microgrid frequency, **b** Diesel generator power

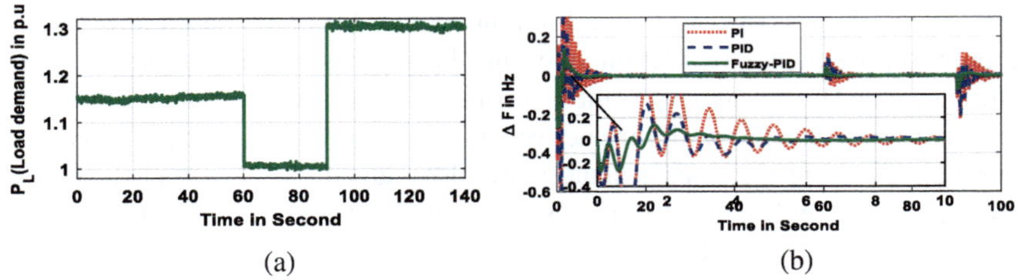

Fig. 5 **a** Stochastic load, **b** Microgrid frequency

5.2 Section 2: Stochastic Load Noise

In the same manner, the ability of the suggested fuzzy PID approach also has been tested in the action of a stochastic load disturbance and has been given in Fig. 5a. In the action of such stochastic noise, the developed response on microgrid frequency is depicted in Fig. 5b. The resulted response clearly identifies improved action of the suggested fuzzy PID approach in related to quick stability and wash-out oscillation as compared to PI and PID approach.

5.3 Section 3: Wind and Solar Energy Dynamics

As the wind velocity and solar ray intensity are weather dependent that's for their magnitude goes on varying and carries large uncertainties. These dynamic issues affect the microgrid performances especially voltage and frequency of the system. This assignment especially highlights on frequency performance of the system. Both the responses of dynamic in wind speed and solar power intensity are given in Fig. 6a, b, respectively. With the action of wind velocity fluctuation, the developed response of vibrant in microgrid frequency is shown in Fig. 6c. In continuation with this, the resulted response of microgrid frequency under solar power uncertainty is given in Fig. 6d. All the responses are resulted in the MATLAB platform with employing SCA-based fuzzy PID, PID and PI controllers. Analysis over both frequency responses concludes outstanding controllability action of the proposed fuzzy PID approach for frequency correction of this AC microgrid system.

5.4 Section 4: Technique Effectiveness Study

This result analysis also examines the effectiveness of the suggested SCA techniques for obtaining optimum parameters of the implemented fuzzy PID approach. To validate the effectiveness of the suggested SCA technique, its optimization property has been compared with basic PSO and GA techniques through few simulated

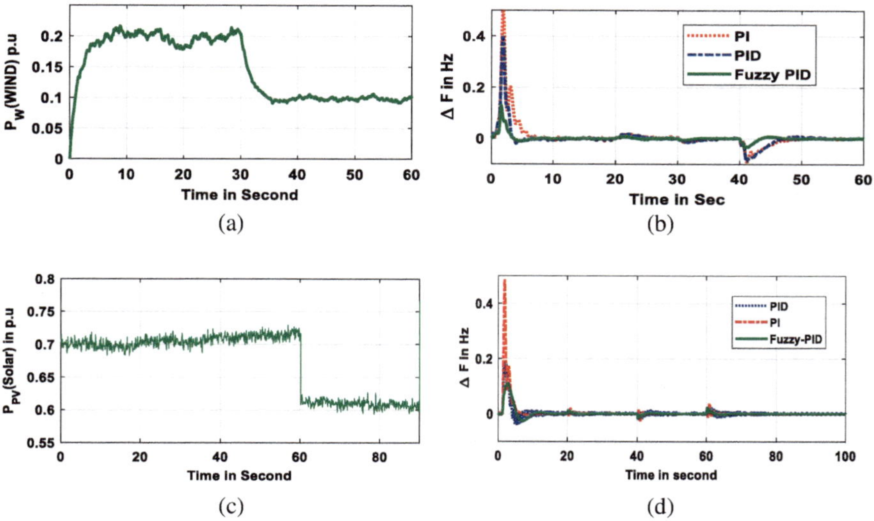

Fig. 6 **a** Wind oscillation, **b** Microgrid frequency, **c** Solar intensity variation, **d** Microgrid frequency under solar

responses. All the techniques are utilized to produce the best parameters of the fuzzy PID controller. The respective gains are utilized to get controlled microgrid frequency response under a step load disturbance. The resulted frequency response is depicted in Fig. 7, which confers super performance of the SCA technique in view of microgrid frequency control.

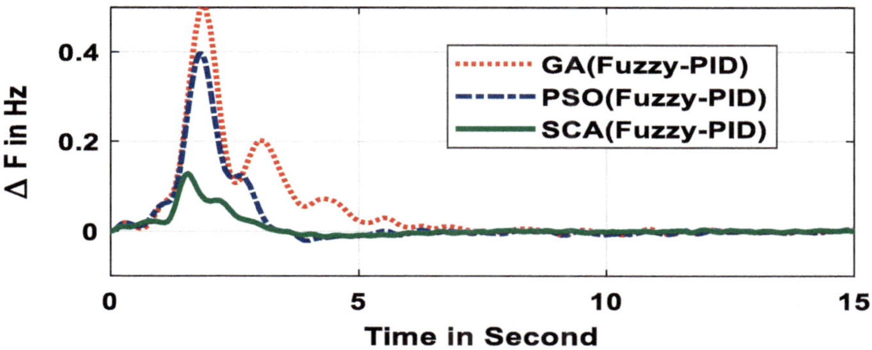

Fig. 7 Microgrid frequency response with different approaches

6 Conclusion

This section highlights the outcomes of the proposed research study on frequency governing of microgrid system. The work has been done in the MATLAB platform and the proposed techniques are well-modeled in this software. The analysis on frequency regulation study of islanded microgrids concludes that the proposed approaches are well in their area of application. In control action concern, the suggested fuzzy PID controller outperforms the basic PID and PI approach in relate to the frequency governing of the microgrid system. The injected fuzzy PID helps to mitigate the frequency error effectively and tries to amend steady situation quickly in the system. The controller not only fits for step load disturbance but also handles the stability of the system under stochastic load disturbances. In a technique point of view, the suggested SCA technique exhibits outstanding performance for fit parameter formulation of the robust fuzzy controller.

Appendix

Microgrid system parameters:

Wind Plant: T_{WTG} (Time constant of WTG)= 1.5s; Solar Plant: T_{PV} (TC of PV system) = 1.8s; Diesel Power Plant: T_{DEG} (TC of DEG) = 2s; Micro-Turbine Set: T_{MT} (TC of MT)= 2s; Fuel-Cell : T_{FC} (TC of FC)= 0.2s; Battery: T_{BES} (TC of battery) = 0.1s; Flywheel unit: T_{FES} (TC of Flywheel)= 0.1s; Microgrid inertia = M =0.2; D (Co-efficient of damping) = 0.012, R (Plant regulation) = 0.05.

References

1. Mustapaevich DK, Mnajatdin o'g'li MD. Properties of coal, processes in coal mining companies, methods of coal mining in the world. JournalNX 7(10):231–236
2. Sahu PC, Prusty RC (2019) Stability analysis in RECS integrated multi-area AGC system with modified-SOS optimized fuzzy controller. Recent Adv Electr Electron Eng (Formerly Recent Patents on Electrical & Electronic Engineering) 12(6):532–542
3. Bhatta SK, Mohapatra S, Sahu PC, Swain SC, Panda S (2021) Load frequency control of a diverse energy source integrated hybrid power system with a novel hybridized harmony search-random search algorithm designed Fuzzy-3D controller. In: Energy sources, Part A: recovery, utilization, and environmental effects, pp 1–22
4. Sun L, Wu G, Xue Y, Shen J, Li D, Lee KY (2017) Coordinated control strategies for fuel cell power plant in a microgrid. IEEE Trans Energy Convers 33(1):1–9
5. Sahu PC, Prusty RC, Panda S (2020) Improved-GWO designed FO based type-II fuzzy controller for frequency awareness of an AC microgrid under plug in electric vehicle. J Ambient Intell Human Comput 1–18

6. Sahu PC, Prusty RC (2020) Frequency and tie-line power awareness in eco-AGC of multi-area power system with SSO-based fractional order controller. Int J Power Energy Convers 11(2):200–221
7. Sahu PC, Prusty RC, Panda S (2020) Approaching hybridized GWO-SCA based type-II fuzzy controller in AGC of diverse energy source multi area power system. J King Saud Univ-Eng Sci 32(3):186–197
8. Sahu PC, Mishra S, Prusty RC, Panda S (2018) Improved-salp swarm optimized type-II fuzzy controller in load frequency control of multi area islanded AC microgrid. Sustain Energy, Grids Netw 16:380–392
9. Pavković D, Lobrović M, Hrgetić M, Komljenović A (2016) A design of cascade control system and adaptive load compensator for battery/ultracapacitor hybrid energy storage-based direct current microgrid. Energy Convers Manag 114:154–167
10. Sahu PC, Prusty RC, Panda S (2020) Frequency regulation of an electric vehicle operated micro grid under WOA tuned fuzzy cascade controller. Int J Ambient Energy 1–18
11. Sahu PC, Prusty RC, Panda S (2017, April) MFO algorithm based fuzzy-PID controller in automatic generation control of multi-area system. In: 2017 international conference on circuit, power and computing technologies (ICCPCT). IEEE, pp 1–6
12. Sahu PC, Prusty RC, Sahoo BK (2020) Modified sine cosine algorithm-based fuzzy-aided PID controller for automatic generation control of multiarea power systems. Soft Comput 1–18
13. Zhang Y, Xu Y, Zhou X, Guo H, Zhang X, Chen H (2019) Compressed air energy storage system with variable configuration for accommodating large-amplitude wind power fluctuation. Appl Energy 239:957–968
14. Sahu PC, Prusty RC, Panda S (2019) A gray wolf optimized FPD plus (1+ PI) multistage controller for AGC of multisource non-linear power system. World J Eng
15. Sahu PC, Prusty RC, Panda S (2020) Optimal design of a robust FO-Multistage controller for the frequency awareness of an islanded AC microgrid under i-SCA algorithm. Int J Ambient Energy 1–13
16. Adu JA, Penaloza JDR, Napolitano F, Tossani F (2019, May). Virtual inertia in a microgrid with renewable generation and a battery energy storage system in islanding transition. In: 2019 1st international conference on energy transition in the mediterranean area (SyNERGY MED). IEEE, pp 1–5
17. Baral KK, Sahu PC, Barisal AK, Mohanty B (2022) Combined analysis on AGC and ELD of a hybrid power system with D-WCA designed Gaussian type-2 fuzzy controller. In: Evolving Systems, pp 1–18
18. Mishra D, Sahu PC, Prusty RC, Panda S (2021) Power Generation Monitoring of a hybrid Power System with I-GWO designed trapezoidal type-II fuzzy Controller. Int J Modell Simul 1–17
19. Mahapatra AK, Samal P, Mohapatra S, Sahu PC, Panda S (2021) Analysis of Gaussian fuzzy logic-sliding model control and flexible AC transmission systems controllers for automatic generation control of hybrid power system under chaotic-water cycle algorithm approach. Int Trans Electr Energy Syst 31(12):e13163
20. Sahu PC, Prusty RC, Panda S (2021) Active power management in wind/solar farm integrated hybrid power system with AI based 3DOF-FOPID approach. Energy Sources, Part A: Recov Utiliz Environ Effects 1–21
21. Mishra D, Sahu PC, Prusty RC, Panda S (2021). Power generation monitoring of a hybrid power system with I-GWO designed trapezoidal type-II fuzzy controller. Int J Modell Simul 1–17
22. Sahu PC, Sahoo S, Prusty RC, Sahu BK (2021) Binary dragonfly algorithm-designed fuzzy cascade controller for AGC of multi-area power system with nonlinearities. In: Green technology for smart city and society. Springer, Singapore, pp 445–456
23. Begum B, Sahu PC, Sahu BK (2021) Comparative performance investigation of fractional order and conventional PID controller implemented for frequency stability. In Green technology for smart city and society. Springer, Singapore, pp 487–496

24. Mohapatra S, Sahu PC, Agrawal R, Prusty RC, Patel NC (2020, July) Superiority of JAYA Optimized 2DOF-IDD control strategy for AGC of hybrid power system. In: 2020 international conference on computational intelligence for smart power system and sustainable energy (CISPSSE). IEEE, pp 1–6
25. Sahu PC, Begum B, Prusty RC, Sahu BK, Devnath MK (2021, January) A grasshopper optimized FO-multistage controller for frequency control of an AC microgrid. In: 2021 1st Odisha international conference on electrical power engineering, communication and computing technology (ODICON). IEEE, pp 1–5
26. Mohapatra S, Sahu PC, Baral KK, Bhoi SK, Prusty RC (2021, October) Performance stability study of a hybrid power system under levy flight CSA based fuzzy controller. In: 2021 international conference in advances in power, signal, and information technology (APSIT). IEEE, pp 1–6
27. Sahu PC, Baliarsingh R, Prusty RC, Panda S (2020) Automatic generation control of diverse energy source-based multiarea power system under deep Q-network-based fuzzy-T2 controller. Energy Sources, Part A: Recov Utiliz Environ Effects 1–22
28. Sahu PC, Baliarsingh R, Prusty RC, Panda S (2022) Novel DQN optimised tilt fuzzy cascade controller for frequency stability of a tidal energy-based AC microgrid. Int J Ambient Energy 43(1):3587–3599
29. Sahu PC, Bhoi SK, Jena NK, Sahu BK, Prusty RC (2021, January) A robust multi verse optimized fuzzy aided tilt controller for AGC of hybrid power system. In: 2021 1st Odisha international conference on electrical power engineering, communication and computing technology (ODICON). IEEE, pp 1–5
30. Jena NK, Sahoo S, Sahu PC, Sahu BK, Mohanty KB (2021, January) Automatic generation control of a renewable energy integrated power system in the presence of fractional order 3DOF controller. In: 2021 1st Odisha international conference on electrical power engineering, communication and computing technology (ODICON). IEEE, pp 1–6

Integrating Solar Photovoltaic Water Pumping System for Paddy Irrigation

Mohamed Salman

Abstract In developing countries, especially India, solar pumping systems are an appealing option for agriculture and water supply systems due to the vast solar potential and a large rural population living in remote areas. This study aimed to design a standalone solar photovoltaic pumping system in the West Godavari district of Andhra Pradesh to meet a paddy field's water requirements. A photovoltaic (PV) water pumping system with a centrifugal pump of 18 kW powered by a PV array of 20 kW was designed. Based on the simulation, the total water pumped yearly was 87,820 m^3, and the performance ratio was 36.7%. In the economic analysis calculations, the energy cost was determined to be 3.77 INR/kWhr. The results of this study showed that solar photovoltaic pumps have a high investment cost, but they have several advantages over traditional water pumps. This study found that solar PV pumps have an increased investment cost but have several advantages over conventional pumps. Results indicate that solar PV water pumps are a viable and attractive solution to meeting agriculture's electricity demands in developing countries.

Keywords Solar · Water pumping system · Photovoltaic · Standalone · Agriculture · Paddy field

1 Introduction

The majority of water pumping worldwide is driven by traditional or diesel-powered electricity. Pumping water with solar energy reduces fossil fuel consumption. In addition to using expensive fuels, diesel and propane-powered pumps produce noise and air pollution. In terms of operation, maintenance, and replacement cost, a diesel pump is two to four times more expensive than a solar photovoltaic (PV) pump [1]. Solar power pumping systems are environmentally friendly, require minimal maintenance, and do not require any fuel [2].

M. Salman (✉)
Heriot-Watt University, Dubai Knowledge Park, United Arab Emirates
e-mail: mmoh803@aucklanduni.ac.nz

J. Kumar et al. (eds.), *Recent Advances in Power Systems*, Lecture Notes in Electrical Engineering 1192, https://doi.org/10.1007/978-981-97-2788-9_9

The lack of grid electricity in rural and remote areas makes solar PV pumping a promising application for solar energy [1]. The technology used in this system is the same as any conventional water pumping system, except that it is powered by solar panels. In recent years, PV water pumping has gained prominence due to electricity shortages and rising diesel prices. PV pumping systems offer significant long-term savings when compared to conventional pumping systems [1]. Rather than batteries for electricity storage, tanks can be used in PV pumping systems to store energy (water) [3]. In developing countries like India, agriculture depends heavily on rain and is greatly impacted by summer droughts. During the summer, solar radiation is highest, so PV pumping systems can pump more water to meet increasing water needs [1].

According to Chandel et al. [1], PV water pumping systems have proven to be an economical and reliable alternative to diesel pumps for irrigation of crops. Several studies have examined the feasibility of PV water pumping systems for agriculture in the past few years. In their research, Narale et al. [4] assessed the feasibility of PV water pumping for banana irrigation. Their study found that using a PV pumping system for irrigating banana fields in remote areas is beneficial, as opposed to diesel pumps. Based on their economic analysis, the life cycle cost of PV pumps is significantly lower than diesel pumps. Senthil Kumar et al. [5], discussed that electricity and water consumption could be minimized by combining sprinklers with green-friendly PV pumps.

Agriculture accounts for almost 20% of India's GDP [6]. Despite this, meeting the water demands of agriculture is still a challenge and an area that needs to be explored more. Due to climate change, the demand for water for agricultural purposes is continuously increasing in India [7]. Water pumping becomes more challenging and expensive in rural areas due to fuel costs and a lack of grid connectivity. A growing number of renewable energy sources are being studied to solve these problems. Solar energy is used across India for various applications owing to its substantial solar resources. Especially in remote areas without grid connectivity, solar water pumps are the most effective solution.

This project aims to design a PV water pumping system (PVWPS) for a paddy field in West Godavari, Andhra Pradesh, India. As a standalone system, it will operate independently. The PV pump system will meet the field's water demands for paddy cultivation. An economic analysis was also conducted as part of the system's evaluation. The results of this study will provide valuable knowledge on PVWPS and contribute to the existing literature.

2 Design Approach

This study aims to design a PVWPS for a paddy field located in the West Godavari district of Andhra Pradesh. The location of the paddy field is shown in Fig. 1.

The paddy field size is 12 acres, and the water requirement per acre is 20 m^3/day. As a result, the paddy field water requirement is 240 m^3/day per day. The deep well

Fig. 1 Paddy field location [8]

Table 1 Resource information

Location	West Godavari, Andhra Pradesh
Latitude	17.15 °N
Longitude	81.62 °E
Altitude (m)	26
Type of water resource	Deep well to storage
Borehole diameter (m)	20
Static level depth (m)	40
Pump level (m)	50
Lower dynamic level (m)	45

to storage system pumping type was considered for this study since it is suited to agricultural applications. Table 1 details the location and the resources available for PVWPS installation.

2.1 *Theoretical Calculations*

Hydraulic power, motor sizing, and PV array sizing were calculated theoretically to pre-size the PVWPS. Based on the following Eq. (1), hydraulic power requirements can be calculated for the system,

$$P_H = \frac{\rho \times g \times H \times Q}{3.6 \times 10^6} \tag{1}$$

where P_H is hydraulic power (kW), g is the gravitational acceleration (m/s), H is the sum of static head and frictional losses (m), ρ is the density of water (kg/m^3), and Q is the water flow rate (m^3/hr). Taking $\rho = 1000$ kg/m^3, $g = 9.81$ m/s, $Q = 80$ m^3/hr, $H = 48$ m (considering 20% frictional losses), the hydraulic power (P_H) required for the system was found to be 10.5 kW. The power required by the motor depends on the efficiency of the pump and can be calculated using Eq. (2),

$$\text{Motor Power Rating} = \frac{P_H}{\eta_p} \tag{2}$$

where P_H is hydraulic power (kW) and η_p is pumping efficiency. Taking η_p to be 80% [9], the required motor power rating was calculated to be 13.1 kW. The required PV array power can be determined using the system efficiency,

$$\text{PV Array Power Rating} = \frac{\text{Motor Power Rating}}{\eta_s} \tag{3}$$

where η_s is system efficiency. Taking η_s to be 70%, the required PV array power rating was determined to be 18.7 kW.

2.2 System Design

The use of simulation software helps the system to be designed in the most efficient manner. Software results enable problems with the system to be corrected before it is installed on a field [10]. PVWPS can be designed using a variety of software packages. The PVWPS system for this study was designed and simulated using PVSyst [11] software. Meteo data was loaded into PVsyst using the location coordinates of the paddy field. A schematic diagram of the proposed PVWPS system for paddy fields is shown in Fig. 2. Solar PV panels or arrays, pump controller (MPPT), surface or submersible pump, storage tank and pipes are the main components of PVWPS.

A solar panel's tilt angle should be optimized to minimize loss and maximize global radiation on the plane. The fixed tiled plane type was selected for the PV array, and the tilt angle of the panels was optimized with respect to the annual irradiation yield of the site. The tilt angle corresponding to the location was determined to be 22° with 0° azimuth. The PV panel orientation is shown in Fig. 3.

In this study, the deep well to storage system was considered. The pump, which is deep inside the reservoir, pumps water into an overhead storage tank. The well characteristics along with storage tank and hydraulic circuit properties were defined in PVsyst to model the pumping hydraulic circuit. The field's daily water requirements (240 m^3/day) were inputted into PVsyst to define the system's water needs.

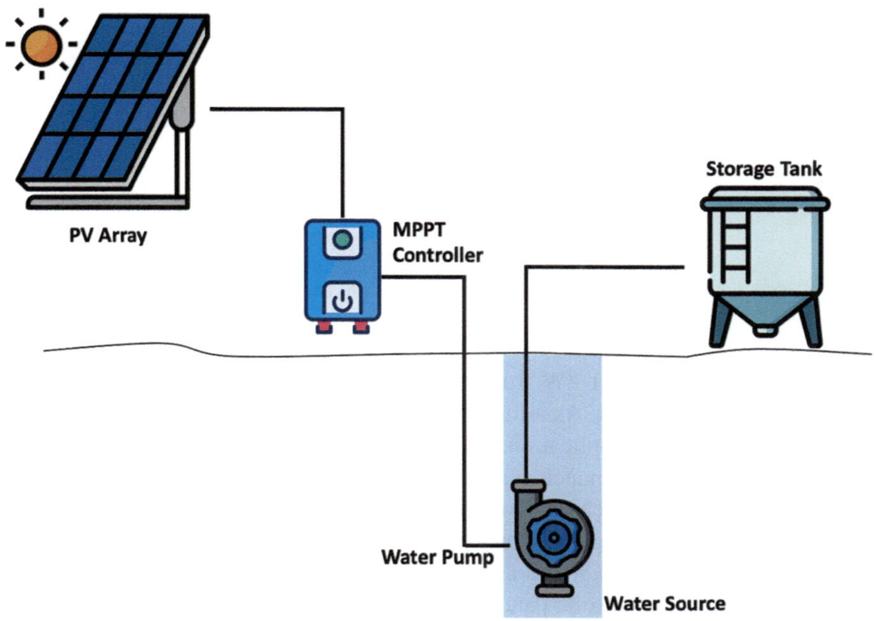

Fig. 2 Schematic diagram of PVWPS

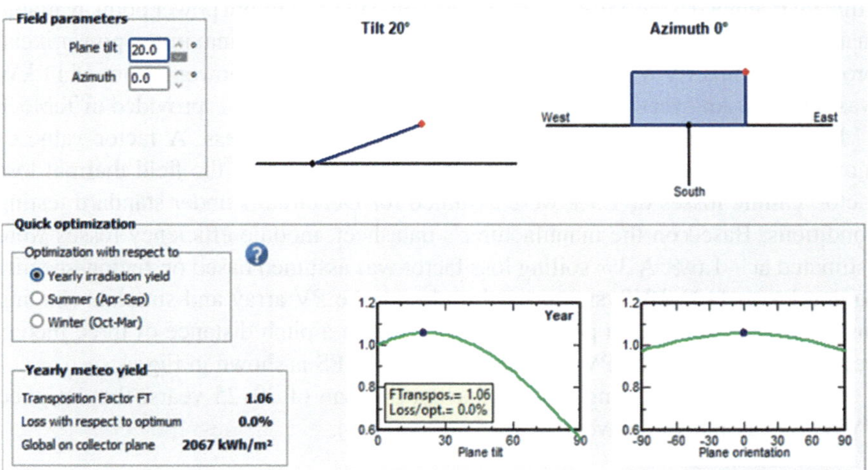

Fig. 3 PV panel orientation

It was assumed that the static head of 40 m would remain constant throughout the year. Table 2 summarizes the pumping hydraulic circuit properties and water needs.

Pumps are defined based on static head, hydraulic energy required to run the system, and other parameters related to the system's requirements. We derived a

Table 2 Pumping hydraulic circuit properties

Tank volume (m^3)	500
Tank diameter (m)	5.7
Water full height (m)	19.5
Feeding altitude (m)	4
Pipe diameter (inches)	3
Pipe length (m)	60
Water requirement (m^3/day)	240
Well static depth variations (mW)	40

motor power rating of 13.1 kW from our theoretical calculations, which matched PVsyst's sizing suggestion. As a result, a pump system with a power output of 13 kW was required for this application. A 13 kW Grundfos multistage centrifugal pump was selected for its performance and cost-effectiveness.

Based on a theoretical calculation, the PV array needed a power rating of 18.7 kW, similar to PVsyst's suggested array size. Considering these two factors, the PV power rating was set at 20 kW. Tata's solar panels are among the most reliable, cost-effective, and widely used solar panels in the country [12]. Therefore, Tata's 320 Wp polycrystalline silicon panels were selected for this PVWPS. A total of 80 panels make up the PV array, including 16 modules in series and five strings of modules.

A PV system should be optimized to maximize its solar power for a load in an efficient manner. Using MPPT, we can determine the maximum power point available in a PV system at a particular irradiance point and extract the maximum power it can provide. For this PVWPS, an ABB MPPT controller with a power output of 11 kW was selected. An overview of pump and PV array information is provided in Table 3.

PVWPS losses were considered during the design process. A factor value of mounted panels with free air circulation was used to model the field thermal loss factor. Ohmic losses of 1.5% were assumed for DC circuits under standard testing conditions. Based on the manufacturer's datasheet, module efficiency losses were estimated at −1.6%. A 3% soiling loss factor was assumed based on region-specific soiling losses [13]. PVSyst was used to design the PV array and simulate shading losses. It was arranged in portrait orientation with a pitch distance of three meters between the panels. The PV array layout for PVWPS is shown in Fig. 4.

In general, PV pumping systems have a lifespan of 20–25 years; the designed PVWPS is assumed to have a 20-year lifespan [14].

2.3 Economic Analysis

Financial analysis can be performed by doing economic calculations, which help calculate water and energy costs. The main factors used in economic calculations are total investment, operating and maintenance costs, and system financing [10]. Online

Table 3 Pump and PV array information

Pump	
Pump type	Centrifugal multistage
Model	Grundfos, SP 95-3
Head (mW)	20 to 62
Rated power (kW)	13
PV array	
Panel type	Si-poly
Model	Tata, TP 320L
Panel power rating (Wp)	320
PV array rated power (kW)	20
Number of modules	84
Number of modules in series	14
Number of strings	6
PV array area (m^2)	163
MPPT controller model	ABB, ACSM1 11kW 440 Vac
MPPT controller rated power (kW)	11

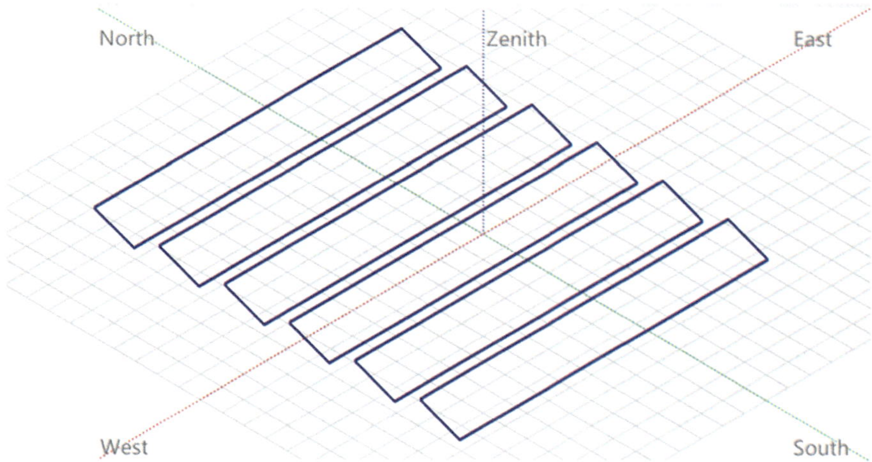

Fig. 4 PV array layout for PVWPS

databases and resources were used to estimate the system component's costs [15, 16]. PVsyst was then used to perform an economic analysis of the PVWS. It was assumed that annual maintenance costs would be 0.5% of the capital costs [17]. The financial analysis takes into account a discount rate of 10%. Table 4 shows the breakdown of PVWS project costs.

Table 4 PVWPS project costs

Investment			
PV modules (Tata TP 320L)	84 units	9,200 INR/unit	772,800 INR
Support for modules			294,000 INR
Pump (Grundfos, SP 95–3)	1 unit	80,000 INR/unit	80,000 INR
Controller (ABB ACSM1 11kW)			130,000 INR
Storage tank			20,000 INR
Hydraulic circuit			2,000 INR
Wirings and other components			1,500 INR
Engineering			30,000 INR
Assembly and transportation			10,000 INR
Well (drilling, building)			10,000 INR
Total investment (without taxes)			1,345,300 INR
Tax (VAT)		18%	242,154 INR
Total investment (with taxes)			1,587,454 INR
Government subsidies		60%	952,472 INR
Net investment (including taxes)			**634,982 INR**
Operating cost			
Maintenance (repair, cleaning, etc.)			67,265 INR/year
Total annual costs			**67,265 INR/year**

The economic analysis is conducted in Indian rupees (INR). The Pradhan Mantri Kisan Urja Suraksha Evam Utthaan Mahabhiyan (PM-KUSUM) scheme was launched in 2019 to provide farmers with solar pumps, grid-connected solar power plants, and other renewable energy sources [18]. PM-KUSUM is designed to ensure farmer's energy security. In addition, it intends to honor India's commitment to increase non-fossil fuel capacity by 40% by 2030 [18]. Under this scheme, the central government will cover 30% of the benchmark cost of standalone solar agriculture pumps. In addition, the government will provide a 30% subsidy, while farmers are responsible for the remaining 40% [19]. As a result, a government subsidy of 60% of the investment in this PVWPS is considered.

3 Results and Discussions

The PVWPS was designed to address the water needs of a paddy field in West Godavari and is built with a 40 m static deep well-to-storage system. PVsyst is used to select a pump with a proper head, a PV module, and an MPPT controller based on the project's needs. A simulation of the final system design was performed using

PVsyst in which all losses were taken into account. Figure 5 shows the normalized energy production.

According to Fig. 5, the highest energy production occurs between February and April. It is possible to determine four critical parameters of PVWPS based on normalized energy production, namely unused energy, effective energy at the pump, converter losses, and PV-array losses. Results reveal that system/convertor and PV-array losses are 0.98 kWhr/kWp/day and 0.64 kWhr/kWp/day, respectively, while unused energy is 1.96 kWhr/kWp/day. It was determined that the effective energy at the pump is 2.07 kWhr/kWp/day. Figure 6 represents the amount of water pumped and used monthly for the paddy field.

The above graph indicates that the PVWPS will pump 87,820 m^3 of water annually. It is estimated that paddy fields require 87,600 m^3 of water each year. As a result, the PVWPS effectively meets the annual water requirement for paddy cultivation, with a yearly excess of 220 m^3 pumped by the plant. The performance ratio of a solar system is the ratio between effective energy and irradiance over a given period of time. The PVWPS performance ratio is shown in Fig. 7.

A performance ratio of 36.7% was found for the PVWPS designed for the paddy field. As part of the Performance ratio, the optical losses are considered (shadings, soiling, etc.), the array losses are considered (aging, module quality, wiring, etc.), and the system losses are taken into consideration (unused losses) [10]. According to Zhao et al. [20], a PVWPS is of high quality if the performance ratio is between 35 and 45%. The simulation results from PVsyst are presented in Table 5.

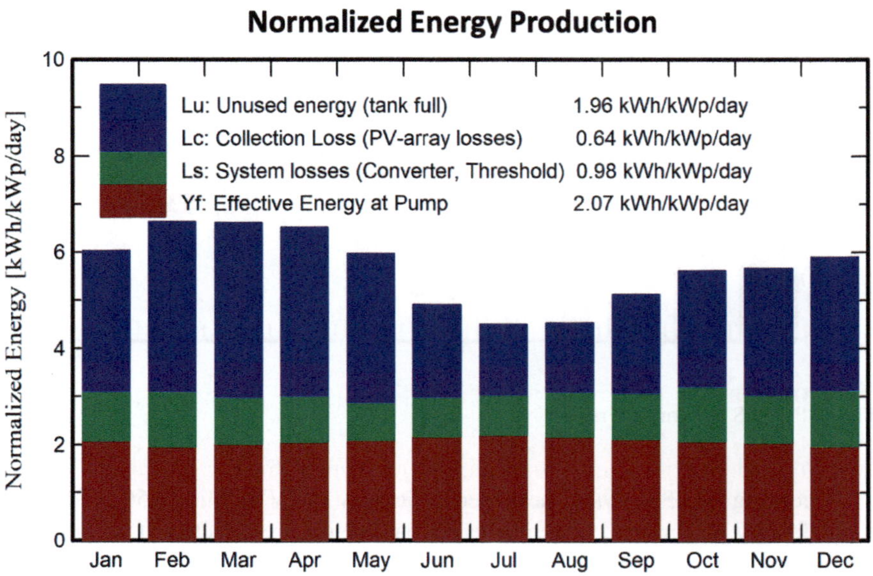

Fig. 5 PVWPS normalized energy production

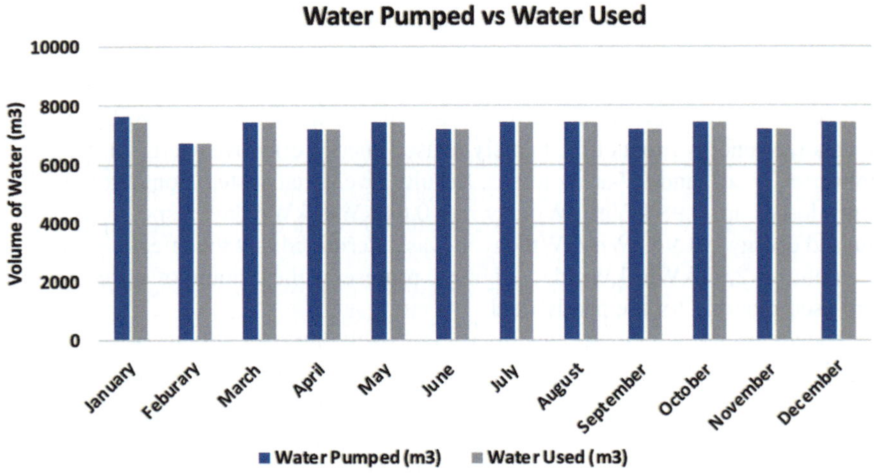

Fig. 6 Paddy field monthly water pumping and water usage

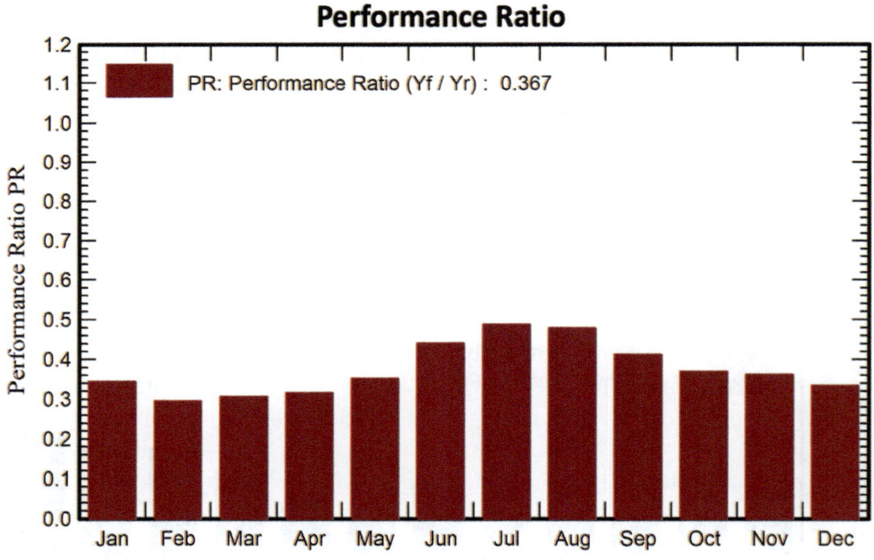

Fig. 7 PVWPS performance ratio

According to the analysis, the system efficiency of PVWPS is 44.5%, indicating a satisfactory performance when different parameters are selected. It is found that the pumping system uses the majority of energy generated by the PV array and that 23.2% of total generated energy is wasted. A summary of PVWPS's economic results is presented in Table 5. The total investment for the system was determined to be 1,587,454 INR, but with a government subsidy of 60% via the PM-KUSUM scheme,

Table 5 PVWPS simulation results

System production	
Water pumped (m^3)	87,820
Specific water pumped (m^3/kWp/bar)	678
Water needs (m^3)	87,600
Missing water (%)	−0.3
Energy at pump (MWhr)	20.35
Specific energy (kWhr/m^3)	0.23
Unused energy (MWhr)	10.60
Unused fraction (%)	23.2
System efficiency (%)	44.5
Economic analysis	
Total investment (including taxes)	1,587,454 INR
Specific investment	59.1 INR/kWp
Annual running cost	67,265 INR/year
Energy cost	3.31 INR/kWhr

the total investment for the paddy field owner will be 634,982 INR. PVWPS annual operating costs were estimated at 67,265 INR.

It is proposed that an indirect field-oriented control strategy be used for this system, which will maximize motor flux while minimizing motor losses. The results indicate that the energy cost is 3.31 INR/kWhr. Water pumping costs per m^3 can be used to compare solar water pumping systems with other pumping systems. According to a study conducted by Purohit [21], the pumping cost of an 18 kW PV pump was 0.69 INR/m^3, which is similar to the findings of this study. In Purohit's [21] study, the water pumping costs of diesel pumps were determined to be 1.13 INR/m^3, which is higher than that of PV pumps. As a result, PV water pumping has economic benefits despite its high upfront investment costs.

4 Conclusions

PV water pumps are an attractive and cost-effective alternative to electric and diesel pumps for irrigation of crops. The purpose of this study was to design a solar PV water pumping system to meet the water demands of a paddy field in West Godavari, Andhra Pradesh. A PVWPS consisting of a 20 kW PV array and a multistage centrifugal pump of 18 kW is proposed. A total of 87,820 m^3 of water is pumped yearly by the PVPWS to satisfy the paddy field's annual water demands of 87,600 m^3. The designed PVWPS has a system efficiency of 44.5% and a performance ratio of 36.7%. According to the economic analysis, the energy cost is 3.31 INR/kWhr. Overall, the PVWPS designed met the paddy fields' water needs.

The system was found to be cost-efficient during the economic analysis. It does, however, require a significant amount of investment at the beginning. The results of this study contribute to the existing literature on solar PV pumping systems. Moreover, similar PV pumping systems can be used to meet the agricultural water needs across the country. Further research is needed on factors influencing the efficiency and use of highly efficient PV modules to lower costs, improve performance and extend pumping system lifespans. Additionally, carbon reductions and system losses could be assessed in a future study.

References

1. Chandel SS, Nagaraju Naik M, Chandel R (2015) Review of solar photovoltaic water pumping system technology for irrigation and community drinking water supplies. Renew Sustain Energy Rev 49:1084–1099. https://doi.org/10.1016/j.rser.2015.04.083
2. Foster R, Ghassemi M, Cota A (2009) Solar energy: renewable energy and the environment. CRC Press
3. Rohit K, Karve G (2013) Solar water pumping system
4. Narale P, Rathore N, Kothari S (2013) Study of solar PV water pumping system for irrigation of horticulture crops. Int J Eng Sci Invent 2(12):54–60
5. Senthil Kumar S, Bibin C, Akash K, Aravindan K, Kishore M, Magesh G (2020) Solar powered water pumping systems for irrigation: a comprehensive review on developments and prospects towards a green energy approach. Mater Today: Proc 33:303–307. https://doi.org/10.1016/j.matpr.2020.04.092
6. Shagun (2021) Agri share in GDP hit 20% after 17 years: economic survey. DownTo Earth, ed
7. Praveen B, Sharma P (2020) Climate change and its impacts on Indian agriculture: an econometric analysis. J Public Aff 20(1):e1972
8. Google (2022) Google Maps, ed
9. Stoffel B (2015) Chapter 4—Physical and technical background of the efficiency of pumps. In: Stoffel B (ed) Assessing the energy efficiency of pumps and pump units. Elsevier, Boston, pp 45–61
10. Sharma R, Sharma S, Tiwari S (2020) Design optimization of solar PV water pumping system. Mater Today: Proc 21:1673–1679. https://doi.org/10.1016/j.matpr.2019.11.322
11. P. SA (2022) PVsyst, ed
12. Joshi A (2022) The Indian solar markets' top players in 2021. Mercom India. https://mercomindia.com/indian-solar-markets-top-players-2021/
13. Ghosh A (2020) Soiling losses: a barrier for India's energy security dependency from photovoltaic power. Challenges 11(1). https://doi.org/10.3390/challe11010009
14. Mohanraj M, Chandramohan P, Sakthivel M, Kamaruzzaman S. Performance of photovoltaic water pumping systems under the influence of panel cooling. Renew Energy Focus 31:31–44. https://doi.org/10.1016/j.ref.2019.10.006
15. Chandra N (2022) Tata solar panel price in India. Loom Solar. https://www.loomsolar.com/blogs/pricelist/tata-solar-panel-price-list-in-india. Accessed 27 Jan 2023
16. Baltic AS (2023) Price list for ABB ACSM1 machinery series variable frequency drives. Advanced Systems https://asb-drives.eu/equipment/price.php?section=255. Accessed 27 Jan 2023
17. Goel S, Sharma R (2021) Economic analysis of solar water pumping system for irrigation. In: Green technology for smart city and society: Springer, pp 157–167
18. Cleartax. Pradhan Mantri Kisan Urja Suraksha Evam Utthaan Mahabhiyan (PM-KUSUM). https://cleartax.in/s/pm-kusum-pradhan-mantri-kisan-urja-suraksha-evam-utthaan-mahabhiyan-scheme. Accessed 3 March 2023

19. G. o. India (2023) PM-KUSUM (Pradhan Mantri Kisan Urja Suraksha evam Utthaan Mahabhiyan) Scheme. https://www.india.gov.in/spotlight/pm-kusum-pradhan-mantri-kisan-urja-suraksha-evam-utthaan-mahabhiyan-scheme. Accessed 3 March 2023
20. Zhao B, Ren Y, Gao D, Xu L (2019) Performance ratio prediction of photovoltaic pumping system based on grey clustering and second curvelet neural network. Energy 171:360–371. https://doi.org/10.1016/j.energy.2019.01.028
21. Purohit P (2007) Financial evaluation of renewable energy technologies for irrigation water pumping in India. Energy Policy 35(6):3134–3144. https://doi.org/10.1016/j.enpol.2006.11.013

Design and Modeling of 0.55 MWp Solar Carport

Santu Hore and Umesh Kumar Sinha

Abstract Installation of solar PV plant and converting solar energy into electrical energy is the most attractive and promising option to reduce power cost followed by production cost, reduce carbon emissions as well as fulfillment of RPO for Indian industries. To install the solar PV plant, larger areas are required due to low efficiency of PV panels. There are numerous challenges to ensure larger lands for solar PV plant but industries are having a larger amount of unused areas in terms of building roofs, industrial roof sheds, car parking areas, scrap yards, material loading and unloading areas, storage areas, etc. In this paper, a car parking area of an industry in the state of Jharkhand has been identified and detailed engineering has been done towards design of grid-connected 550 kWp solar carport under which vehicles can easily be parked and green energy can also be generated for use. This paper also describes the technical parameters of critical PV plant components, interconnection diagrams, projected energy generation from this plant, losses against different parameters and performance ratio of the plant. The design and simulation of the plant were done through PVsyst software.

Keywords Solar carport · Performance Ratio (PR) · Loss factor

1 Introduction

The vision of India is to become Net Zero Emission by 2070 and to achieve this, India has taken initiatives to increase capacity of renewable energy installation to 500 GW till 2030 [1]. To fulfill the 500 GW renewable energy target, 300 GW solar energy has also been planned and can be achieved through 100 GW rooftop solar energy installations [2].

S. Hore (✉) · U. K. Sinha
National Institute of Technology, Jamshedpur 831014, India
e-mail: horeaec@gmail.com

U. K. Sinha
e-mail: uksinha.ee@nitjsr.ac.in

J. Kumar et al. (eds.), *Recent Advances in Power Systems*, Lecture Notes in Electrical Engineering 1192, https://doi.org/10.1007/978-981-97-2788-9_10

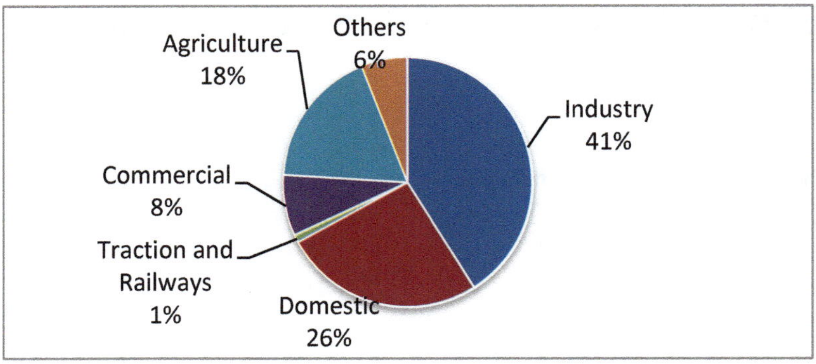

Fig. 1 Sector-wise electricity consumption in India during FY 2020–21. *Source* MOSPI (2022)

Consumption of electricity is increasing day by day across the world. In India, consumption of electricity increased from 7.85 × 105 GWh to 12.27 × 105 GWh from 2011 to 2021, at CAGR of 5.09%. While analyzing the electricity consumption for 20–2021, industry sector is found responsible for the maximum use–41.09%, while domestic sector is in 2nd position with approx. 25.67% use [3] (Fig. 1).

Considering the current availability and exhaustion of conventional energy resources, it is utmost essential to move forwards with non-conventional energy sources like wind, solar, tide energy, small hydro plants, biogas, etc.

The estimated potential of RE resources as on 31 March 2021 indicates that Rajasthan has the highest potential of approx. 18.2% followed by Gujarat with 12.1%. Maharashtra and Karnataka are also having very good potential of renewable energy and the mentioned four states are having 50% or more of the total potential of Renewable Power in India. When we talk about Jharkhand, this state is having approx. 1% potential in terms of renewable energy [3] (Fig. 2).

2 Overview

When we talk about low production cost, conventional power cost is found as major challenge for industries and when we talk about sustainability, high carbon emission is also one of the major concerns for industries. After the recent pandemic of COVID-19, industries are taking different initiatives to reduce manufacturing cost and carbon emission. To control the power cost and carbon emission, every industry is looking to resource power from green and cheapest sources or to setup their own renewable power generation systems. Among all renewable energy resources, grid-connected solar plant is found the most easy and cost-effective solution [4].

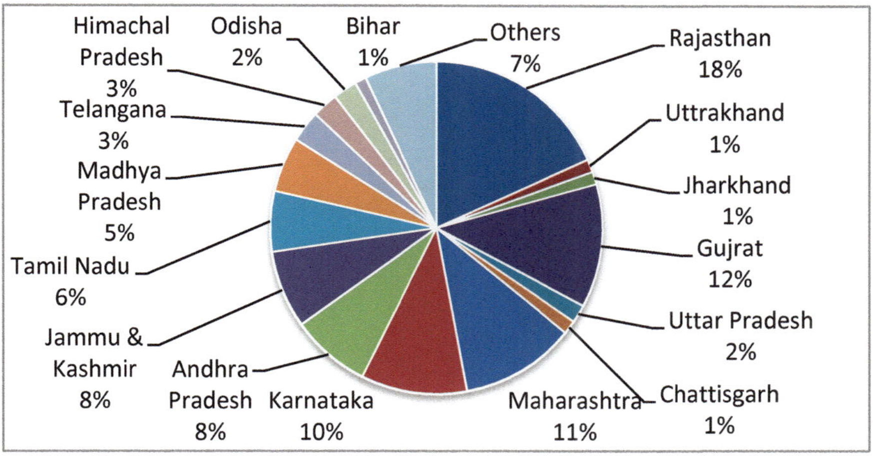

Fig. 2 State-wise estimated potential of renewable power in India as on March 2021. *Source* MOSPI (2022)

Solar energy is the most attractive and viable renewable energy source. To convert the solar energy into the electrical energy, PV module is the main component and therefore performance of PV modules must be assessed in real outdoor conditions to ensure its actual output [5]. Solar energy is fluctuating in nature and is very unreliable on cloudy days and or night hours. Grid-connected solar plant or solar plant with storage arrangement can resolve this problem in most efficient way but storage arrangement is also not at all cost-effective [6]. Therefore, to ensure cost effectiveness, grid-connected 550 KWp solar carport is planned and vehicles can also be parked easily under the plant area [7].

3 Site Assessment and Site Selection

The proposed site selected for installation of solar carport is located at Jamshedpur in the state of Jharkhand. The temperature of Jamshedpur varies from 12 to 40 °C with an annual average temperature range of 19–33 °C [8] (Fig. 3).

The daylight hours are one of the critical parameters to estimate generation from the plant as it varies from month to month. In Jamshedpur, monthly average daylight hours vary from 10.8 h to 13.5 h in the month of December and June, respectively [8] (Fig. 4).

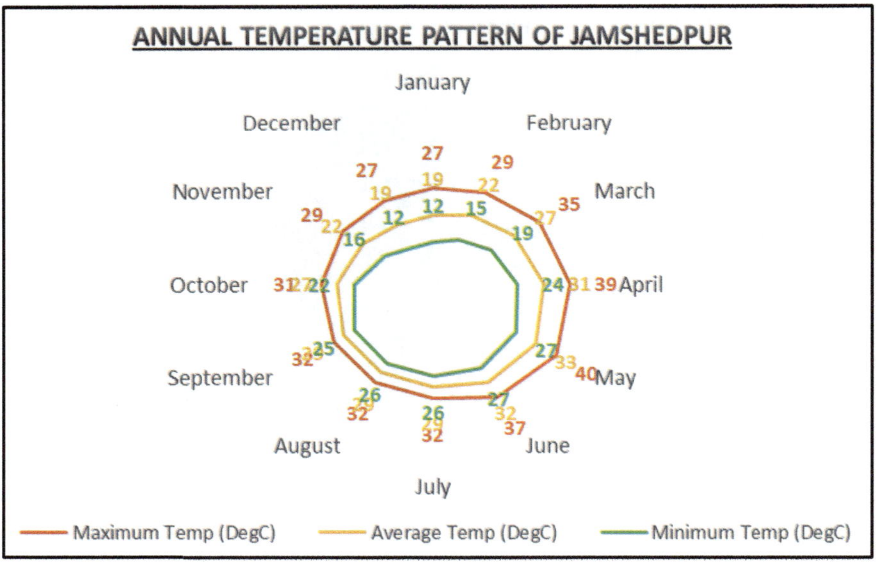

Fig. 3 Temperature pattern of Jamshedpur

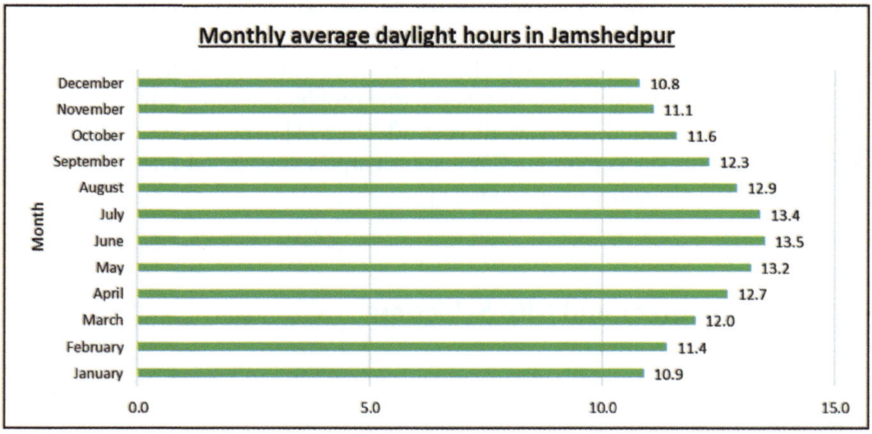

Fig. 4 Average daylight hours in Jamshedpur

The location for solar carport was finalized at latitude 22.96° N and longitude 86.52° E without any shedding effect from the surrounding buildings, sheds or any other structures, etc (Fig. 5; Table 1).

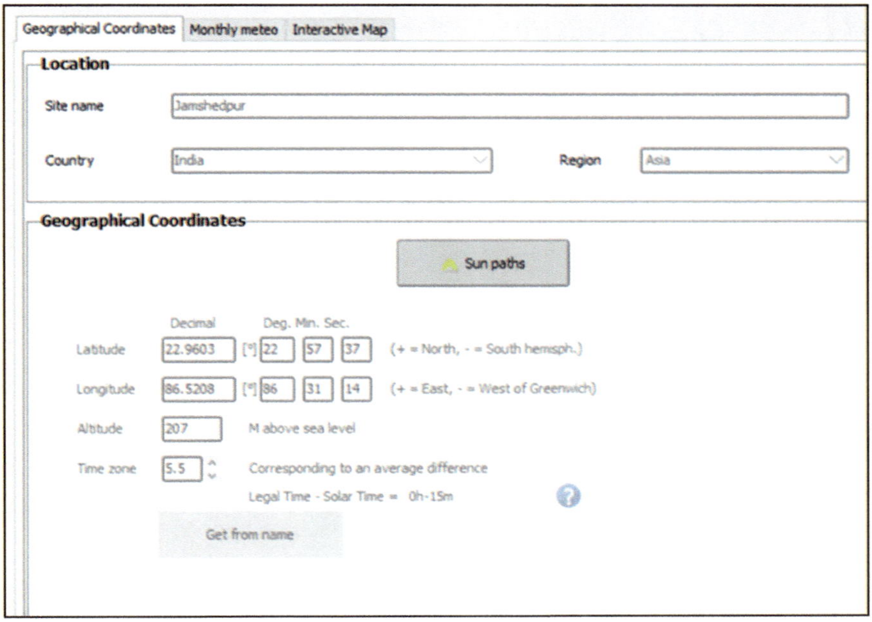

Fig. 5 Geographical location of the solar PV plant

Table 1 Location details of the proposed solar carport

Serial number	Parameters	Values
1	Location	Jamshedpur, Jharkhand
2	Latitude/longitude	22.96° N/86.52° E
3	Tilt angle	5°
4	Interconnection voltage	0.44 kV

The tilt angle and azimuth angle of the proposed plant are considered as 5° and 0°, respectively (Fig. 6).

4 Design of Grid-Connected Solar Carport

4.1 Proposed System Layout

To design a solar carport plant, the designer must know the different parameters of site selection, data related to solar radiation, load data, metering topology, the latest available and suitable components specifications, etc. [9]. To ensure the optimum generation as per the available area, size and capacity of the PV modules need to

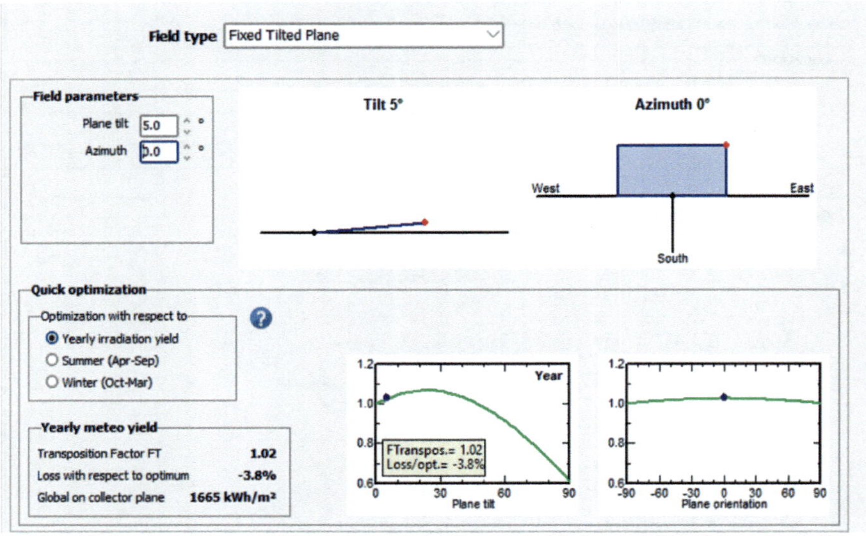

Fig. 6 Tilt angle and azimuth angle considered for the proposed plant

be decided very accurately. After selection of PV module, capacity of the inverters, cables, etc. must be decided considering the efficiency, peak and lean load conditions, and peak generation and lowest generation conditions also [10] (Fig. 7).

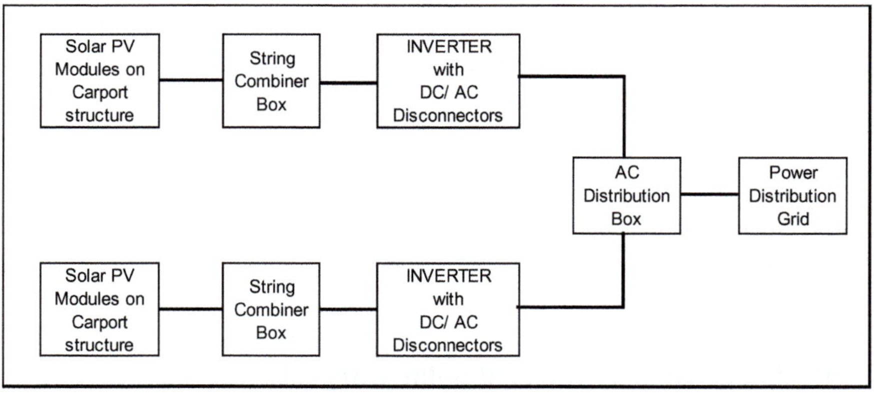

Fig. 7 Block diagram of grid-connected solar PV plant on Carport structure

Table 2 Technical parameters of PV module

Serial number	Parameters	Values
1	PV panel	Monocrystalline
2	Cells per panel	110 numbers
3	Maximum power (Pmax)	550 Wp
4	Maximum power voltage (Vmp)	31.6 V
5	Maximum power current (Imp)	17.40 A
6	Open Ckt. Volt. (Voc)	37.9 V
7	Short Ckt. current (Isc)	18.52 A
8	Power tolerance	0 to +5 W
9	Junction box	IP 68
10	Module efficiency	21.00%

4.2 System Components

4.2.1 Solar PV Module

Considering cost and efficiency, PV panels required to design carport have been decided. For designing of this PV plant, several options have been identified and evaluated considering Mono and Polycrystalline panels of different make and models in terms of lifecycle, performance, warranty, and other critical parameters [5]. Finally, 550 Wp Monocrystalline PV module has been selected and adopted for this project [11]. Detailed technical parameters of PV modules are tabulated in Table 2.

Based on the available space and module size, 1008 number, PV modules have been considered for the plant and the detailed array layout was worked out as shown in Fig. 8 to ensure optimum use of the available area.

4.2.2 Grid-Connected Inverters

Six numbers 80 KW each, three-phase grid-connected inverter has been finalized for this carport. 40% overloading has been taken into account for the inverters [12]. Technical details of inverters are tabulated in Table 3.

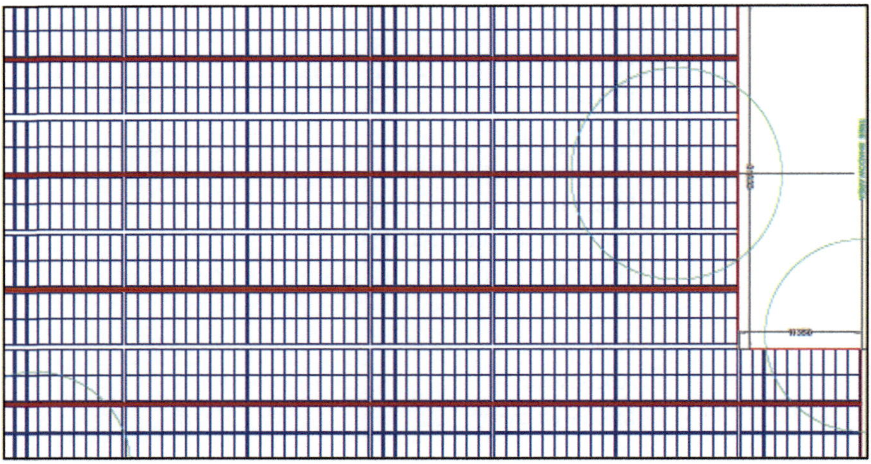

Fig. 8 Proposed array layout of carport area

Table 3 Technical details of inverters

Serial number	Parameters	Values
1	Type	Grid-connected PV inverter
2	Vmax PV/Isc PV (abs. maximum)	1100 V/9 × 40 A
3	MPPT voltage range D.C	180–1000 V
4	Max. input current	9 × 26 A
5	Overvoltage category	II (PV); III(MAINS)
6	Rated output power of each inverter	80,000 W
7	Rated grid voltage/frequency	3/N/PE 230/ 400 V; 50 Hz
8	Max. AC output active power	88,000 W
10	Max. continuous output current	133.7 A
11	Power factor	[−0.8.....1......0.8]

5 Design of Plant by PVsyst Software

Subsequent paragraphs, however, are indented. The complete solar PV plant has been designed with 1008 number solar PV module of 550 Wp monocrystalline panels, and six numbers 80 kW inverter. The total number of string made 42 numbers and module connected in each series connection is 24 numbers. Zero shedding hours have been considered and tilt angle is considered 5° where azimuth angle is considered as 0° [13] (Figs. 9 and 10).

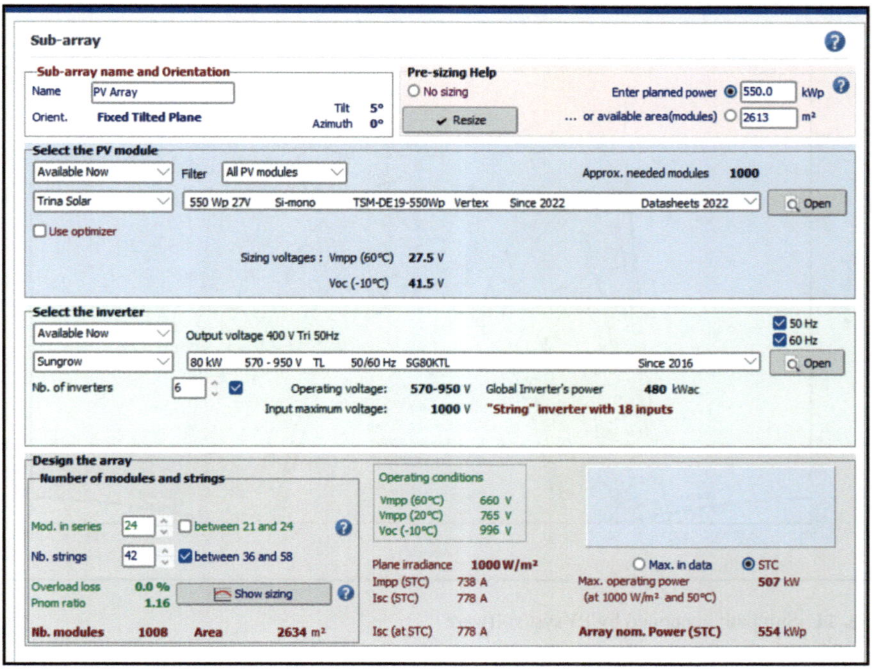

Fig. 9 Design input parameters of solar PV plant

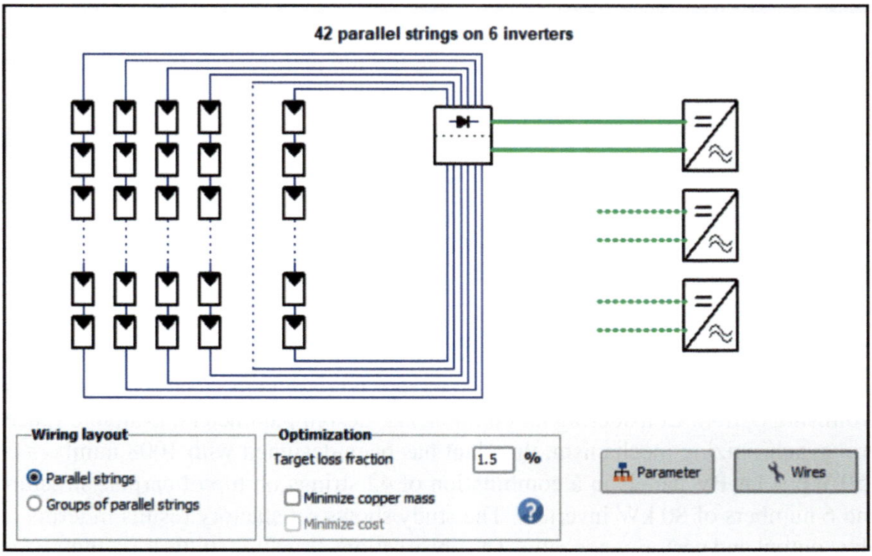

Fig. 10 Configuration of solar PV plant

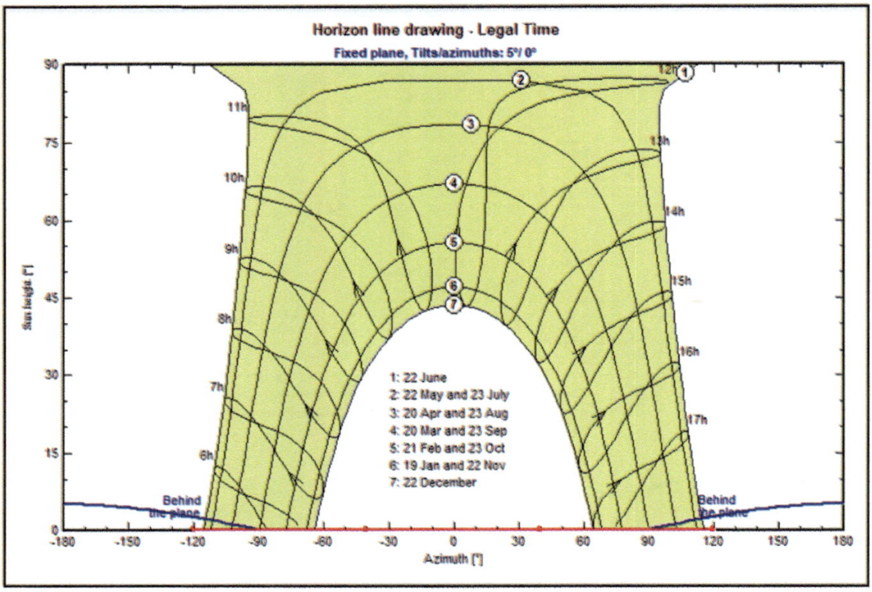

Fig. 11 Sun path generated by PVsyst software

6 Simulation Results from PVsyst Software

A simulation study of the designed solar carport has been conducted through PVsyst software and observed no shading effects in the sun path generated by the software [14]. Total energy generated, total energy injected into the grid, loss factors, performance ratio, CO_2 emission throughout the plant life have also been simulated through the software (Fig. 11).

7 Conclusion

To meet the target of low production cost and low carbon emission, Indian industries are looking at different renewable sources and among them solar PV is the most promising option. Considering all components, design parameters, available space, grid synchronizing mechanism, the plant has been designed with 1008 numbers of 550 Wp solar PV panels in a combination of 42 strings on top of carport structure and 6 numbers of 80 kW inverters. The study shows satisfactory results in terms of plant output and performance ratio. The observations from this study are summarized below:

- The PV plant can deliver 720,130 kWh into the grid in a year with maximum power generation in May and minimum in December (Ref. Fig. 12).

New simulation variant
Energy use and User's needs

	E_Grid kWh	PR ratio
January	55151	0.808
February	55669	0.793
March	71428	0.776
April	72746	0.764
May	76540	0.764
June	62776	0.772
July	56229	0.780
August	54543	0.782
September	56031	0.782
October	55528	0.785
November	53330	0.797
December	50158	0.807
Year	720130	0.782

Fig. 12 PVsyst report of total energy injected into grid and performance ratio (PR) of the PV plant

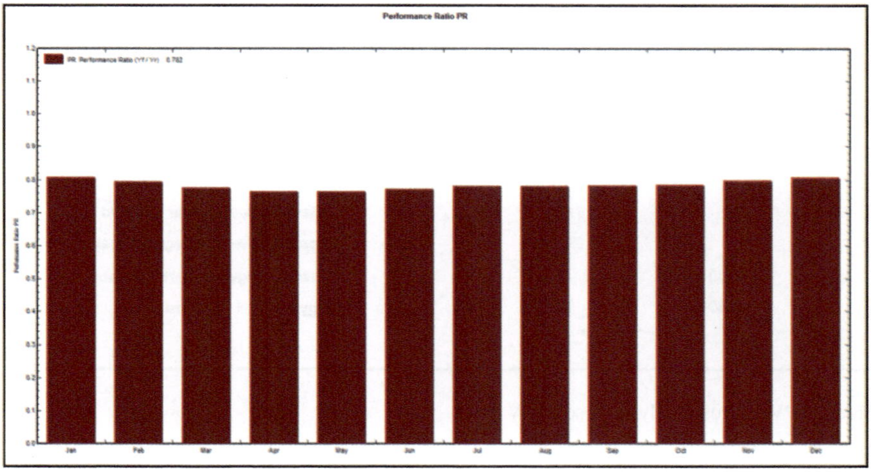

Fig. 13 PVsyst report of performance ratio (PR) of the PV plant

- Total energy generated annually by the plant is 729174 kWh with a yearly system loss figure of 9044 kWh (Ref. Fig. 14).
- Overall performance ratio varies from 76 to 81% with yearly average of 78% (Ref. Fig. 13).
- This plant can save 16,529.101 tons of CO_2 in its projected 30-year lifecycle (Ref. Fig. 15).

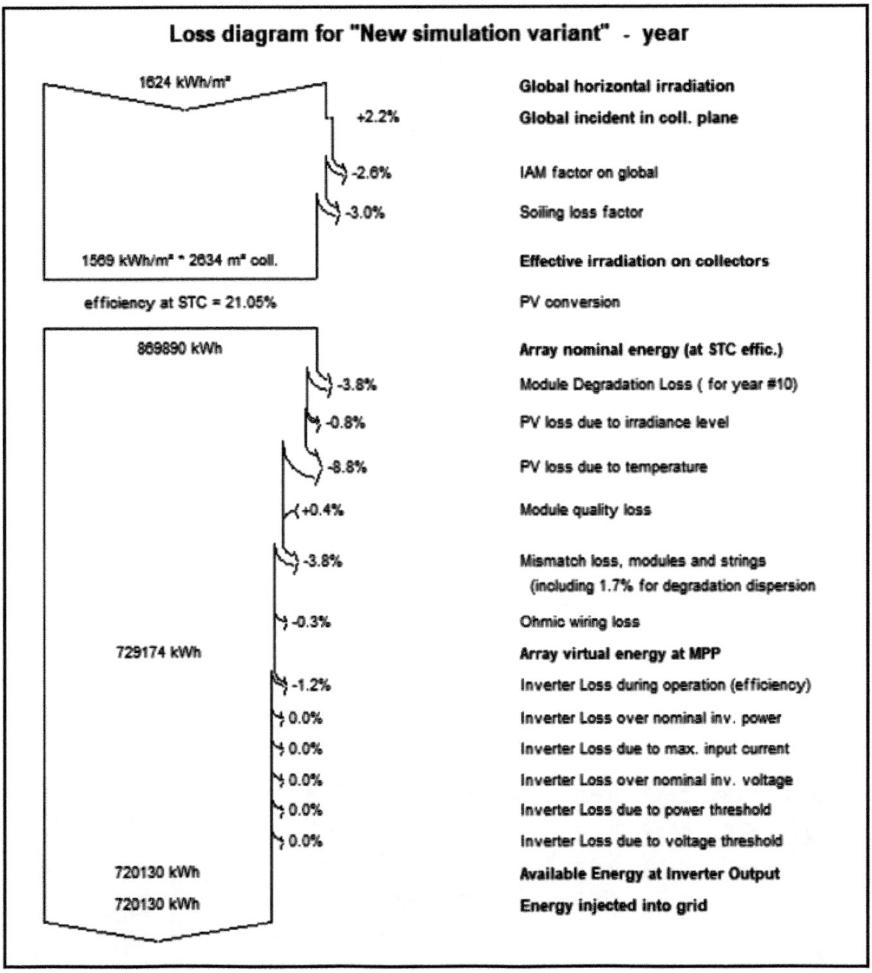

Fig. 14 PVsyst loss analysis of solar PV plant

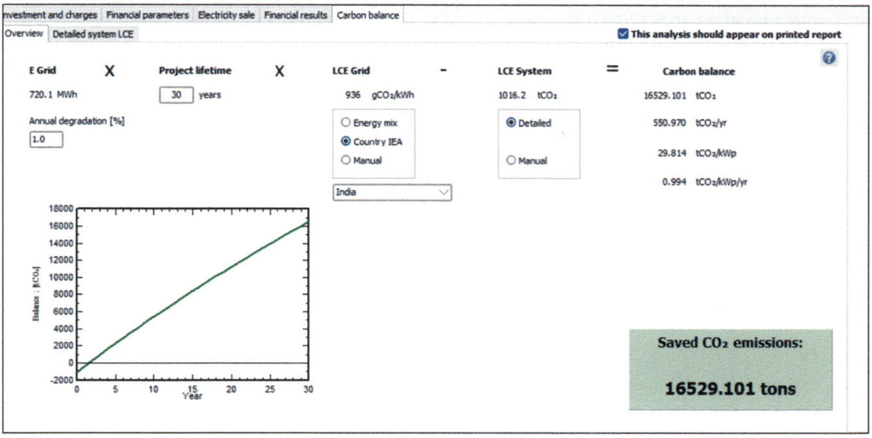

Fig. 15 PVsyst CO_2 emission analysis of solar PV plant

References

1. International Energy Agency. https://www.iea.org/commentaries/india-s-clean-energy-transition-is-rapidly-underway-benefiting-the-entire-world
2. Ministry of New and Renewable Energy. https://www.mnre.gov.in/solar/current-status/
3. Energy Statistics 2022, Ministry of Statistics and Programme Implementation (MOSPI). http://www.mospi.gov.in
4. Hore S, Sakile R, Kumar Sinha U. Design and implementation of 1.43MWp grid connected rooftop solar PV plant. Lecture Notes in Electrical Engineering (LNEE), vol 812. ISBN No: 978-981-16-6970-5. https://doi.org/10.1007/978-981-16-6970-5_41
5. Vasanthkumar, Kumarappa S, Naganagouda H (2017) Design and development of 5MW solar PV grid connected power plant using PVsyst. Int Res J Eng Technol (IRJET) 4(8)
6. Hore S, Sakile R, Kumar Sinha U. Monitoring of grid connected 1.43 MWp rooftop solar PV plant by Internet of Things. Lecture Notes in Electrical Engineering (LNEE), vol 926. ISBN No: 978-981-19-4970-8. https://doi.org/10.1007/978-981-19-4971-5_39
7. Umer F, Shehzad Aslam M, Shoaib Rabbani M, Javed Hanif M, Naeem N, Tafseer Abbas M. Design and optimization of solar carport canopies for maximum power generation and efficiency at Bahawalpur. Hindawi, Int J Photo Energy 2019. Article ID 6372503. https://doi.org/10.1155/2019/6372503
8. Climate and monthly weather forecast Jamshedpur, India. https://www.weather-atlas.com/en/india/jamshedpur-climate
9. Hore S, Sakile R, Kumar Sinha U. Solar power generation and utilization-policies in India. Lecture Notes in Electrical Engineering (LNEE), vol 926. ISBN No: 978-981-19-4974-8. https://doi.org/10.1007/978-981-19-4971-5_66
10. Nirwan D, Thakur T (2017) Performance evaluation of grid connected solar PV plant using PVsyst. Int Res J Eng Technol (IRJET) 4(5). e- ISSN: 2395-0056. p-ISSN: 2395-0072
11. Chandra Rout K, Kulkarni PS. Design and performance evaluation of proposed 2 kW solar PV rooftop on grid system in Odisha using PVsyst. In: 2020 IEEE international students' conference on electrical, electronics and computer science. ISSN: 978-7281-4862-5

12. Hindocha K, Shah S (2020) Design of 50 MW grid connected solar power plant. Int J Eng Res Technol (IJERT) 9(4). ISSN: 2278-0181
13. Alnoosani A, Oreijah M, Alhazmi M, Samkari Y, Faqeha H (2019) Design of 100MW solar PV on-grid connected power plant using (PVsyst) in Umm Al-Qura University. Int J Sci Res (IJSR) 8(11). ISSN: 2319-7064
14. Barua S, Arun Prasath R, Boruah D (2017) Rooftop solar photovoltaic system design and assessment for the academic campus using PVsyst software. Int J Electron Electr Eng 5(1)

Transactive Energy Systems: Current Trends and Future Perspectives

Vidura Bandara Wijekoon and Neeraj Gupta

Abstract As the power grid evolves with the integration of renewable energy sources and advanced communication technologies, Transactive Energy Systems (TES) emerge as a vital solution for managing the energy transition. This paper examines the key components, technologies, and market mechanisms driving TES implementation, focusing on distributed energy resources, smart grid advancements, and secure energy transactions. We also discuss the role of regulatory frameworks and policies, highlighting successful TES deployments and their benefits. Lastly, we explore future trends, challenges, and opportunities in scaling TES, aiming to foster a sustainable and robust energy future (As shown in Fig. 4).

Keywords Transactive energy systems · Distributed energy resources · Smart grid · Artificial intelligence · Regulatory frameworks · Energy storage systems

1 Introduction

The energy sector's landscape is swiftly transforming, with the accelerated integration of renewable energy resources, the transition to decentralized power generation, and the assimilation of groundbreaking technologies like electric vehicles and energy storage systems. This metamorphosis redefines our methodologies for generating, distributing, and utilizing energy, presenting unique hurdles and prospects. Transactive Energy Systems (TES) have been introduced to address these shifts as a potent approach, offering a more streamlined, dependable, and eco-friendly energy framework [1].

V. B. Wijekoon (✉)
Faculty of Engineering, Univeristy of Ruhuna, Galle, Sri Lanka
e-mail: eg20183507@engug.ruh.ac.lk
URL: https://www.ruh.ac.lk/index.php/en/

N. Gupta
National Institute of Technology, Srinagar, JK, India
e-mail: neeraj5822@gmail.com

J. Kumar et al. (eds.), *Recent Advances in Power Systems*, Lecture Notes in Electrical Engineering 1192, https://doi.org/10.1007/978-981-97-2788-9_11

This academic paper offers a thorough analysis of TES, studying contemporary developments, future projections, and the essential components required for the effective execution and comprehensive expansion of TES. It investigates how TES can help accommodate an increasing share of renewable and distributed generation, manage peak demand, improve grid reliability, and enable consumer participation in energy markets.

The paper begins with a comprehensive overview of TES fundamentals, detailing the key principles that underpin transactive energy transactions. This segment elucidates the role of TES in streamlining the active, market-driven transfer of energy, enhancing the equilibrium between energy supply and consumption, and fostering the efficient distribution of resources.

The document investigates the cutting-edge technologies underpinning TES, including blockchain, artificial intelligence (AI), and sophisticated communication infrastructures. It elaborates on utilizing these technologies to fortify energy exchanges, fine-tunes grid functionalities, and handle the intricacies of a dispersed energy environment.

Subsequently, the paper discusses the regulatory frameworks and policies supporting the development and deployment of TES. It underscores the need for adaptive and supportive environments that foster innovation, facilitate investment, and encourage the transition toward a more decentralized, intelligent, and consumer-centric energy system [2].

Furthermore, the paper explores the future trajectories of TES, focusing on the continued integration of emerging technologies, innovative solutions, and new market designs. It also addresses the trends and challenges anticipated to shape the energy sector in the coming years, including the role of prosumers, the impact of digitalization, and the integration of multi-energy systems [3].

Finally, the paper examines the critical aspects of scaling TES, touching upon interoperability, standardization, cybersecurity, and data privacy issues. It identifies the opportunities and barriers that could influence TES deployment, including technical, economic, and regulatory aspects.

This study aims to present precious insights and an all-encompassing comprehension of TES to facilitate a more proficient, ecologically friendly, and robust energy future. It encourages additional investigation and creativity in this area, contributing significantly to the progression of our energy domain.

2 Transactive Energy Systems Fundamentals

Transactive Energy Systems (TES) is an innovative approach to managing energy resources rooted in critical principles to achieve a more efficient, reliable, and sustainable energy landscape. The fundamentals of TES revolve around the following transaction energy principles [11].

1. Highly automated, coordinated self-optimization: TES utilizes advanced algorithms and real-time data to enable the automated and coordinated management of energy resources, ensuring optimal performance and efficiency.
2. Non-discriminatory participation: TES promotes an inclusive energy marketplace by allowing qualified participants, including utilities, prosumers, and third-party providers, to participate on equal terms, fostering competition and innovation.
3. Accountability for performance standards: In TES, transacting parties are responsible for meeting established performance and reliability standards, ensuring high-quality service and a stable energy infrastructure.
4. Observable and auditable interfaces: TES ensures transparency and trust by facilitating the observability and auditability of energy transactions, promoting secure and reliable energy exchanges.
5. Maintaining system reliability and control: TES prioritizes preserving system reliability and control while seamlessly integrating distributed energy resources, such as renewables, energy storage systems, and electric vehicles.
6. Scalability, adaptability, and extensibility: TES is designed to be easily scaled, adapted, and extended across various devices, participants, and geographic areas, allowing for the widespread deployment and integration of transactive energy solutions.

2.1 Distributed Energy Resources

Distributed Energy Resources (DERs) are essential components of Transactive Energy Systems (TES), as they provide the necessary flexibility, resilience, and efficiency to optimize energy supply and demand. Four main categories of DERs play a crucial role in sustaining effective TES, which include distributed generation, energy storage systems, electric vehicles, and smart loads. This section provides a detailed description of each category and explains their significance in maintaining a robust transactive energy level [5].

2.1.1 Distributed Generation

Distributed generation (DG) refers to producing electricity from small-scale, decentralized energy sources near the consumption point. Examples of DG include solar photovoltaic panels, wind turbines, and combined heat and power (CHP) systems. DG reduces the reliance on traditional centralized power plants, minimizes transmission and distribution losses, and contributes to a more resilient and environmentally friendly energy infrastructure. In TES, DG allows for real-time energy transactions between prosumers (producers and consumers), promoting efficient energy utilization and enabling demand-side management.

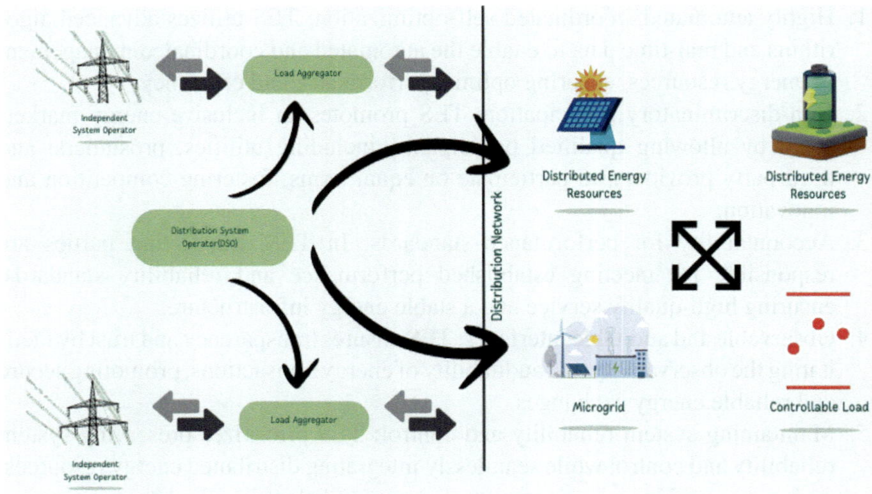

Fig. 1 Transactive energy architecture in distribution networks

2.1.2 Energy Storage Systems

Energy storage systems (ESS) store surplus energy during periods of low demand or high production and release it during periods of high demand or low production. Examples of ESS include batteries, pumped hydro storage, and compressed air energy storage. ESS plays a critical role in TES by providing grid stability, enabling the integration of intermittent renewable energy sources, and allowing for peak-shaving and load-shifting. The integration of ESS in TES enables more efficient energy management and enhances the reliability and resilience of the overall energy system (Fig. 1). Transactive energy architecture in distribution networks (adapted from [1])).

The transactive energy structure in distribution networks showcases the role of Independent System Operators (ISOs) in guiding load aggregators connected to the distribution network. The Distribution System Operator (DSO) oversees two simultaneous pathways involving load aggregators and their relationship with the distribution network. Furthermore, the DSO manages distributed energy resources, microgrids, and controllable loads connected to the system. This configuration emphasizes the significance of coordination and communication among ISOs, DSOs, load aggregators, and other associated parties in refining energy distribution and utilization in a transactive energy framework (Fig. 1. Transactive energy architecture in distribution networks (adapted from [1]).

2.1.3 Electric Vehicles

Electric vehicles (EVs) are increasingly becoming an essential component of modern energy systems, as they contribute to reducing greenhouse gas emissions and serve as mobile energy storage units. In TES, EVs can participate in vehicle-to-grid (V2G) and vehicle-to-building (V2B) schemes, selling stored energy back to the grid or buildings during peak demand periods or charging during off-peak periods. This bidirectional energy flow enhances the flexibility and efficiency of TES and further supports the integration of renewable energy sources.

2.1.4 Smart Loads

Smart loads are energy-consuming devices with communication and control technologies that respond dynamically to grid conditions, price signals, and user preferences. Examples of smart loads include smart thermostats, smart appliances, and demand-responsive industrial processes. In TES, smart loads play a vital role in demand-side management, as they can adjust their consumption patterns based on real-time energy prices, grid conditions, and user-defined settings. This demand flexibility contributes to the transactive energy system's efficiency, resilience, and sustainability.

In conclusion, DERs, including distributed generation, energy storage systems, electric vehicles, and smart loads, are essential for sustaining a robust transactive energy level. Their integration into TES enables flexible and efficient energy management, enhances the resilience and reliability of the energy infrastructure, and supports the transition toward a more sustainable and decentralized energy landscape.

2.2 Smart Grid Technologies

Smart grid technologies form the backbone of Transactive Energy Systems (TES), enabling the efficient integration of distributed energy resources, advanced communication systems, and market-based approaches to optimize energy supply and demand. These technologies provide the necessary intelligence, control, and communication capabilities to facilitate TES's secure, reliable, and real-time energy transactions, as shown in Fig. 2[2]. This section describes key smart grid technologies, focusing on their role in enabling efficient and resilient transactive energy systems [3].

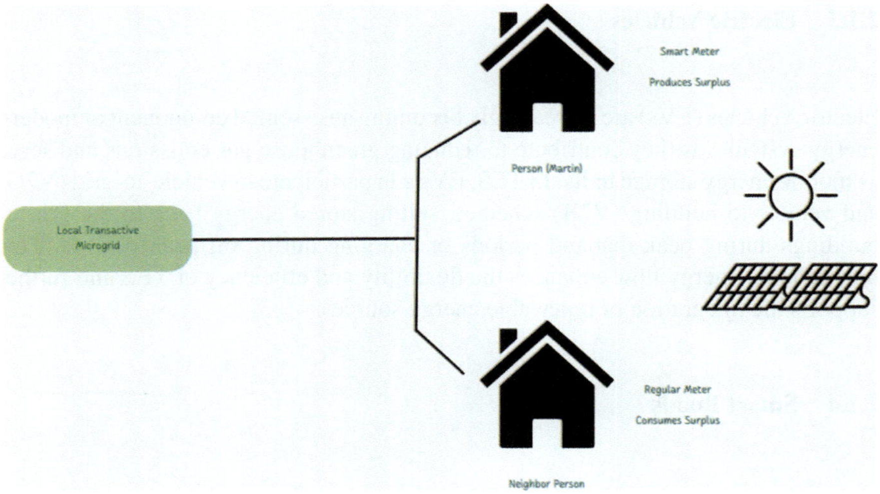

Fig. 2 TransActive grid

2.2.1 Advanced Metering Infrastructure (AMI)

Advanced Metering Infrastructure (AMI) comprises smart meters, communication networks, and data management systems that allow for the remote and real-time monitoring, control, and analysis of energy consumption data. AMI enables two-way communication between utilities and consumers, providing accurate and timely information on energy usage patterns and price signals. In TES, AMI is essential for demand-side management. It allows consumers and prosumers to make informed decisions based on real-time energy prices and grid conditions, thus facilitating efficient energy transactions and grid optimization.

Figure 2 portrays a neighborhood transactive microgrid connecting two nearby homes, such as Martin's residence and the house next door. Both homes are equipped with solar panel systems, and Martin's home generates excess energy that the adjacent residence consumes. Each dwelling incorporates smart meters as an Advanced Metering Infrastructure (AMI) component. This arrangement emphasizes the significance of AMI in promoting efficient energy transfers within the transactive grid by enabling the sharing of surplus power between the households, thereby contributing to a more robust and environmentally friendly energy network (Fig. 2. TransActive Grid (adapted from [3])).

2.2.2 Distribution Automation (DA)

Distribution Automation (DA) refers to deploying sensors, control devices, and communication systems throughout the distribution network to monitor, control, and optimize the grid's performance. DA technologies enhance the distribution system's reliability, efficiency, and resilience by enabling real-time fault detection, isolation, and restoration, as well as voltage and reactive power control. In TES, DA is crucial for maintaining grid stability and ensuring the seamless integration of distributed energy resources, thereby supporting efficient energy transactions and grid optimization [6].

2.2.3 Demand Response (DR)

Demand Response (DR) is a smart grid strategy that encourages consumers to adjust their energy consumption patterns in response to grid conditions, price signals, or incentive programs. DR can be achieved through various means, such as direct load control, time-varying pricing, and behavioral demand response programs. In TES, DR plays a pivotal role in demand-side management by enabling consumers and prosumers to actively participate in energy transactions and optimize their consumption based on real-time energy prices and grid conditions. This demand flexibility contributes to the transactive energy system's efficiency, resilience, and sustainability.

2.2.4 Cybersecurity and Communication Technologies

Cybersecurity and communication technologies are essential for ensuring TES's secure, reliable, and efficient operation. Advanced communication technologies, such as wireless networks, fiber optics, and power line communications, provide the necessary connectivity and data exchange between various smart grid components, facilitating real-time energy transactions and grid optimization. Meanwhile, cybersecurity measures are vital for protecting the confidentiality, integrity, and availability of critical grid infrastructure and sensitive consumer data. In TES, robust cybersecurity and communication technologies enable secure and transparent energy transactions, safeguard grid stability, and ensure consumer trust.

In summary, smart grid technologies, including Advanced Metering Infrastructure, Distribution Automation, Demand Response, and cybersecurity and communication technologies, are integral to successfully implementing Transactive Energy Systems. These technologies enable efficient energy management, enhance the resilience and reliability of the energy infrastructure, and support the integration of distributed energy resources, ultimately contributing to a more sustainable and decentralized energy landscape.

2.3 Market-Based Approaches

Market-based approaches play a crucial role in implementing and operating Transactive Energy Systems (TES) by providing the necessary incentives, price signals, and mechanisms to encourage efficient and flexible use of energy resources. These approaches create a competitive environment where energy producers, consumers, and prosumers can actively participate in energy transactions, optimizing energy supply and demand in real time. This section describes key market-based approaches used in TES, focusing on their significance in promoting efficient, reliable, and sustainable energy management.

2.3.1 Dynamic Pricing

Dynamic pricing is a market-based approach that reflects the real-time cost of energy production, transmission, and distribution and other factors such as congestion, environmental impact, and scarcity. Examples of dynamic pricing schemes include time-of-use (TOU) rates, critical peak pricing (CPP), and real-time pricing (RTP). In TES, dynamic pricing encourages consumers and prosumers to adjust their energy consumption and production patterns in response to price signals, ultimately contributing to efficient energy transactions, peak load reduction, and grid optimization.

2.3.2 Demand-Side Bidding

Demand-side bidding (DSB) allows consumers and prosumers to submit bids for energy consumption, production, or curtailment based on their preferences, willingness to pay, or generation capabilities. DSB fosters active consumer participation in TES, enabling them to influence energy prices and grid conditions. DSB contributed to the transactive energy system's overall efficiency, resilience, and sustainability by incentivizing demand-side management and distributed energy resource integration.

2.3.3 Capacity Markets and Ancillary Services

Capacity markets and ancillary services are market-based approaches to ensure the power system's long-term adequacy, stability, and reliability. Capacity markets incentivize investments in generation, storage, and demand-side resources to meet future demand. At the same time, ancillary services maintain the short-term balance and stability of the grid through frequency regulation, voltage control, and operating reserves. In TES, capacity markets and ancillary services create additional revenue streams for distributed energy resource owners and enable their integration into the grid, thus promoting efficient energy management and enhancing grid resilience.

2.3.4 Peer-to-Peer Energy Trading

Peer-to-peer (P2P) energy trading is a decentralized market-based approach that allows energy producers, consumers, and prosumers to trade energy directly, bypassing traditional intermediaries such as utilities or energy retailers. P2P energy trading can be facilitated through platforms like blockchain, which ensures secure, transparent, and automated transactions. In TES, P2P energy trading fosters local energy exchange, community engagement, and demand-side management, ultimately contributing to the energy system's overall efficiency, resilience, and sustainability [7].

In conclusion, market-based approaches, including dynamic pricing, demand-side bidding, capacity markets and ancillary services, and peer-to-peer energy trading, are essential for successfully implementing Transactive Energy Systems. These approaches incentivize the efficient and flexible use of energy resources, promote active consumer participation, and support the integration of distributed energy resources, ultimately contributing to a more efficient, reliable, and sustainable energy landscape.

3 Enabling Technologies for Transactive Energy Systems

Enabling technologies play a pivotal role in successfully implementing and operating Transactive Energy Systems (TES) by providing the necessary tools, platforms, and intelligence to facilitate efficient, secure, and real-time energy transactions. These technologies enhance the coordination, optimization, and integration of distributed energy resources, market mechanisms, and smart grid technologies in TES. This section describes key enabling technologies for TES, focusing on their significance in promoting efficient, reliable, and sustainable energy management [8].

3.1 Blockchain for Secure Energy Transactions

Blockchain is a distributed ledger technology that enables secure, transparent, and tamper-proof data storage and transactions without centralized intermediaries. In TES, blockchain can facilitate peer-to-peer (P2P) energy trading, ensuring secure, automated, and transparent energy transactions between energy producers, consumers, and prosumers. By providing a trustless and decentralized platform, blockchain promotes local energy exchange, community engagement, and demand-side management, ultimately contributing to the energy system's overall efficiency, resilience, and sustainability [8].

3.2 Artificial Intelligence and Machine Learning for TES Coordination and Optimization

Artificial Intelligence (AI) and Machine Learning (ML) are advanced computational techniques that enable machines to learn, adapt, and make decisions based on data analysis. AI and ML can optimize energy transactions, coordinate distributed energy resources, and enhance demand-side management in TES. Examples of AI and ML applications in TES include forecasting energy generation and consumption, optimizing energy storage operations, and enabling real-time energy pricing. By providing advanced analytics and decision-making capabilities, AI and ML contributed to the transactive energy system's overall efficiency, reliability, and resilience [9].

3.3 Advanced Communication Technologies

Advanced communication technologies are essential for ensuring the seamless exchange of data and information between various smart grid components, facilitating real-time energy transactions, and grid optimization in TES. These technologies include wireless networks (e.g., 5G), fiber optics, power line communications, and Internet of Things (IoT) devices. In TES, advanced communication technologies enable two-way communication between utilities, consumers, and distributed energy resources, providing the necessary connectivity and data exchange for efficient energy management. Additionally, these technologies support the integration of advanced metering infrastructure, distribution automation, and demand response strategies, further enhancing the resilience and reliability of the energy infrastructure.

In summary, enabling technologies, including blockchain, artificial intelligence, machine learning, and advanced communication technologies, are integral to successfully implementing Transactive Energy Systems. These technologies enable secure and transparent energy transactions, enhance coordination and optimization of distributed energy resources, and provide the necessary connectivity and data exchange for efficient energy management. Ultimately, these enabling technologies contribute to a more efficient, reliable, and sustainable energy landscape.

4 Regulatory Frameworks and Policies in Transactive Energy Systems

Regulatory frameworks and policies play a crucial role in the development, implementation, and operation of Transactive Energy Systems (TES) by setting the rules, standards, and incentives that guide the behavior of market participants and ensure the

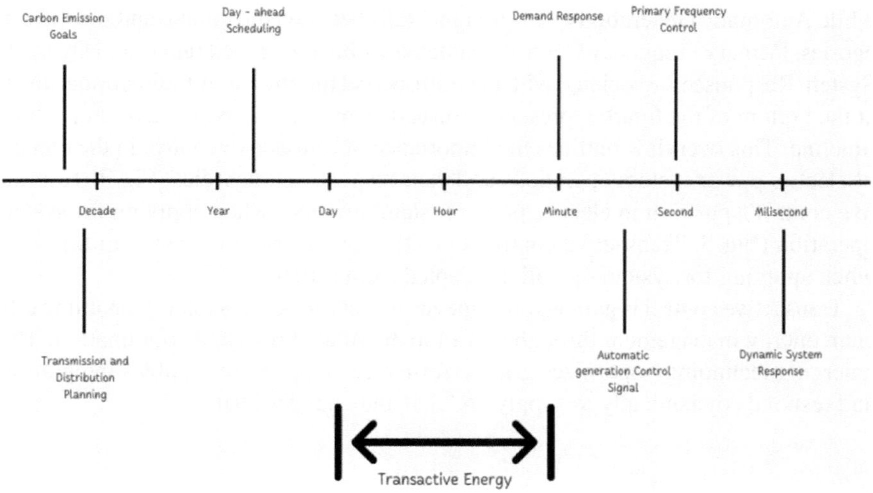

Fig. 3 Transactive control's position in electric power system timelines when applying for system operation

overall stability, reliability, and fairness of the energy system. An adaptive and supportive regulatory environment fosters innovation, investment, and consumer engagement in TES. This section provides a detailed description of the role of regulation in TES development, the need for an adaptive and supportive regulatory environment, and case studies of successful TES implementations [2].

4.1 The Role of Regulation in TES Development

Regulation plays a pivotal role in TES development by shaping the market structure, incentives, and requirements for integrating distributed energy resources, smart grid technologies, and market-based approaches. Key regulatory aspects that influence TES development include grid access and interconnection rules, technical standards for equipment and communication, data privacy and security regulations, and market rules for energy transactions and pricing. By providing a clear and stable regulatory framework, policymakers can encourage investment in TES, support the adoption of innovative technologies and practices, and ensure the equitable distribution of benefits among different stakeholders [2].

Figure 3 presents a timeline displaying various electric power system operations on a scale ranging from milliseconds to decades. It highlights the position of transactive control within these operations. Carbon emission goals and Transmission and Distribution Planning span over a decade, while Day-ahead Scheduling occurs between the year and day marks. Demand response occurs within the minute range, while Automatic Generation Control Signal falls between the minute and second categories. Primary Frequency Control is situated within the second range, and Dynamic System Response is associated with the millisecond timeframe. A bidirectional arrow at the bottom of the figure represents transactive energy, extending across the entire timeline. This overview outlines the importance of transactive control in the context of electric power system operations and their respective timescales (Fig. 3. Transactive control's position in electric power system timelines when applying for system operation (Fig. 3. Transactive control's position in electric power system timelines when applying for system operation(adapted from [10])).

Transactive control is gaining prominence in electric power systems, enabling efficient energy management through real-time distributed resource coordination. This emerging technology optimizes grid performance, supports renewables integration, and responds dynamically to supply and demand changes [10].

4.2 Adaptive and Supportive Regulatory Environment

An adaptive and supportive regulatory environment is essential for fostering the growth and success of TES, as it enables the energy sector to respond to emerging trends, challenges, and opportunities. Adaptive regulation involves continuously evaluating and modifying policies, rules, and standards based on technological advancements, market developments, and lessons from pilot projects and case studies. Supportive regulation, on the other hand, entails the provision of incentives, grants, and other policy tools that encourage adopting and scaling TES solutions. By creating an adaptive and supportive regulatory environment, policymakers can facilitate the transition to a more decentralized, intelligent, and sustainable energy landscape as shown in Fig. 3 [2].

5 Current Scenario in Transactive Energy Systems

Transactive Energy Systems (TES) are currently concentrating on developing and deploying novel technologies and methodologies to manage the power grid effectively and assimilate distributed energy resources (DERs) efficiently. TES aims to establish a market-driven structure that facilitates various energy stakeholders' participation in transparent and efficient energy exchange, optimizing grid operations and improving system resilience.

The power grid faces new challenges in managing supply and demand, ensuring grid stability, and avoiding congestion due to the rapid expansion of renewable energy sources such as solar and wind, as well as the rising adoption of energy storage systems, electric vehicles, and demand response programs. By harnessing advanced technologies like blockchain, artificial intelligence, and the Internet of Things (IoT),

TES offers a solution to these challenges by enabling secure and seamless energy transactions among diverse participants.

Furthermore, regulatory frameworks and policies adapt to support TES deployment by fostering an environment conducive to widespread adoption. These policies frequently emphasize promoting DER usage, increasing grid flexibility, and encouraging market-oriented mechanisms for efficient energy exchange.

In a nutshell, the current landscape of TES revolves around:

1. Integrating DERs into the power grid.
2. Utilizing state-of-the-art technologies for secure and efficient energy transactions.
3. Developing suitable regulatory frameworks and policies that support TES adoption.
4. Investigating innovative business models and market designs that facilitate TES implementation.
5. Evaluating the potential advantages and challenges associated with large-scale TES deployment.

5.1 Implementation A—Brooklyn Microgrid Project

The Brooklyn Microgrid project in New York, USA, where a blockchain- enabled P2P energy trading platform to facilitate local energy transactions among prosumers and consumers, supported by regulatory waivers and pilot program provisions. This project covers the current landscape of Transactive Energy landscape [11].

1. Integrating DERs into the power grid: The project connected local prosumers (producing and consuming energy) and consumers within a microgrid. The energy generated by solar panels and other DERs was traded within the community, promoting local energy use and increasing the integration of renewable energy sources into the power grid.
2. Utilizing state-of-the-art technologies for secure and efficient energy transactions: The project employed a blockchain-enabled P2P energy trading platform, which ensured secure and transparent energy transactions among participants. Blockchain technology provides an immutable record of transactions and facilitates real-time energy trading.
3. Developing suitable regulatory frameworks and policies that support TES adoption: The Brooklyn Microgrid project operated under regulatory waivers and pilot program provisions, allowing for experimentation and testing of the P2P energy trading model. This support from regulatory authorities helped create a favorable environment for TES adoption.
4. Investigating innovative business models and market design that facilitate TES implementation: The project explored a new business model for local energy transactions among prosumers and consumers. By enabling P2P energy trading, the Brooklyn Microgrid created a new market design that allowed participants

to buy and sell locally produced energy, fostering community engagement and energy independence.

5. Evaluating the potential advantages and challenges associated with large-scale TES deployment: While the Brooklyn Microgrid project served as a successful pilot, scaling up the TES model to larger areas would require addressing challenges such as grid stability, cybersecurity, and regulatory support. The project provided valuable insights into the advantages of local energy trading and the potential obstacles to be overcome for broader TES adoption

5.2 Implementation B—PowerMatcher Project in the Netherlands

The PowerMatcher project in the Netherlands applied the following functionalities [12]:

1. Integrating DERs into the power grid: The project used a multi-agent system to coordinate and optimize energy consumption and production among distributed energy resources. This approach facilitated the integration of various DERs, such as solar panels and energy storage systems, into the grid by efficiently managing energy flows.
2. Utilizing state-of-the-art technologies for secure and efficient energy transactions: The multi-agent system employed advanced algorithms and communication protocols to enable real-time coordination and optimization of energy consumption and production. This technological approach ensured efficient energy transactions and improved grid management.
3. Developing suitable regulatory frameworks and policies that support TES adoption: The PowerMatcher project operated within a regulatory sandbox, which provided flexibility in market rules and grid tariffs. This supportive regulatory environment allowed for the experimentation and testing of innovative TES concepts, promoting their adoption and development.
4. Investigating innovative business models and market design that facilitate TES implementation: The project explored a new market design by leveraging a multi-agent system to coordinate and optimize energy transactions. This innovative approach enabled more efficient energy management and the potential for new business models supporting DER integration.
5. Evaluating the potential advantages and challenges associated with large-scale TES deployment: The PowerMatcher project offered insights into the benefits of coordinated energy management, such as increased grid efficiency and improved DER integration. However, challenges to be addressed for large-scale TES deployment could include ensuring cybersecurity, maintaining grid stability, and gaining widespread regulatory support. The project's experience within the regulatory sandbox provided valuable lessons for addressing these challenges in future TES implementations.

5.3 Implementation C—Simulation-Based Valuation of Transactive Energy Systems

The Simulation-Based Valuation of Transactive Energy Systems project encompasses the following functionalities [13]:

1. Integrating DERs into the power grid: The project's focus on Transactive Energy (TE) promotes the integration of responsive loads and distributed energy resources, thus advancing grid modernization. The comprehensive simulation-based TE valuation method helps assess the impact of DERs within the grid.
2. Utilizing state-of-the-art technologies for secure and efficient energy transactions: The open-source simulation platform integrates transmission, distribution, and building simulators with TE and non-TE agents through the Framework for Network Co-Simulation (FNCS). This advanced approach ensures the secure and efficient analysis of energy transactions in the TE context.
3. Developing suitable regulatory frameworks and policies that support TES adoption: The project's methodical approach in defining valuation scenarios, use cases, baseline, and valuation metrics provides a structured foundation for assessing the potential benefits of TE systems. This process can inform the development of supportive regulatory frameworks and policies for TES adoption.
4. Investigating innovative business models and market design that facilitate TES implementation: The separation of simulation and valuation and base and final valuation metrics enhances the transparency and flexibility of the project's valuation. This allows for a more in-depth investigation of innovative business models and market designs that can facilitate TES implementation.
5. Evaluating the potential advantages and challenges associated with large-scale TES deployment: The developed TE simulation program and valuation method allow for systematic analysis of the benefits and challenges of TE systems. The project's example, based on the Smart Grid Interoperability Panel (SGIP) Use Case 1, demonstrates the potential value of TE and highlights the considerations for large-scale deployment.

In conclusion, current trends in TES highlight its potential to transform the power sector by creating a more adaptable, efficient, and eco-friendly energy environment.

6 Future Perspectives of Transactive Energy Systems

As the energy sector continues to evolve, Transactive Energy Systems (TES) are poised to play an increasingly important role in shaping the future of energy management, production, and consumption. Advancements in technology, coupled with supportive regulatory frameworks and the integration of new resources, will further enhance the potential of TES to create a more efficient, resilient, and sustainable energy landscape. This section provides a detailed description of the future perspectives of TES, focusing on integrating electric vehicles, microgrids, and energy storage systems, and emerging trends and challenges.

6.1 Integration of Electric Vehicles

The widespread adoption of electric vehicles (EVs) presents both opportunities and challenges for TES. On the one hand, EVs can act as flexible loads and distributed energy resources, contributing to demand-side management and grid optimization through vehicle-to-grid (V2G) and vehicle-to-building (V2B) technologies. On the other hand, the increased demand for electricity due to EV charging can strain the grid, necessitating the development of smart charging strategies, dynamic pricing schemes, and the integration of renewable energy sources. In the future, TES can facilitate EVs' efficient and sustainable integration by enabling real-time energy transactions, coordinating charging and discharging schedules, and incentivizing clean and distributed energy resources [14].

6.2 Microgrids and Energy Storage Systems

Microgrids and energy storage systems are essential to the future energy landscape, providing localized energy management, resilience, and autonomy. TES can support the operation and integration of microgrids and energy storage systems by coordinating the dispatch of generation, storage, and demand-side resources and facilitating energy transactions within and across microgrids. Moreover, TES can optimize energy storage operations, considering real-time energy prices, grid conditions, and customer preferences. As the deployment of microgrids and energy storage systems continues to grow, TES will play a vital role in unlocking their full potential, enhancing grid resilience and sustainability.

6.3 Emerging Trends and Challenges

Several emerging trends and challenges will shape the future of TES, including the increasing penetration of distributed energy resources, the growing complexity of energy markets, and the need for enhanced cybersecurity and data privacy. To address these challenges and capitalize on emerging opportunities, TES must continuously adapt and evolve, leveraging advancements in enabling technologies such as AI, ML, and blockchain and fostering innovation through adaptive and supportive regulatory frameworks. In addition, TES must tackle the issue of social equity, ensuring that the benefits of decentralized and intelligent energy systems are accessible to all members of society [15].

In conclusion, integrating electric vehicles, microgrids, and energy storage systems and emerging trends and challenges strongly influence the future perspectives of Transactive Energy Systems. By leveraging technological advancements and fostering supportive regulatory environments, TES can play a pivotal role in shaping a more efficient, resilient, and sustainable energy future, ultimately contributing to the global transition toward a decentralized, intelligent, and equitable energy landscape.

7 Scaling Transactive Energy Systems

As Transactive Energy Systems (TES) continue to evolve and gain traction, it is essential to address the challenges and opportunities associated with scaling these systems to maximize their potential in transforming the energy landscape. Successfully scaling TES requires addressing interoperability and standardization, cybersecurity concerns, and identifying opportunities and barriers in TES deployment. This section describes these critical aspects of scaling TES, emphasizing their significance in promoting efficient, reliable, and sustainable energy management.

7.1 Interoperability and Standardization

Interoperability and standardization are crucial to successfully scaling TES, ensuring seamless integration and communication among various technologies, components, and stakeholders. As TES involves diverse devices, systems, and market participants, establishing common protocols, data formats, and technical standards is essential to facilitate real-time energy transactions, coordination of distributed energy resources, and data exchange. Collaborative efforts among industry, academia, and policymakers must develop and adopt comprehensive and flexible standards to accommodate the continuous evolution of TES technologies and practices, ultimately promoting large-scale deployment and market growth [1].

7.2 Cybersecurity Concerns

TES's increasing interconnectedness and complexity raises critical cybersecurity concerns that must be addressed to ensure the energy infrastructure's reliability, resilience, and security. Cyber threats, such as unauthorized access, data breaches, and malicious attacks, can compromise the integrity and availability of TES, potentially leading to disruptions in energy supply and demand and compromising sensitive information. Implementing robust cybersecurity measures, such as encryption, authentication, and intrusion detection, is essential to protect TES from potential threats. Moreover, fostering a culture of cybersecurity awareness among stakeholders, and developing comprehensive cybersecurity policies and standards, will play a vital role in safeguarding TES as they scale [16].

8 Opportunities and Barriers in TES Deployment

Various opportunities and barriers exist in deploying Transactive Energy Systems (TES), influencing their large-scale adoption and impact. Opportunities include the increasing penetration of distributed energy resources [17] advancements in enabling technologies such as artificial intelligence (AI), machine learning (ML), blockchain, advanced communication, and supportive regulatory frameworks that encourage innovation and investment. These factors provide a strong foundation for the widespread deployment of TES and for realizing their full potential in transforming the energy landscape [18].

Conversely, barriers to TES deployment may include technical challenges, such as integration of legacy systems, grid stability, and power quality; economic factors, such as high upfront costs and lack of financial incentives; and social issues, such as consumer awareness, acceptance, and participation. Identifying and addressing these barriers through targeted policy interventions, technological advancements, and stakeholder engagement will be critical in facilitating the scaling of TES and ensuring its long-term success [19].

In conclusion, scaling Transactive Energy Systems requires addressing critical aspects such as interoperability and standardization, cybersecurity concerns, and opportunities and barriers in TES deployment. By addressing these challenges and capitalizing on emerging opportunities, TES can play a pivotal role in transforming the energy sector, ultimately contributing to a more efficient, reliable, and sustainable energy future.

Based on the research pipeline, below graphical abstract can be created as follows (Fig. 4).

9 Conclusion

With this detailed review concerning Transactive Energy Systems: Current Trends and Future Perspectives, following observations, have been obtained.

1. Transactive Energy Systems have the potential to revolutionize the energy sector by enabling flexible, scalable, and secure energy management. By leveraging distributed energy resources, smart grid technologies, and market-based approaches, TES can contribute to increased efficiency, resilience, and sustainability.

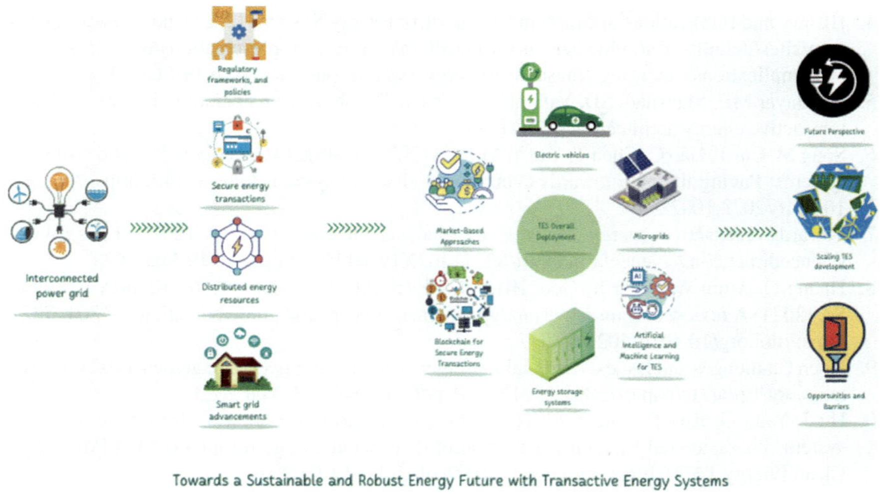

Fig. 4 Paper graphical abstract of the research development

2. To fully realize their potential, it is crucial to address interoperability, standardization, and cybersecurity challenges and establish a supportive and adaptive regulatory environment. This paper provides a foundation for understanding TES's current trends and future perspectives, encouraging further research and innovation to pave the way for a more sustainable and resilient energy future.
3. The future of TES lies in the continued integration of emerging technologies and innovative solutions, such as electric vehicles, microgrids, and energy storage systems, as well as in addressing the trends and challenges that will shape the energy sector in the coming years.

Acknowledgements I would like to express my sincere gratitude to Prof. Neeraj Gupta, my mentor, for his invaluable guidance, support, and encouragement throughout the development of this research paper on Transactive Energy Systems: Current Trends and Future Perspectives. His insights and expertise have been instrumental in shaping my understanding of the topic and its significance in the broader context of the energy sector.

References

1. What Is Transactive Energy? https://blockchain.ieee.org/verticals/transactive-energy/topics/what-is-transactive-energy. Accessed 21 Mar 2023
2. Zhou H, Li B, Zong X, Chen D (2022) Transactive energy system: concept, configuration, and mechanism. Journal 10. https://doi.org/10.3389/2022.1057106
3. Lilic J (2023) TransActive Grid 1–13

4. History and Implications of Emerging Transactive Energy Systems (2023). https://www.psma.com/sites/default/files/uploads/tech-forumsalternative-energy/presentations/is84-history-and-implications-emerging-transactive-energy-systems.pdf. Accessed 19 Mar 2023
5. Khodayar ME, Manshadi SD, Vafamehr A (2016) The short-term operation of microgrids in a transactive energy architecture 2(9):41–48
6. Song M, Cai Y, Gao C, Chen T, Yao Y, Ming H (2022) Transactive energy in power distribution systems: Paving the path towards cyber-physical-social system. Journal 142. https://doi.org/10.1016/2022.108289
7. Towards transactive energy systems: an analysis on current trends (2023) https://www.sciencedirect.com/science/article/pii/S2211467X19301105. Accessed 19 Mar 2023
8. Huang Q, Amin W, Umer K, Gooi HB, Eddy FYS, Afzal M, Shahzadi M, Khan AA, Ahmad SA (2021) A review of transactive energy systems: concept and implementation. 7:7804–7824. https://doi.org/10.1016/2021.05.037
9. Open Challenges and Issues: Artificial Intelligence for Transactive Management (2023) https://arxiv.org/ftp/arxiv/papers/2001/2001.03238.pdf. Accessed 19 Mar 2023
10. Hu J, Yang G, Kok K, Xue Y (2016) Transactive control: a framework for operating power systems characterized by high penetration of distributed energy resources. Mod Power Syst Clean Energy J 5(3). https://doi.org/10.1007/s40565-016-0228-1
11. The Brooklyn microgrid: blockchain-enabled community power. In: Power technology. https://www.power-technology.com/features/featurethe-brooklyn-microgrid-blockchain-enabled-community-power-578356. Accessed 1 Feb 2023
12. Bliek F, van den Noort A, Roossien B, Kamphuis R, de Wit J, van der Velde J, Eijgelaar M (2020) PowerMatching City, a living lab smart grid demonstration. In: Conference: innovative smart grid technologies conference Europe (ISGT Europe), 2010 IEEE PES. https://doi.org/10.1109/ISGTEUROPE.2010.5638863
13. Huang Q, McDermott T, Tang Y, Makhmalbaf A, Hammerstrom D, Fisher A, Marinovici L, Hardy T : Simulation-based valuation of transactive energy systems. In: IEEE Transactions on Conference: Power Systems, pp 99. https://doi.org/10.1109/TPWRS.2018.2838111
14. Lopes JAP, Soares F, Almeida PR (2011) Integration of electric vehicles in the electric power system. Journal https://doi.org/10.1109/JPROC.2010.2066250
15. Transactive energy platform: trends, benefits, and challenges of grid modernization. https://blockchain.ieee.org/verticals/transactive-energy/topics/transactive-energy-platform-trends-benefits-and-challenges-of-grid-modernization. Accessed 21 Mar 2023
16. Dasgupta R, Sakzad A, Rudolph C (2021) Cyber attacks in transactive energy market-based microgrid systems. Journal 14(4). https://doi.org/10.3390/en14041137
17. Liua G, Jiangb T, Ollisa TB, Zhangc X, Tomsovicd K (2016) Distributed energy management for community microgrids considering network operational constraints and building thermal dynamics. Journal. https://doi.org/10.1016
18. Transactive energy management in the smart grid. https://blockchain.ieee.org/verticals/transactive-energy/topics/transactive-energy-management-in-the-smart-grid. Accessed 19 Mar 2023
19. Lee D, Hess DJ, Neema H (2020) The challenges of implementing transactive energy: a comparative analysis of experimental projects 3(3). https://doi.org/10.1016/1040-6190

Approach for Islanding Detection in a PV Grid-Connected System Using Negative Sequence Impedance

Kumaresh Pal, A. K. Akella, K. Namrata, and Anshuman Bhuyan

Abstract Photovoltaic (PV) power systems that are linked to the grid have the advantage of being quick and dependable sources of electricity. The power industry has been obliged to transition over to more PV-penetrated distributed generation as a result of solar energy's favorable environmental effects in order to keep up with rising load demand. In grid-tied, PV systems, the issue of unintentional islanding in grid connectivity still presents a barrier. Therefore, quick islanding detection is required for efficient and trustworthy functioning of the system. This paper proposes a passive islanding detection method based on negative sequence impedance computation at the Point of Common Coupling (PCC) for a PV interconnected grid system. The approach is simulated and tested under several test scenarios. By determining the reciprocal of square root of the obtained Negative Sequence Impedance (NSI) at the PCC, which is calculated from the ratio of negative sequence voltage and negative sequence current found at PCC, the islanding mode is identified. By identifying the Non-Detection Zone (NDZ), that fulfils the IEEE requirements of execution time of 2 s, the results of this technique have been achieved and confirmed. A 100 kW of PV integrated grid system is used for the study using MATLAB/Simulink (version 2018a).

Keywords Photovoltaic (PV) system · Point of Common Coupling (PCC) · Unintentional islanding · Negative Sequence Impedance (NSI) · Islanding mode

K. Pal (✉) · A. K. Akella · K. Namrata
Department of Electrical Engineering, NIT, Jamshedpur, India
e-mail: kumaresh.pal@rediffmail.com

K. Namrata
e-mail: namrata.ee@nitjsr.ac.in

A. Bhuyan
Department of Electrical Engineering, Gandhi Institute for Education and Technology, Bhubaneswar, India

J. Kumar et al. (eds.), *Recent Advances in Power Systems*, Lecture Notes in Electrical Engineering 1192, https://doi.org/10.1007/978-981-97-2788-9_12

1 Introduction

Sustainable energy sources have exploded in popularity during the last decades. It produces less emissions and releases fewer greenhouse gases than fossil fuels and nuclear power [1]. Examples of distributed generation such as landfill gases, photovoltaic (PV) energy, wind energy, micro-hydro turbines and biomass categorised as renewable energy [2]. Because of the world's fast expansion, distributed generation (DG) technology is advancing at a rapid pace, and as scientific, technological and economic progress advances, people are placing greater demands on the environment. Including these DGs in the distribution network provides several advantages, including improved reliability, increased energy efficiency, lower line losses and improved power quality. Distributed generating technology is not only safe, clean, environmentally friendly, cost-effective and efficient, but it can also assess the whole power system's dependability and efficiency [3–5].

A PV system becomes "islanded" when a section of the utility system that includes load as well as distributed resources, such as a PV inverter, stays operational even though the utility system is disconnected [6]. Islanding mode can result in a number of challenges, including safety concerns, load damage, reclosing issues and more [7]. Grid voltage and frequency may fluctuate because a grid-connected inverter cannot regulate the grid's voltage and frequency when it is in islanding mode. Loads may sustain irreparable damage as a result of this. Grid synchronisation is also impossible in islanding mode due to the lack of a standard grid voltage. Additionally, it is challenging to synchronise an isolated grid with the utility grid during recloser operation due to the isolated grid's voltage being out of phase with the electrical utility grid, and the grid might trip again due to an overcurrent fault.

For grid-connected PV inverters, Anti-Islanding Detection (AID) is a necessary function since islanding might pose a hazard to the operation of the grid. When an island is detected, the PV inverter must stop energising the grid within the allotted period. A number of AID algorithms have been commercialised and have been developed to prevent islanding. These algorithms could be classified into passive techniques and active techniques [8]. Any significant system disturbance, a grid failure, or intentional islanding for a planned or unexpected event might cause the islanding to happen. An islanding situation can be an unintentional or intentional. Unintentional islanding situation is a threat to the electrical grid's protection, posing a risk to utilities, equipment and maintenance personnel. As a result, following island formation, the islanding system must be disconnected in 2 s in accordance with different islanding standards such as IEEE Std. 929-2000, IEEE Std. 1547-2003, IEC 62116 and distributed generating operations must be stopped [9, 10].

Islanding detection can be categorised as communication-based, passive, active, or hybrid, depending on the methodology. Monitoring of grid parameters is the foundation of passive approaches. By using a relay of frequency and voltage, islanding can be easily detected, relays such as UOV (Under or Over Voltage)/UOF (Under or Over Frequency). Measuring voltage as well as frequency that are not within the suitable range can avoid islanding. The detection of an islanding operation can also be done using a phase jump or a change in harmonics [11]. Active approaches disrupt the grid using a PV inverter and monitor how the grid's parameters respond. Two categories can be made for this. The first active approach uses positive feedback, which causes the grid to become unstable when islanding situation occurs. When an islanding condition arises, a positive feedback loop to power references or current phase angle starts and voltage or frequency instability would be tracked [12–14]. The frequency drift techniques (FDTs) are representative techniques, like active frequency drift (AFD), slip-mode frequency shift (SMS) and Sandia-frequency shift (SFS) [15, 16]. The second active method type injects a signal into a grid and monitors how system parameters respond [17]. The injection signal could be a current or a voltage. It is possible to feel the voltage generated by the injected components after islanding due to the injection of current. As high-frequency voltage is injected into the grid, the impedance of the grid can be measured and used as a possible indicator of islanding. This paper looks into a dilemma that arises at the point where a distributed generation facility meets the rest of the power system. The dilemma could be described by the power systems' recognition of an islanding condition. In recent years, a great number of contemporary methods for determining islanding scenarios have been described.

2 Suggested Technique

Out of several detection methods, the essential requirement for the existence of any disturbances in the voltage signal and the current signal detected at the PCC is the negative-sequence component approach [18]. Therefore, to be able to effectively identify islanding and notch perturbance caused by islanding occurrences, this research has analysed the negative-sequence component of voltage signals and the current signals discovered at PCC. The symmetrical component of analysis can be used to describe the negative, positive and zero sequence components of the voltage signal at the PCC as [19].

$$V_n = \frac{1}{3}(V_a + \lambda^2 V_b + \lambda V_c) \tag{1}$$

$$V_p = \frac{1}{3}(V_a + \lambda V_b + \lambda^2 V_c) \tag{2}$$

$$V_z = \frac{1}{3}(V_a + V_b + V_c) \tag{3}$$

where V_a, V_b and V_c represent the three-phase voltages, while V_n, V_p and V_z represent the negative, positive and zero sequence voltages, respectively. The complex operator $\lambda = 1\angle 120^\circ$ is also included in the analysis [20].

The symmetrical component of analysis can be used to indicate the zero, negative and positive sequence components of the current signal at PCC as

$$I_n = \frac{1}{3}(I_a + \lambda^2 I_b + \lambda I_c) \tag{4}$$

$$I_p = \frac{1}{3}(I_a + \lambda I_b + \lambda^2 I_c) \tag{5}$$

$$I_z = \frac{1}{3}(I_a + I_b + I_c) \tag{6}$$

where I_a, I_b and I_c represent the three-phase currents, while In, I_p and I_z are the negative, positive and zero sequence currents, respectively. The complex operator $\lambda = 1\llcorner 120°$ is also included in the analysis.

The negative-sequence impedance may be expressed by the symmetrical component of analysis given as.

$$Z_n = \frac{V_n}{I_n} \tag{7}$$

The reciprocal of square root of negative-sequence impedance can be expressed by the symmetrical component of analysis given as

$$Z_x = \frac{1}{\sqrt{Z_n}} \tag{8}$$

The PCC-collected three-phase voltage and current measurements serve as inputs to the corresponding MATLAB block that performs sequence analysis. The negative-sequence component is taken into account from among the three sequential components. In this paper, negative sequence impedance-based passive technique is used because it emulates the information following perturbance condition. Evaluation of the reciprocal of square root of negative-sequence impedance at the point of common coupling provides excellent noise shielding for islanding detection, therefore it performs well. The waveform output of reciprocal of square root of negative-sequence impedance is being denoised and smoothed by signal analyser tool for further analysis. Variations in magnitude of reciprocal of the square root of NSI can be used to identify the islanding mode [21]. It is possible to determine the magnitude of the negative sequence impedance (NSI) by dividing the negative-sequence voltage by the negative-sequence current that is measured at the point of common connection (PCC) (Fig. 1).

The technique is examined on a test system that includes a distributed generation unit powered by photovoltaic system and connected to a distribution network system through 100 KVA, 0.260/25 kV transformers. The total generation capacity of PV system is 100 KW. The distribution network system is operated at 25 kV. The total

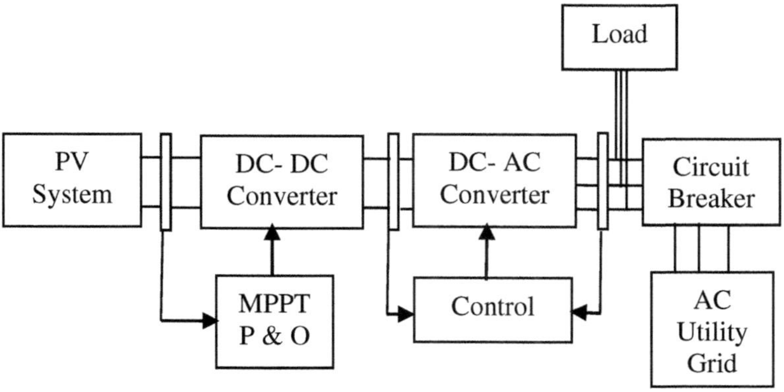

Fig. 1 Block diagram of islanding in the system

capacity of PV connected grid system is 200 kW. Loads are rated at 100 KW, 30 KVAR and 50 KW, 40 KVAR (Fig. 2).

3 Result and Discussion

The method is tested for various scenarios like symmetrical fault, unsymmetrical fault, disconnection and connection of heavy load, capacitor bank switching and islanding. The system is simulated for 1.5 s. By opening the circuit breaker' contact for three conductors at 1 s with both 0% and 50% active power mismatch circumstances, the microgrid region is islanded and hence detected.

3.1 Proposed Method Determined for 50% Active Power Mismatch and 0% Active Power Mismatch

A test scenario was simulated for the system with a 50% power mismatch, causing an islanding event at 1 s. Under these circumstances, it is discovered that the approach produces satisfactory results. In Figs. 3 and 4, the simulated waveforms are shown.

In Figs. 3 and 4, the consequence of anti-islanding detection using the suggested methodology is displayed. At 1 s, the grid side's input current is blocked and then reciprocal of sqrt of NSI increased. The simulation results show that the suggested technique, using negative sequence components, effectively estimates islanding.

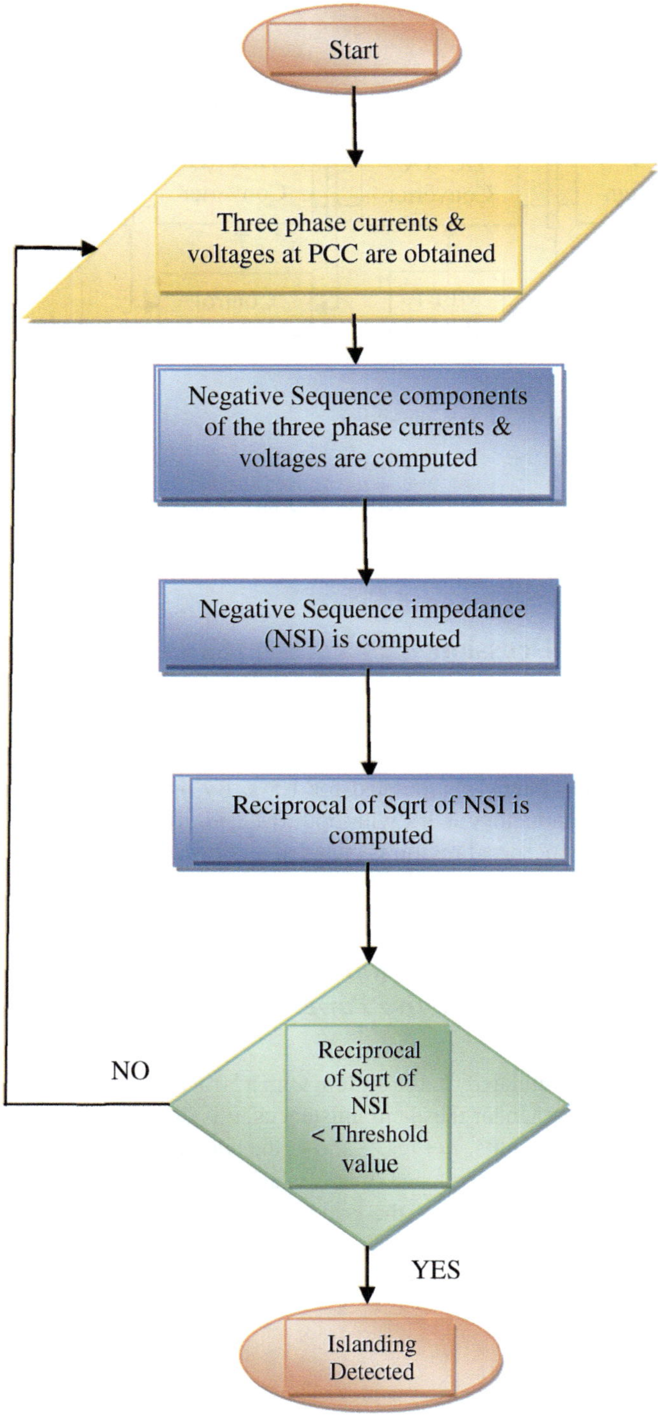

Fig. 2 Flowchart for the method described

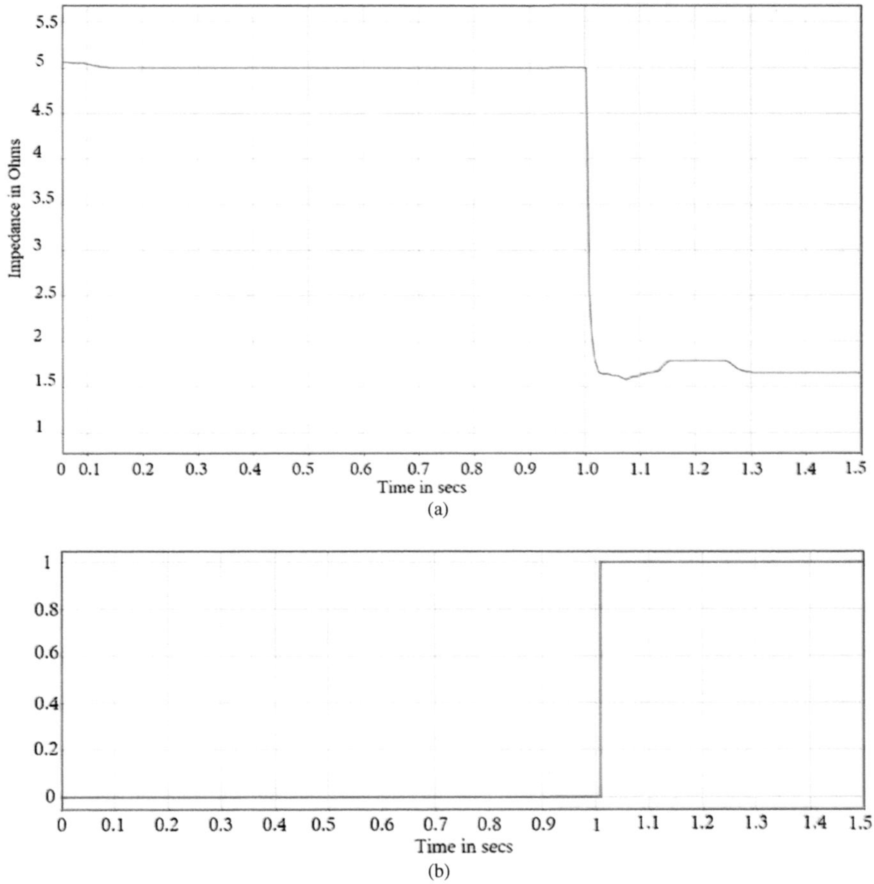

Fig. 3 Simulation result for 50% mismatch in active power with islanding at 1 s. **a** output signal, **b** decision signal

3.2 *Proposed Method Examined for Fault Occurring at PCC*

The simulation of the system involves the introduction of several types of faults occurring at the point of common connection (PCC) within a time range of 0–1.5 s. From the simulation results, the method is resistant to these perturbations and does not generate a false decision signal for these disturbances. Figures 5 and 6 display the simulation's outcome for symmetrical and unsymmetrical short circuit fault conditions.

In Figs. 5 and 6, the results are obtained for the load side short circuit faults, and the method is discovered to work satisfactorily. All other kinds of short circuit faults have similar results.

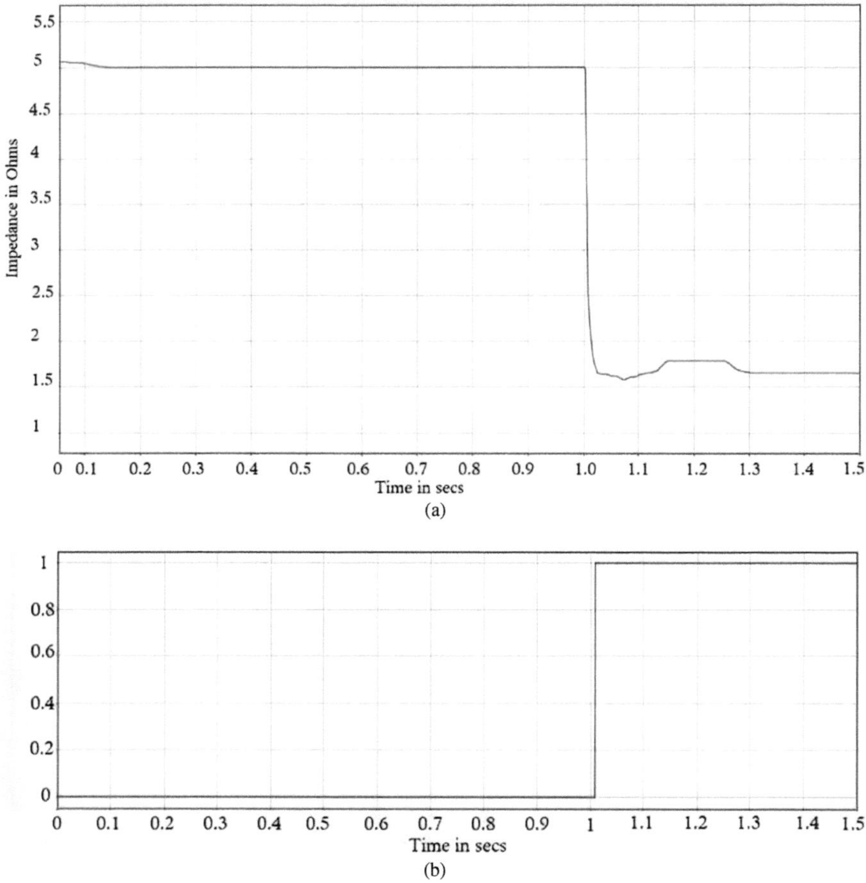

Fig. 4 Simulation result for 0% mismatch in active power with islanding at 1 s. **a** output signal, **b** decision signal

3.3 Proposed Method Tested for Disturbances Caused Due to Disconnection and Reconnection of Heavy Loads

The method is examined to see how it responds to disruptions brought on by large loads entering and exiting the system. From 1 and 1.3 s, the situation is simulated. For 1.5 s, the system is simulated. Figure 7 displays the simulation outcome.

Figure 7 clearly indicates that the method is resistant to these perturbations and does not generate a false decision signal for these disturbances. The method is also tested in DG terminals, where it is discovered to behave consistently.

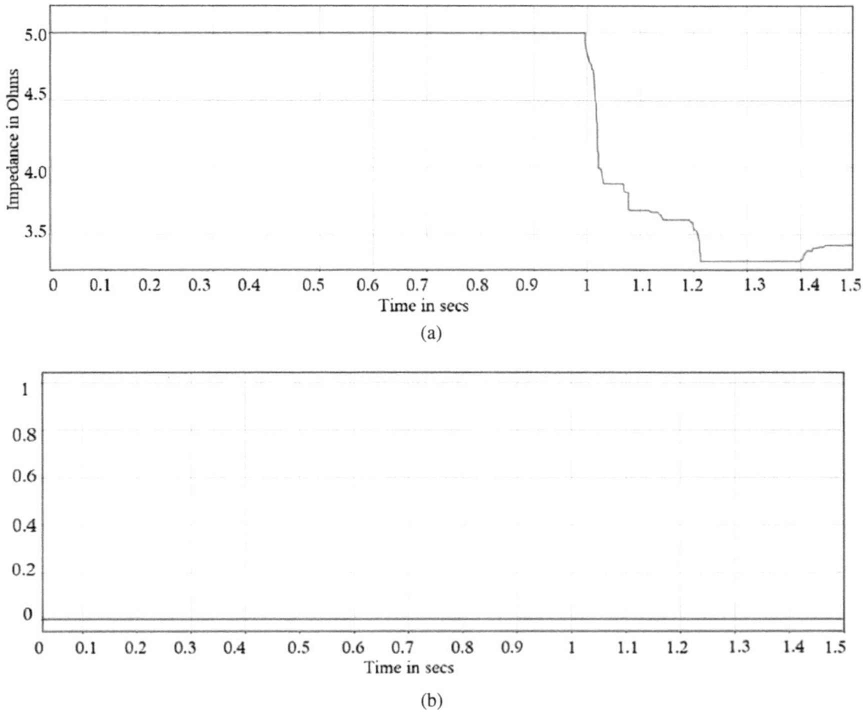

Fig. 5 Simulation result for LG fault at 1 s. **a** output signal, **b** decision signal

3.4 Proposed Technique Observed for Capacitor Bank Switching

The effect of these interruptions on the behaviour of the technique is investigated by activating a capacitor bank at 1 s and reconnecting it at 1.3 s. There is a 1.5-s simulation of the system. Figure 8 illustrates the result that had been actually observed.

In the case of capacitor bank switching, as shown by the waveforms in Fig. 8, the suggested technique does not generate any erroneous decision signals. Only under the islanding scenario the decision signal is initiated.

Detection of islanding events by analysing negative sequence impedance had a high detection rate and low non-detection zone, making it a reliable and effective method for detecting islanding [3].

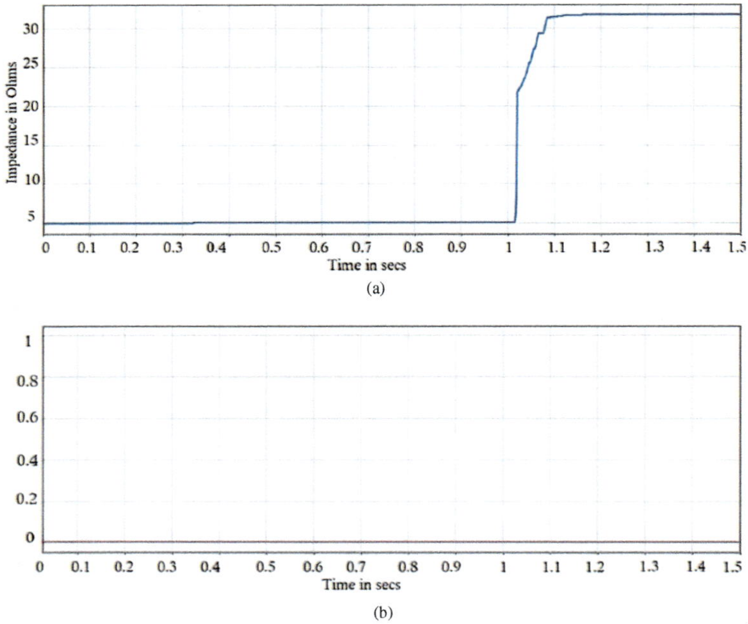

Fig. 6 Simulation result for LLLG fault at 1 s. **a** output signal, **b** decision signal

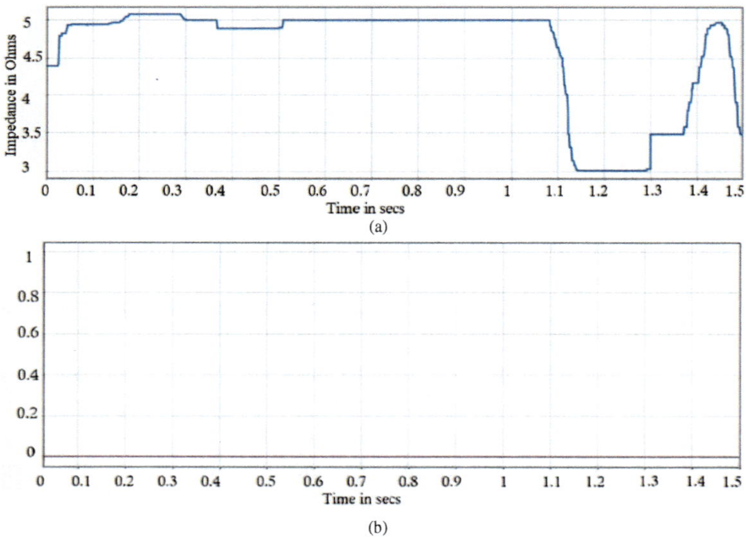

Fig. 7 Simulation result for load disconnection at 1 s and reconnection at 1.3 s. **a** output signal, **b** decision signal

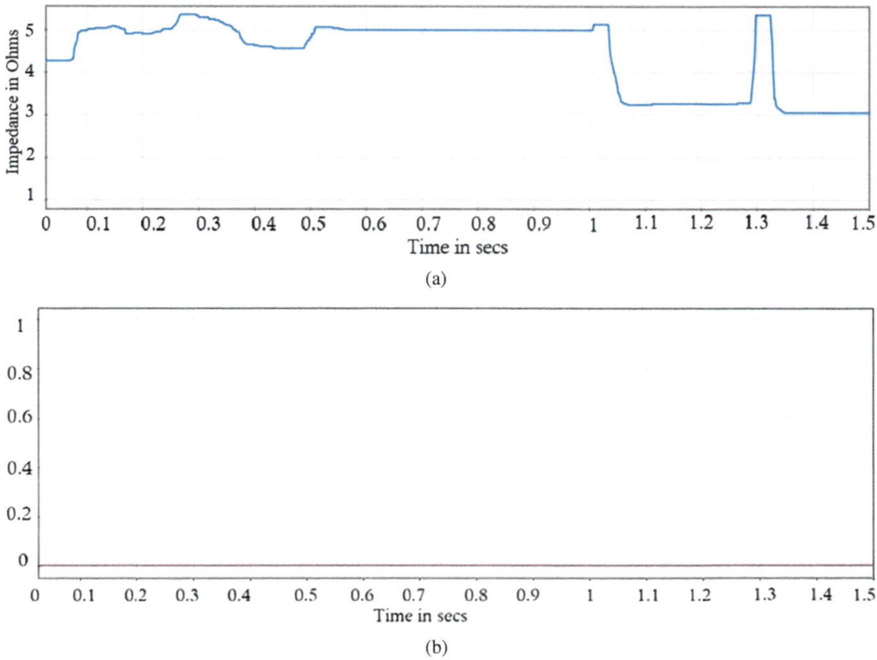

Fig. 8 Simulation result for capacitor bank connection at 1 s and disconnection at 1.3 s. **a** output signal, **b** decision signal

4 Conclusion

A method that is able to identify the islanding situation depending on the computation of the magnitudes of reciprocal of square root of NSI from negative sequence components of current and voltage found at PCC is discussed in this paper. This approach can identify islands when there is no mismatch in active power and 50% mismatch in active power. The method has proven to work satisfactorily in a variety of circumstances. The magnitude of the modified negative sequence impedance (NSI) is significant enough to differentiate between regular operation and islanding operation. It is possible to identify islanding based on this variation. Variations in modified NSI magnitude can be used to identify the islanding mode. The method also performs well for measurements taken at DG side and at point of common connection.

References

1. Swarnkar NK, Mahela OP, Khan B, Lalwani M (2021) Identification of islanding events in utility grid with renewable energy penetration using current based passive method. IEEE Access 9:93781–93794. https://doi.org/10.1109/ACCESS.2021.3092971

2. Panigrahy N, Ilamparithi T, Kashinath MV, Prakash R (2016) Comparison and review of islanding detection techniques for power distribution studies. Int J Adv Res Electr, Electron Instrum Eng 5(7):6485–6492
3. Fadzil NA, Hairi MH, Hanaffi F, Kamarudin MN, Isira AS, Packeer Mohamed MF (2019) A research of islanding detection method for distributed generation: mechanism, merits and demerits. Int J Innov Technol Explor Eng (IJITEE) 8(12S2). ISSN: 2278-3075
4. Dutta S, Sadhu PK, Jaya Bharata Reddy M, Mohanta DK (2018) Shifting of research trends in islanding detection method—A comprehensive survey. Prot Contr Mod Power Syst
5. Dash PK, Padhee M, Panigrahi TK (2012) A hybrid time–frequency approach based fuzzy logic system for power island detection in grid connected distributed generation. Int J Electr Power Energy Syst 42(1):453–464
6. Ahmad KNEK, Selvaraj J, Rahim NA (2013) A review of the islanding detection methods in grid-connected PV inverters. Renew Sustain Energy Rev 21:756–766
7. Soumya AV, Edward JB (2016) Review of islanding detection techniques for distributed energy sources. Int Adv Res J Sci Eng Technol 3(3). National Conference on Emerging Trends in Engineering and Technology (NCETET'16). Lourdes Matha College of Science & Technology
8. Kumar N, Namrata K, Samadhiya A (2023) Bi-level decision making in techno-economic planning and probabilistic analysis of community-based sector-coupled energy system. Appl Intell 53:6604–6628. https://doi.org/10.1007/s10489-022-03794-9
9. Panigrahi BK, Bhuyan A, Shukla J, Ray PK, Pati S (2021) A comprehensive review on intelligent islanding detection techniques for renewable energy integrated power system. Int J Energy Res 45(10):14085–14116
10. Ahmadipour M, Hizam H, Othman ML, Radzi MAM, Chireh N (2019) A novel islanding detection technique using modified Slantlet transform in multi-distributed generation. Int J Electr Power Energy Syst 112:460–475
11. Bhuyan A, Panigrahi BK, Pal K, Pati S (2022)Convolutional neural network based fault detection for transmission line. In: 2022 international conference on intelligent controller and computing for smart power (ICICCSP), Hyderabad, India, pp 1–4. https://doi.org/10.1109/ICICCSP53532.2022.9862446
12. Gupta D, Kumari N, Samadhiya A (2020)Photovoltaic modeling using single diode model in MATLAB. In: 2020 IEEE international conference on computing, power and communication technologies (GUCON), Greater Noida, India, pp 734–739. https://doi.org/10.1109/GUCON48875.2020.9231165
13. Khamis A, Shareef H, Bizkevelci E, Khatib T (2013) A review of islanding detection techniques for renewable distributed generation systems. Renew Sustain Energy Rev 28:483–493
14. Pal K, Akella AK, Namrata K, Pati S (2022)Classification of fault using artificial neural network and power quality improvement using DVR in a PV integrated hybrid power system. In: 2022 international conference on intelligent controller and computing for smart Power (ICICCSP), Hyderabad, India, pp 1–4. https://doi.org/10.1109/ICICCSP53532.2022.9862499
15. Namrata K, Sharma SP, Seksena SBL (2014)Determining regression constants for calculating global solar radiation at Jharkhand (India) region. In: 2014 international conference on renewable energy research and application (ICRERA), Milwaukee, WI, USA, pp 795–797. https://doi.org/10.1109/ICRERA.2014.7016494
16. Pal K, Kumar Panigrahi B, Mohapatra S, Mohapatra A (2017)Impact of STATCOM on voltage profile in a DG penetrated grid connected system. In: 2017 international conference on circuit, power and computing technologies (ICCPCT), Kollam, India, pp 1–5. https://doi.org/10.1109/ICCPCT.2017.8074213
17. Prasanna SL, Pal K, Mandal D (2022) Predicting the monthly average incident shortwave solar energy for Hubli, India by using training functions in ANN. In: 2022 international conference on intelligent controller and computing for smart power (ICICCSP), Hyderabad, India, pp 1–4. https://doi.org/10.1109/ICICCSP53532.2022.9862510
18. Kumar Panigrahi B, Nandi R, Mahanta B, Pal K (2016) Islanding detection in distributed generation. In: 2016 international conference on circuit, power and computing technologies (ICCPCT), Nagercoil, India, pp 1–5. https://doi.org/10.1109/ICCPCT.2016.7530295

19. Hartmann NB, dos Santos RC, Grilo AP, Vieira JCM (2017) Hardware implementation and real-time evaluation of an ANN-based algorithm for anti-islanding protection of distributed generators. IEEE Trans Ind Electron 65(6):36, 5051–5059
20. Pal K, Kumar Panigrahi B, Mohapatra S, Mohapatra A (2017) Impact of STATCOM on voltage profile in a DG penetrated grid connected system. In: 2017 international conference on circuit, power and computing technologies (ICCPCT), Kollam, India, pp 1–5. https://doi.org/10.1109/ICCPCT.2017.8074213
21. Pal K, Akella AK, Namrata K, Pati S (2022) Face detection using artificial neural network and wavelet neural network. In: International conference on intelligent controller and computing for smart power (ICICCSP), Hyderabad, India, pp 1–6. https://doi.org/10.1109/ICICCSP53532.2022.9862349

Investigating the Effectiveness of Reconfiguration Technique in Overall Improvement of the System Performance: An All-Encompassing Methodology for Management of Voltage Profile and Reduction of Power Loss in Power Distribution System

Farishta Rehman, Neeraj Gupta, and Nasir Rehman

Abstract This paper provides a detailed overview of reconfiguration and its applications in power distribution networks. Specifically, the study uses the IEEE 33 bus network as a test system to demonstrate the impact of reconfiguration on reliability and voltage deviation indices. The objective of this paper is to compare different methodologies employed to minimize active power loss within a system. Initially, the study focuses on the optimal positioning and sizing of capacitors, followed by the combined implementation of capacitor placement and reconfiguration techniques. The results demonstrate that employing both techniques simultaneously leads to more effective power loss minimization. The results of the study highlight the effectiveness of reconfiguration in enhancing the performance of distribution networks and addressing their operational challenges.

Keywords Reconfiguration · Voltage profile improvement · Reliability indices · Voltage deviation indices · Distribution system

1 Introduction

In a normal power system, electricity travels via the transmission and distribution networks from power generation facilities, like power plants or offshore wind farms, to consumers. Large amounts of power can be transported over great distances with little loss because the transmission system (*TS*) usually works at high voltages. Electricity is transported along high-voltage transmission lines from power plants to substations, where a lower voltage is stepped down for delivery to end users. The distribution system (*DS*) provides electricity to nearby residences, businesses, and other customers while operating at lower voltages [1]. The degree of control over

F. Rehman (✉) · N. Gupta · N. Rehman
Department of Electrical Engineering, National Institute of Technology, Srinagar, India
e-mail: farishtarehman.fr@gmail.com

J. Kumar et al. (eds.), *Recent Advances in Power Systems*, Lecture Notes in Electrical Engineering 1192, https://doi.org/10.1007/978-981-97-2788-9_13

electricity flows is one important distinction between the distribution infrastructure and the *TS*. Power flows in the *TSs* are carefully controlled to maintain the system stability and guarantee effective power delivery. However, in the *DS*, power flows can vary greatly and are less predictable based on things like customer demand and the accessibility of nearby generators. The *DS* management and supervision may become more difficult as a result. The majority of customer interruptions, it has been observed, are caused by flaws in the distribution infrastructure. This is because the *DS* is more prone to interruptions than the *TS* because it serves a large range of customers directly and runs at lower voltages [2]. Additionally, because the *DS* is frequently exposed to more diverse and regular stresses, such as disruptions from construction or weather-related events.

A decent distribution network is one where customers do not suffer frequent interruptions and get a reliable supply of power. Because both consumers and energy concessionaires may suffer large financial losses as a result of power outages. Appliances, machinery, and sensitive electronic equipment are all susceptible to damage from power outages. Consumers may incur repair or replacement expenditures as a result of equipment failure brought on by sudden power surges or voltage fluctuations during restoration. Industrial facilities with sophisticated and specialized equipment may sustain significant losses as a result of such harm. Following interruptions in electricity, businesses that depend on refrigeration, cold storage, or controlled environments may face financial losses because of the deterioration of perishable commodities. For sectors like food manufacturing, medicines, and healthcare, this is especially important. Due to product spoiling or deterioration, retailers and eateries may also experience inventory losses. Power interruptions can lead to customer dissatisfaction particularly if they happen frequently or at crucial moments. As a result, energy concessionaires may be subjected to compensation demands from customers who experience lengthy or frequent power outages. Concessionaires could have to pay more for services like customer assistance, emergency response, and repair and maintenance.

To figure out if a network has satisfactory or poor reliability, we can make use of a few measures of reliability. Reliability indices are mathematical parameters that are used to measure the performance and dependability of a radial distribution system [3]. Similarly, voltage deviation indices are commonly used to evaluate the voltage profile of a power system with n number of buses [4]. Voltage variations can have serious consequences for the electrical system and its components [5]. These deviations can damage or harm equipment and decrease system reliability. As a result, it is critical to keep the voltage within a certain range to guarantee the proper operation of the power system and to avoid damage to equipment. These indices can help utility companies and system operators to evaluate the performance of the distribution system and identify areas for improvement [6].

In general, the *DS* is an important part of the power system, and disruptions to this system can have a big effect on customers. To guarantee dependable customer support, we, therefore, need careful planning and efficient procedures. Reconfiguring the *DS* is one method that can help by reducing the length and frequency of power outages to increase the reliability of the power delivery to customers [7]. A well-designed *DS* in addition to ensuring reliability, should also be able to maintain the voltage profile and minimize the power losses [8].

2 Distribution-Network Reconfiguration (DNRC)

The aim of reconfiguration is to ensure that power is distributed efficiently and reliably to all customers while minimizing power losses [9]. Reconfiguration can involve changing the way that power flows through the *DS* by adjusting the connections between feeders, switches, and other devices. By doing so, the power flow can be redistributed to ensure that it is balanced and that voltage drops are minimized. This can help to improve the voltage profile and reduce power losses, leading to a more efficient and reliable electricity supply. Reconfiguration of the *DS* involves the use of sectionalizing and tie switches. To optimize the performance of the *DS* without requiring major upgrades or investment, reconfiguration is a practical technique.

Sectionalizing switches are used to divide the *DS* into sections or zones. These are used to control the flow of power to different sections of the *DS* and to isolate faults or problems. Tie switches, on the other hand, are used to connect two or more sections of the *DS*. These switches allow power to flow between different sections of the system, which can help to balance the load and improve reliability. Tie switches are used to provide redundancy in the *DS*, which helps to ensure that power can flow through alternate routes if there is an outage or fault in one section of the system. Reconfiguration of the *DS* involves changing the status of these sectionalizing and tie switches to optimize the operation of the system [10]. By opening or closing these switches, power can be rerouted around problem areas or redistributed to ensure a more balanced load. This can help to improve the efficiency and reliability of the system while minimizing power losses.

For example, during periods of high demand, the *DS* may experience voltage drops due to overload. Reconfiguration can be used to open sectionalizing switches and reroute power to less loaded feeders or to close tie switches and allow power to flow between different sections of the system [11]. This can help to balance the load and prevent voltage drops and other issues. Similarly, if a fault occurs in a particular section of the *DS*, reconfiguration can be used to isolate the fault and reroute power around the faulty area [12].

3 Objective

DNRC is a complex problem that typically involves non-linear optimization techniques, as the relationships between the system parameters (such as voltage, current, and power flows) are non-linear and, the problem involves many interdependent variables [13].

Apparent power, S, is:

$$S = P + jQ \tag{1}$$

where P is active power demand, and Q is reactive power demand.

The power loss (kW) is caused by the flow of current through the resistance of the conductor. According to Ohm's law, total power loss for n_b branches can be given by:

$$P_{loss} = \sum_{i=0}^{n_b} I_i^2 * R_i \tag{2}$$

where i = 1, 2,..., n_b The aim of this paper is to minimize the total power loss ($\mathcal{F} = P_{loss}$) of the system, improve reliability indices, and reduce voltage deviation.

Constraints

1. Radiality Constraint: For performing reconfiguration, we have to ensure that the distribution system is radial.
2. Voltage at each bus i, (where $i = 1, 2, \ldots, n$) should be within limits that can be allowed by the system design. $V_{min} \leq V_i \leq V_{max}$
3. Each line has a maximum threshold for current magnitude, beyond which it can be subjected to overheating. Therefore, current through each branch i, (where $i = 1, 2, \ldots, n_b$) should be within limit. $I_i \leq I_{max}$

3.1 Performance Evaluation Indices

Some of the commonly used reliability indices in a radial distribution system include:

1. SAIDI
2. SAIFI
3. CAIDI
4. ASAI

System Average Interruption Duration Index (SAIDI)

$$SAIDI = \frac{\sum_{i=1}^{n} \Delta_i C_i}{C_t} \tag{3}$$

where C_i represents the number of customers served by node at the end of branch i, and C_t represents the total number of customers in the system [14]. Δ_i is average interruption duration (in hr/yr) for serving point at the end of ith branch, taking into account both time to detect and repair the fault. It represents the average time that customers connected to that node are without power during each interruption [15].

$$\Delta_i = \frac{\sum_{j=1}^{N_{int}} D_j}{N_{int}} \tag{4}$$

where D_j represents duration of jth Interruption, N_{int} represents total number of interruptions at ith node.

System Average Interruption Frequency Index (SAIFI)

$$SAIFI = \frac{\sum_{i=1}^{n} \lambda_i C_i}{C_t} \tag{5}$$

In this equation, λ_i represents the failure rate $(failures/yr)$ for the node at the end of branch i, in the system. The failure rate is a measure of the frequency at which a node experiences an interruption or outage over a certain period. It is typically expressed in units of failures per year or failures per hour. $\sum_{i=1}^{n} \lambda_i C_i$ calculates the total number of customer hours lost due to outages in the system [16].

$$\lambda_i = \frac{F_i}{T_i} \tag{6}$$

where F_i is the number of failures (or outages) that have occurred during a given time period on node, i.e., at the end of branch i, and T_i is the total time during which the node or component was operational during that period.

Customer Average Interruption Duration Index (CAIDI)

$$CAIDI = \frac{\sum_{i=1}^{n} \Delta_i C_i}{\sum_{i=1}^{n} \lambda_i C_i} \tag{7}$$

A lower CAIDI value indicates that the distribution system is more reliable and has shorter outages.

Average Service Availability Index (ASAI)

$$ASAI = \frac{T_a}{T_t} \tag{8}$$

$$T_a = T_t - \sum_{i=1}^{n} \Delta_i \tag{9}$$

where T_a is the total time in which power is available to customers and T_t is total time.

This value is expressed as a percentage and indicates the percentage of time that customers have access to power. A higher ASAI value indicates better reliability and customer satisfaction.

In a radial distribution system, there are various indices used to measure voltage deviation, for example:

1. Voltage deviation index (VDI)
2. Voltage profile enhancement index ($VPEI$)
3. Voltage profile index (VPI)
4. Voltage regulation index (VRI)

Table 6 gives expressions to calculate these voltage deviation indices for a radial DS.

These indices give us an estimate of how far the actual voltage profile is from the nominal [17]. The rated voltage ($1\ p.u$) is the nominal voltage at which the system is designed to operate, and any deviation from this voltage can lead to various problems such as voltage instability, equipment damage, and power losses. Voltage profile optimization can help us to reduce this deviation [18].

4 Optimization Algorithm

Kennedy and Eberhart [20] made a significant contribution to the field of Particle Swarm Optimization (PSO). They expanded the applicability of PSO to binary spaces by introducing a sigmoid transformation to the velocity component, which constrained the velocities to a range [0 1]. This adaptation replaced the original equation for updating particle positions with a new equation (Eq. 11) that better suited binary spaces.

$$\text{sigmoid}\left(v_{id}^{k+1}\right) = \frac{1}{1 + e^{-v_{id}^{k+1}}} \tag{10}$$

$$x_{id}^{k+1} = \begin{cases} 1, & \text{if } rand < \text{sigmoid}(v_{id}^{k+1}) \\ 0, & \text{else} \end{cases} \tag{11}$$

Just like in the standard PSO, the selective PSO also requires a fitness function $\mathcal{F}$ to be defined. However, in this scenario, $\mathcal{F}$ maps each dimension from dn positions of the selective space S_d, where the particle's position has been transformed from a real-valued space point to a point in the selective space. Selective Particle Swarm Optimization ($S - PSO$) is a variant of the conventional PSO algorithm that functions on a search space S_d in every d-dimension. This search space is composed of dn positions, where dn signifies the quantity of chosen positions in that specific dimension. The sigmoid transformation used in basic PSO is replaced with Eq. (12), and the i_{th} coordinate of each particle's position at a dimension d is a selective value that is updated by Eq. (13). The conversion process allows the $S - PSO$ algorithm to effectively tackle binary optimization problems by constraining the search space to chosen positions and transforming particle positions into selective values [21].

$$\text{sigmoid}(v_{id}^{k+1}) = dn \frac{1}{1 + e^{-v_{id}^{k+1}}} \tag{12}$$

$$x_{id}^{k+1} = \begin{cases} S_{d1}, & \text{if sigmoid}(v_{id}^{k+1}) < 1 \\ S_{d2}, & \text{if sigmoid}(v_{id}^{k+1}) < 2 \\ S_{d3}, & \text{if sigmoid}(v_{id}^{k+1}) < 3 \\ . & \\ . & \\ . & \\ S_{dn}, & \text{if sigmoid}(v_{id}^{k+1}) < dn \end{cases} \quad (13)$$

where $S_{d1}, S_{d2}, S_{d1}, \ldots, S_{dn}$ represent the opted values in the d-dimension. Using Eq. (14), the values of velocity are confined to a certain minimum and maximum values $[Vmin, Vmax]$.

$$v_{id}^{k+1} = \begin{cases} V_{max}, & \text{if } v_{id}^{k+1} > V_{max} \\ v_{id}^{k+1}, & \text{if } |v_{id}^{k+1}| \leq V_{max} \\ V_{max}, & \text{if } v_{id}^{k+1} < V_{min} \end{cases} \quad (14)$$

To prevent particle i's velocity in dimension d from becoming fixed at maximum or minimum values and oscillating between these extremes, Eq. (15) is utilized to guarantee that each particle moves across the entire search space [22].

$$v_{id}^{k+1} = \begin{cases} rand * v_{id}^{k+1}, & \text{if } |v_{id}^{k+1}| = |v_{id}^{k}| \\ v_{id}^{k+1}, & \text{else} \end{cases} \quad (15)$$

4.1 Implementation of DNRC on Test System Using S-PSO

The algorithm was executed in a step-by-step manner.

- Initially, each loop was considered as a dimension. However, it should be noted that a switch can only occur once within a single loop and cannot be placed in any other loop. The allocation of switches to dimensions is done randomly.
- Regarding the IEEE 33 bus test system (Fig. 1), it is evident that five loops can be formed after closing tie switches S33, S34, S35, S36, and S37.
- For running the $S - PSO$ algorithm discussed in Sect. 4, over the test system: the dimensions or loops needed are given. Search space depends on individual dimensions. Sets of branches or switches in each dimension qualify as elements of search space (Table 1).
- The branches that do not belong to any of these five loops have been excluded from the search space in order to simplify the process. Since our test system has a total of five dimensions, d5 is used in place of dn [19].

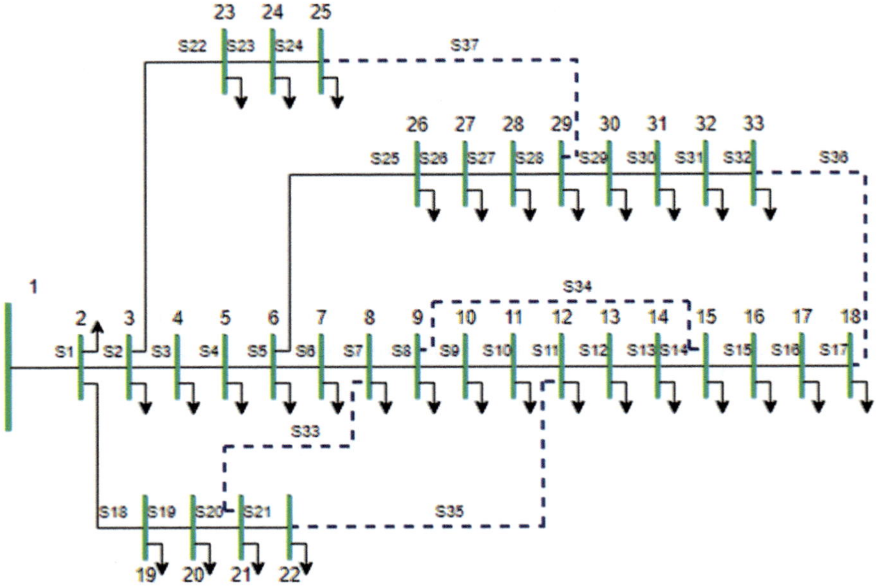

Fig. 1 Standard IEEE 33 bus system

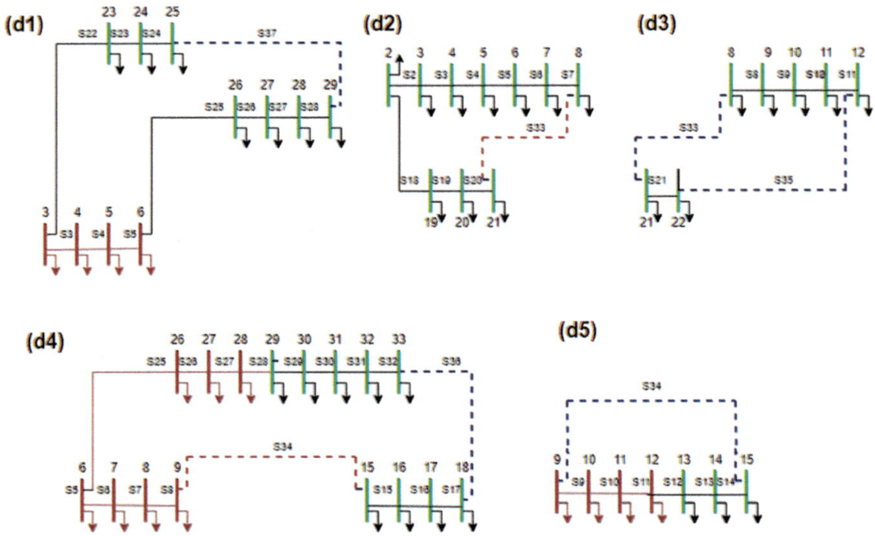

Table 1 Search space for each dimension

Dimension (*d*)	Search space
1	22, 23, 24, 25, 26, 27, 28, 37
2	2, 3, 4, 5, 6, 7, 18, 19, 20
3	8, 9, 10, 11, 21, 33, 35
4	15, 16, 17, 29, 30, 31, 32, 36
5	12, 13, 14, 34

5 Results

Standard IEEE 33 bus system [23] has been used to test the algorithm and compare the results (Fig. 1). In this paper, a method for both reconfiguring a *DS* and computing indices for reliability and voltage deviation has been presented. Four cases have been considered: Base case load flow analysis, DNRC, Optimal positioning and sizing of capacitors (C), and DNRC along with optimal positioning and sizing of capacitors (C).

1. **Standard configuration without any C or DNRC**
   Due to the limitations of using conventional load flow analysis for distribution networks, a revised version of the Newton–Raphson load flow method was employed to conduct load flow analysis on the base case (i.e., configuration without the integration of any C or DNRC). This analysis is important because it helps to ensure that the *DS* can deliver power to customers within the required voltage limits and that the equipment in the system is not overloaded [24].
2. **Configuration with the incorporation of C**
   A pair of capacitors has been used. The range available for use in genetic algorithm is 0 to 2000 KVAr. Through optimization, it was determined that the ideal locations for the capacitors are at nodes 12 and 30, with corresponding sizes of 470.3 *KVAr* and 1057.1 *KVAr*, respectively.
3. **Configuration after DNRC**
   The optimization algorithm utilized in this paper achieves the objective given in Sect. 3.
4. **Configuration with simultaneous C incorporation and DNRC** It is possible to achieve even higher levels of performance by combining the optimization of capacitor positioning and sizing with the reconfiguration of the distribution system. *DS*' performance can be enhanced by finding the best overall solution that strikes a balance between the advantages of capacitors and reconfiguration (Tables 2 and 3).

It can be seen from the results that after DNRC and C installation, the reliability indices and voltage indices have improved (Tables 4 and 5). Total power Loss has also been reduced, as can be seen from Table 2 and Fig. 2 depicting voltage profiles for these four cases. The results indicate that the use of DNRC, either alone or with

Table 2 Active power loss and tie lines

CASE	$P_{loss}(KW)$	$Q_{loss}(KVAr)$	Tie switches	V_{min}
Base case	202.52	135.13	33 34 35 36 37	0.9133
DNRC by S-PSO	139.55	102.3	7 9 14 32 37	0.9378
With C	135.62	90.53	33 34 35 36 37	0.9364
DNRC with C	96.53	71.35	7 9 14 36 37	0.9591

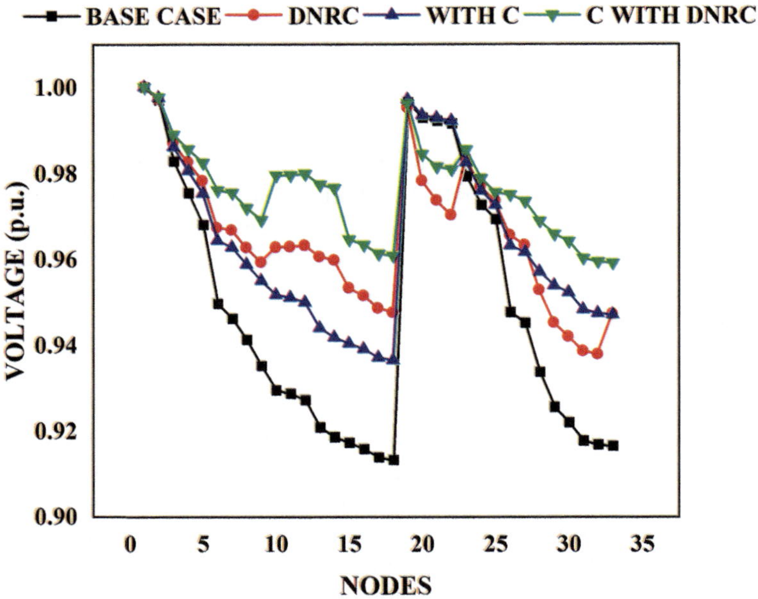

Fig. 2 Voltage profile

C, leads to an improvement in voltage profiles and reactive power management. Similarly, Fig. 3 illustrates voltage regulation indices. It shows that VRI begins to decrease immediately after introducing DNRC to the test system.

DNRC can also help in balancing the loads across the different lines [28, 29]. It reduces the voltage drop across the lines with higher loads and improves the voltage deviation indices.

Tables 4 and 5 provide the values for reliability and voltage deviation indices, which can be calculated using the analytical approaches outlined in Sect. 3.1. Furthermore, upon examining the results obtained from S-PSO and those of other techniques presented in Table 3, it becomes apparent that S-PSO surpasses the other methods in terms of performance. This superiority stems from its capability to deliver optimal solutions while achieving a lower power loss compared to the remaining techniques.

Table 3 Comparison of reconfiguration techniques

CASE	$P_{loss}(KW)$	Tie switches
DNRC by harmony search algorithm [25]	142.67	7 10 14 36 37
DNRC by heuristic-OPF algorithm [26]	140.2735	7 10 14 32 37
DNRC by PSO [27]	139.97	7 9 14 28 32
DNRC by selective-PSO	139.55	7 9 14 32 37

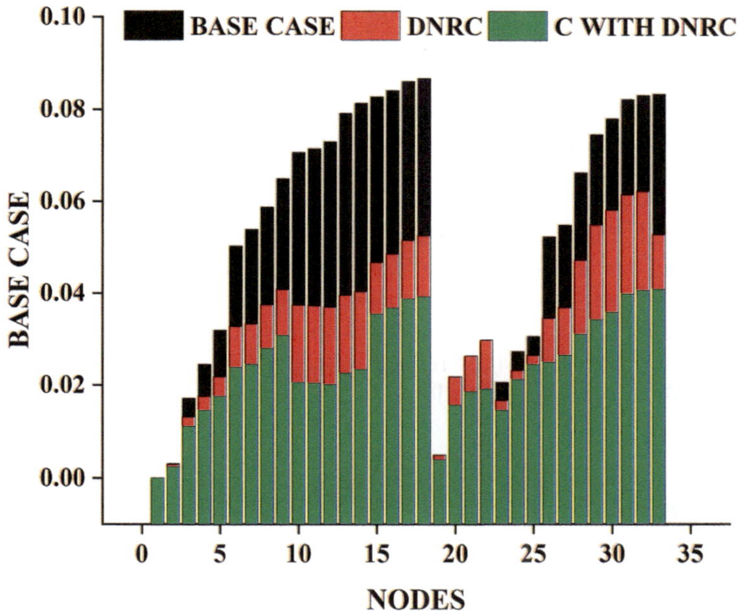

Fig. 3 Voltage regulation index

Table 4 Reliability indices

CASE	SAIDI	SAIFI	CAIDI	ASAI
Base case	0.505	0.0982	5.138	99.88
DNR	0.458	0.0967	4.740	99.89
DNR and C	0.467	0.0972	4.800	99.89

Table 5 Voltage deviation indices

CASE	VDI	VPEI	VPI
Base Case	0.1168	0.0035	0.0867
DNR	0.0487	0.0015	0.0622
DNR and C	0.0234	0.0007	0.0409

6 Conclusion

Reconfiguration entails changing the connections between various sections of the system, such as adding or removing lines or switches, to optimize the flow of electricity and ensure that power is delivered effectively and reliably to customers. Furthermore, reconfiguration can be used to aid in the integration of renewable energy sources into the power distribution system by allowing the system to adjust in real time to changes in power production and demand. This can help to increase the system's overall efficiency and sustainability while also lowering greenhouse gas emissions and other environmental effects. The implementation of the optimization algorithm in this article demonstrates that reconfiguration is a crucial technique for enhancing the performance of power distribution systems. By utilizing reconfiguration, the algorithm successfully reduces voltage deviation at system nodes while simultaneously minimizing power loss.

7 Annexure

See Tables 6 and 7.

Table 6 Expressions for various voltage deviation indices

No.	Voltage deviation index	Expression
1	VDI	$\sum_{i=1}^{n}(V_{rated_i} - V_i)^2 \quad (17)$
2	VPEI	$\frac{\sum_{i=1}^{n}(V_{rated_i} - V_i)^2}{n} \quad (18)$
3	VPI	$max(\|1 - V_i\|), 1 = 1, 2, \ldots, n \quad (19)$
4	VRI	$\frac{\|V_{rated_i} - V_i\|}{V_{rated_i}}, 1 = 1, 2, \ldots, n \quad (20)$

Table 7 Data for calculation of reliability indices

Branch number	Sending bus	Receiving bus	λ	Number of customers	Δ
1	1	2	0.0500	26	0.3000
2	2	3	0.0400	23	0.3000
3	3	4	0.0600	31	0.3000
4	4	5	0.0300	16	0.2000
5	5	6	0.0300	16	0.2000
6	6	7	0.0900	52	0.6000
7	7	8	0.0300	52	0.6000
8	8	9	0.0300	15	0.2000
9	9	10	0.0200	15	0.2000
10	10	11	0.0300	12	0.1000
11	11	12	0.0300	16	0.2000
12	12	13	0.0600	16	0.2000
13	13	14	0.0300	31	0.3000
14	14	15	0.0300	16	0.2000
15	15	16	0.0300	16	0.2000
16	16	17	0.0300	16	0.2000
17	17	18	0.0400	23	0.2000
18	2	19	0.0400	23	0.2000
19	19	20	0.0400	23	0.2000
20	20	21	0.0400	23	0.2000
21	21	22	0.0400	23	0.2000
22	3	23	0.0400	23	0.2000
23	23	24	0.1900	109	1.1000
24	24	25	0.1900	109	1.1000
25	6	26	0.0300	16	0.2000
26	26	27	0.0300	16	0.2000
27	27	28	0.0300	16	0.2000
28	28	29	0.5400	31	0.3000
29	29	30	0.0900	25	0.5000
30	30	31	0.0700	39	0.4000
31	31	32	0.1000	35	0.6000
32	32	33	0.0300	16	0.2000
33	8	21	0.0200	33	0.1000
34	9	15	0.0200	33	0.1000
35	12	22	0.0500	33	0.1000
36	18	33	0.0500	33	0.1000
37	25	29	0.0500	33	0.1000

References

1. Alonso FR, Oliveira DQ (2015) Zambroni de Souza AC (2015) Artificial immune systems optimization approach for multiobjective distribution system reconfiguration. IEEE Trans Power Syst 30(2):840–847
2. Wu H, Dong P (2020) Liu M (2020) Distribution network reconfiguration for loss reduction and voltage stability with random fuzzy uncertainties of renewable energy generation and load. IEEE Trans Ind Inform 16(9):5655–5666
3. López-Prado JL, Vélez JI, Garcia-Llinás GA (2020) Reliability evaluation in distribution networks with microgrids: review and classification of the literature. Energies 13(23):6189
4. Ismail B, Abdul Wahab NI, Othman ML, Radzi MAM, Naidu Vijyakumar K, Mat Naain MN (2020) A comprehensive review on optimal location and sizing of reactive power compensation using hybrid-based approaches for power loss reduction, voltage stability improvement, voltage profile enhancement and loadability enhancement. IEEE Access 8:222733–222765
5. Anastasiadis AG, Kondylis GP, Vokas GA (2020) Effect of augmented distributed generation in distribution networks. Energy Rep 177–187
6. Song Y, Zheng Y, Liu T, Lei S, Hill DJ (2020) A new formulation of distribution network reconfiguration for reducing the voltage volatility induced by distributed generation. IEEE Trans Power Syst 35(1):496–507
7. Shaheen AM, El-Sehiemy RA, Kamel S, Elattar EE, Elsayed AM (2021) Improving distribution networks' consistency by optimal distribution system reconfiguration and distributed generations. IEEE Access 9:67186–67200
8. Montoya F, Baños R, Gil C, Espín A, Alcayde A, Gómez J (2010) Minimization of voltage deviation and power losses in power networks using Pareto optimization methods. Eng Appl Artif Intell 23:695–703
9. Gomes FV, Carneiro S, Pereira JLR, Vinagre MP, Garcia PAN (2006) De Araujo LR (2006) A new distribution system reconfiguration approach using optimum power flow and sensitivity analysis for loss reduction. IEEE Trans Power Syst 21(4):1616–1623
10. Abed A, Alwash S, Ibrahim S (2022) Distribution system reconfiguration with soft open point for power loss reduction in distribution systems based on hybrid water cycle algorithm. Energies 16
11. Dechgummarn Y, Fuangfoo P, Kampeerawat W (2022) Reliability assessment and improvement of electrical distribution systems by using multinomial monte carlo simulations and a component risk priority index. IEEE Access 10:111923–111935
12. Lakshmi G, Jayalaxmi A, Veeramsetty, Venkataramana (2020) Optimal placement of distributed generation using firefly algorithm. In: IOP conference series: materials science and engineering, vol 981, p 042060
13. Alanazi M, Alanazi A, Almadhor A et al (2022) Multiobjective reconfiguration of unbalanced distribution networks using improved transient search optimization algorithm considering power quality and reliability metrics. Sci Rep 12:13686
14. Nick M, Cherkaoui R, Paolone M (2018) Optimal planning of distributed energy storage systems in active distribution networks embedding grid reconfiguration. IEEE Trans Power Syst 33(2):1577–1590
15. Kahouli O, Alsaif H, Bouteraa Y, Ben Ali N, Chaabene M (2021) Power system reconfiguration in distribution network for improving reliability using genetic algorithm and particle swarm optimization. Appl Sci 11(7)
16. Karafotis PA, Bazionis IK, Georgilakis PS (2020) Optimal distribution network reconfiguration for reliability improvement in presence of DG. In: The 12_{th} mediterranean conference on power generation, transmission, distribution and energy conversion (MEDPOWER 2020), pp 354–360
17. Ogunboyo PT, Tiako R, Davidson IE (2018) Effectiveness of dynamic voltage restorer for unbalance voltage mitigation and voltage profile improvement in secondary distribution system. Can J Electr Comput Eng 41(2):105–115
18. Essallah S, Khedher A (2019) Optimal distribution system reconfiguration for loss minimization using BPSO algorithm. In: 10th international renewable energy congress (IREC), pp 1–6

19. Salau AO, Gebru YW, Bitew D (2020) Optimal network reconfiguration for power loss minimization and voltage profile enhancement in distribution systems. Heliyon 20:6(6)
20. Kennedy J, Eberhart R Particle swarm optimization. In: Proceedings of ICNN'95—international conference on neural networks, Perth, WA, Australia, vol 4, pp 1942–1948
21. Daud JG, Kondoj M, Patabo M (2018) Reconfiguration distribution network with ant colony. International Conference on Applied Science and Technology (iCAST). Manado, Indonesia, pp 349–353
22. Khalil TM, Gorpinich AV (2012) Selective particle swarm optimization. Int J Multidiscip Sci Eng (IJMSE) 3(4):2045–7057
23. Baran ME, Wu FF (1989) Network reconfiguration in distribution systems for loss reduction and load balancing. IEEE Trans Power Del 4(2):1401–1407
24. Venkatesan C, Kannadasan R, Alsharif M, Kim MK, Nebhen J (2021) A novel multiobjective hybrid technique for siting and sizing of distributed generation and capacitor banks in radial distribution systems. Sustainability 13:3308
25. Rao RS, Narasimham SVL, Ramalinga Raju M, Srinivasa Rao A (2011) Optimal network reconfiguration of large-scale distribution system using harmony search algorithm. IEEE Trans Power Syst 26(3):1080–1088
26. Gomes FV, Carneiro S Jr, Pereira JLR, Vinagre MP, Garcia PAN, Araujo LR (2006) A new distribution system reconfiguration approach using optimum power flow and sensitivity analysis for loss reduction. IEEE Trans Power Syst 21(4):1616–1623
27. Wu WC, Tsai MS (2011) Application of enhanced integer coded particle swarm optimization for distribution system feeder reconfiguration. IEEE Trans Power Syst 26(3):1591–1599
28. Mahdavi M, Sanjeevikumar P, Siano P, Haes A (2021) Genetic algorithm application in distribution system reconfiguration. Act Electr Distrib Netw: Smart Approach 479–515 (2021)
29. Lima FN, Santos MM, Benetti MA, Milke T, Sperandio M (2021) Power distribution network reconfiguration considering the transmission system usage. IEEE Lat Am Trans 19(12):2113–2121

Optimal Bidirectional Home Energy Management System Considering Photovoltaic and Battery Storage Energy

Lovekush and Sandeep Kakran

Abstract In this paper, a home energy management system (HEMS) architecture with an energy storage system and photovoltaic is proposed for the buying/selling of electricity from/to the main grid. This paper suggests a price-based demand response (DR)-integrated binary particle swarm optimization gravitational search algorithm (BPSOGSA) for the optimization of the HEMS. The suggested method minimizes the home energy cost and schedules the home appliances based on the real-time price-based DR signals received from the utility while meeting the limitations imposed by the users. According to the simulated results, the proposed methodology also reduces the peak-to-average ratio (PAR) and hence improves the load curve significantly. The proposed methodology is proven to find a better solution for the same number of functions examined when compared to the existing method like binary particle swarm optimization (BPSO). In the considered HEMS system, the hybrid optimization algorithm BPSOGSA reduces the home energy cost by 10.22% and PAR by 2.88% compared to the BPSO algorithm.

Keywords Home energy management system · Demand response · Hybrid of BPSOGSA · Smart grid

1 Introduction

Energy consumption has been on the rise globally over the last few decades. A large portion of electricity produced in the past used fossil fuels. However, scientists have been working on new techniques of energy generation along with the renewable energy sources (RES) to satisfy the rising demand for electricity while emitting as few greenhouse gases as possible. However, the spread of RES significantly raised the complexity and dynamics of the power system. [1]. RES, like wind and solar power, are uncertain due to their changing nature. As a result, it is harder to predict and manage the supply and demand for electricity, especially renewable generation,

Lovekush (✉) · S. Kakran
National Institute of Technology, Kurukshetra, India
e-mail: kushlove108@gmail.com

J. Kumar et al. (eds.), *Recent Advances in Power Systems*, Lecture Notes in Electrical Engineering 1192, https://doi.org/10.1007/978-981-97-2788-9_14

which makes energy shortages or surpluses more common [2]. Further, if RES and other distributed generation (DG) are integrated on a big scale, the present electrical system cannot continue to keep its stability. One of the current solutions is to upgrade the electricity infrastructure to a smart grid (SG) using the most advanced information and communication technologies. (ICTs) [3]. Demand side energy management especially home energy management system (HEMS) can play an important role in reshaping of load curve and in maintaining the load-generation balance even in the presence of intermittent RES.

Home Energy Management Systems arise as a crucial tool for tracking, controlling, and optimizing the flow and utilization of energy to support demand-side management programs, e.g., demand response (DR) [4, 5]. DR initiatives include several kinds of measures. DR refers to changes in end-user electricity usage from their usual consumption patterns in response to change in the price of electricity over the time or incentive payments intended to encourage lower electricity usage during periods of high wholesale market price [6]. They provide customers with a method to monitor how their devices are being used to reduce the cost of energy with the smallest possible compromises to their comfort [7]. Overall, a HEMS could be considered a system that provides energy management solutions to efficiently track and control power production, storage, or usage in smart homes [8].

From a household perspective, certain appliances, like washing machines, refrigerator, water heater, etc. can create versatility because their operating profiles can be changed to fit the operator's schedule without sacrificing user comfort [9]. Therefore, this paper introduces a HEMS architecture for the scheduling of washing machines, refrigerator, water heater, clothe dryer and dishwashers in the presence of energy storage system and photovoltaic for the buying/selling of electricity from/to the main grid. The optimized results of energy scheduling with day-ahead real-time price-based DR program are achieved using hybrid BPSO-GSA algorithm [7, 10]. The used hybrid algorithm gives significant decrement in the energy use cost and peak to average ratio after energy scheduling as compared to the other similar algorithms already used in the literature.

The paper is structured as follows: Sect. 1 presents introduction and literature review of similar research articles. In Sect. 2, the architecture of the HEMS is introduced. Section 3 presents the details of home appliances used. Section 4 discusses the methodology considered for getting the solution. Section 5 describes the problem formulation. Section 6 gives the objective function, and results have been discussed in Sect. 7. Section 8 presents the general conclusion of this study.

2 Home Energy Management System (HEMS)

The primary components of the suggested HEMS design are shown in Fig. 1. The ideal house should have an advanced metering infrastructure (AMI), a main controller (MC), an energy storage system (ESS), and a PV system as a renewable energy source (RES). A collection of systems, including smart meters (SM), sophisticated

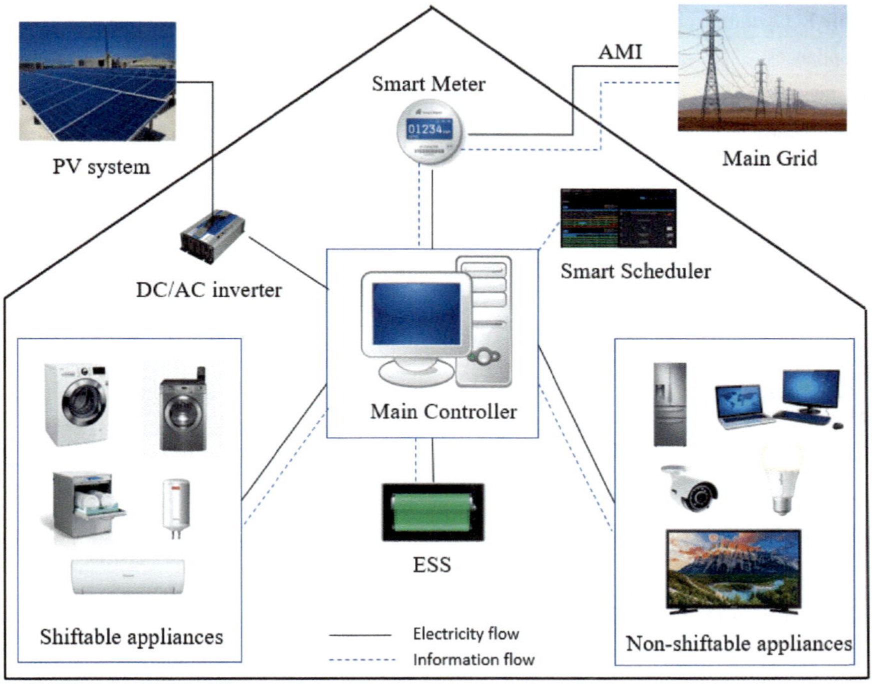

Fig. 1 Diagram of the HEMS [12]

connectivity, and data management systems, are collectively referred to as "AMI." AMI serves as a backbone connecting energy suppliers and customers. SM results as a conduit for information between the home and the suppliers [11]. SM is in charge of receiving, processing, and transmitting AMI data from the home to the providers regarding the energy consumption. AMI is also used to deliver crucial infrastructure data from energy suppliers to customers. This kind of data could be comparable to market data, weather forecasts, and sun irradiance. The PV system is utilized due to its low cost and simple implementation. The DC from the PV system is changed into AC electricity using a DC/AC converter.

3 Home Appliances

Various appliances considered for HEMS are shown in Table 1. Their power ratings with respect to their daily usage are also shown in the table. * Represents shiftable appliances in the below table and the rest are non-shiftable appliances [13].

Table 1 Operating characteristics of appliances [12]

Appliances	Power rating (KW)	Daily usage (Hours)	Start time
Dishwasher*	0.2	3	
Water heater*	1	8	
Air conditioner*	1.3	10	
Clothes dryer*	0.7	4	
Washing machine*	0.8	5	
Refrigerator	0.9	20	2 A.M
Security camera	0.1	24	0 A.M
Personal computers	0.2	18	7 A.M
Television	0.2	8	4 P.M
Microwave oven	0.5	7	3 P.M
Lights	0.1	6	6 P.M

4 Methodology

HEMS are widespread effects of high electric power usage at home that have drawn the attention of today's scientists. Due to the opulent lifestyle that the current generation leads, where daily tasks are completed by electrical appliances and new, innovative products are regularly introduced to the market, the demand for consistent energy flow is greater than the supply [14, 15]. HEMS is used in the presented design to plan and monitor each considered electrical item. As the initial steps of the solution methodology, HEMS specifies initial parameters like cost as startup and control parameters, PV parameters, GSA parameters, e.g., population and number of iteration. Then, it results in generation of a random set of control factors, associated fitness function and finds a suitable fitness value to find the best optimum solutions using the hybrid approach of BPSO-GSA.

Assume that there are two distinct sets of devices in the proposed scheme: M shiftable appliances and N non-shiftable appliances. The group of shiftable devices also includes devices that could operate at any moment $M = a_1, a_2, a_3, \ldots, a_m$, and they can be shift to cheaper timeslots to reduce costs. Users have designated fixed operating time slots for the non-shiftable devices $N = b_1\ b_2, b_3, \ldots, b_n$. to assess the principal amount of the HEMS system and also to compare the current method in use and the suggested method. As a result, the suggested program's fitness function is given in the equations below. The flowchart for the methodology is given in Fig. 2.

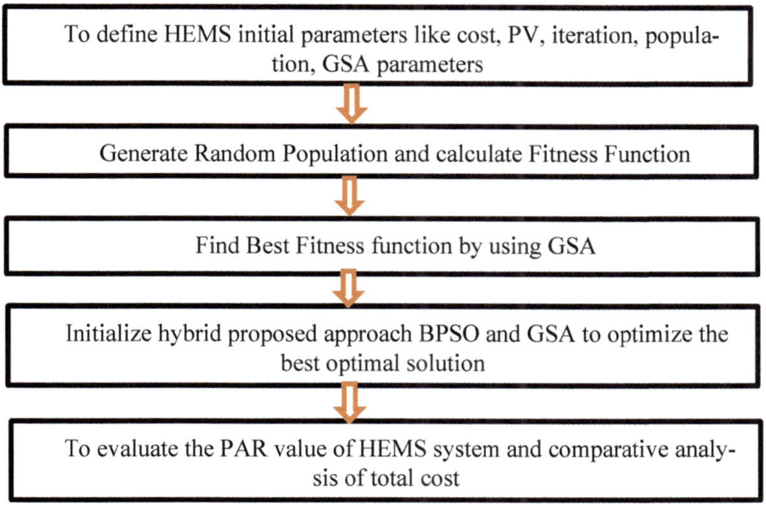

Fig. 2 Flowchart of proposed work

5 Problem Formulation

The mathematical models are created and constraints are applied for all components to reduce the energy use cost and PAR. From 12 a.m. to the next day's 12 a.m., mathematical models for the cost function, PV, and ESS appliances have been modeled. The day is split into T = 24 time periods, with each slot for t = 1 h.

5.1 Flow of Electricity

The flow of electricity block diagram is shown in Fig. 3. It represents the flow of electricity among energy storage system, main grid and PV system in accordance with home appliances.

5.1.1 Energy Generation by PV System

The PV system's generating power (P_{out} (t)) is calculated as

$$P_{out}(t) = GHI.S.\eta^{PV} 7 \leq t \leq 18 \tag{1}$$

where *GHI* is the horizontal irradiation (kW/m^2) of solar panels.

η^{PV} is the efficiency of the PV system and S being a total area of the solar panel (m^2). Generation of electrical energy by PV system:

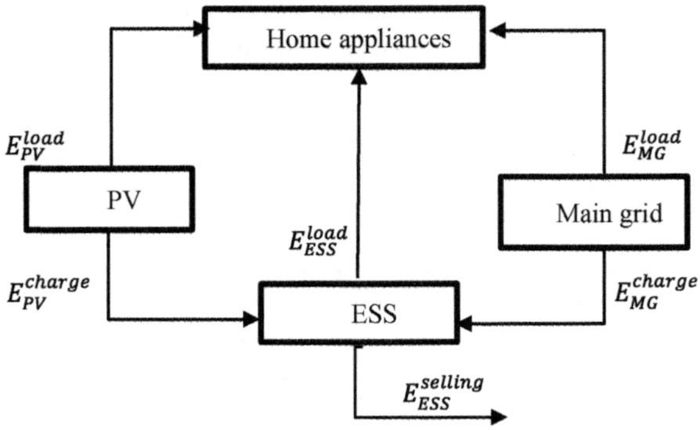

Fig. 3 Flow of electricity [12]

$$E_{PV}(t) = P_{OUT}(t) \times \Delta t \text{ and } \Delta \text{t is } 1 - \text{hour time} \tag{2}$$

ESS for charging as well as household equipment use this energy. E_{PV} (t) is the energy generated by the PV

$$E_{PV}(t) = E_{PV}^{load}(t) + E_{PV}^{charge}(t) \tag{3}$$

whereas E_{PV}^{Load} (t) and $E_{PV}^{Charge}(t)$ the PV generated energy transferred to the load and charging of ESS, respectively.

5.1.2 Energy Storage System

A battery has been used as an energy storage system that is being charged by either the main grid or PV. The energy from PV is more economical. The storage system stores energy at low prices which are used for home loads at the time of higher prices. $E_{ESS}^{Discharged}$ (t) is discharging energy of energy storage system.

$$E_{ESS}^{Discharged}(t) = \left(E_{ESS}^{load}(t) + E_{ESS}^{selling}(t)\right) \cdot \left(1 - Mode^{ESS}(t)\right) \tag{4}$$

whereas E_{ESS}^{Load} (t) & $E_{ESS}^{Selling}(t)$ are the storage energy transferred to the load and sold to the main grid.

$$E_{ESS}^{Charge}(t) = \left(E_{PV}^{charge}(t) + E_{MG}^{charge}(t)\right) \cdot Mode^{ESS}(t) \tag{5}$$

$$Mode^{ESS}(t) = \begin{cases} 1 \text{ } if \text{ } ESS \text{ } charging \\ 0 \text{ } if \text{ } ESS \text{ } charging \end{cases} \tag{6}$$

whereas

E_{ESS}^{Charge} (t) is charging energy of energy storage system from the main grid and PV.

E_{MG}^{Charge} (t) represents as charging from the main grid to ESS.

$$E_{ESS}^{level}(t) = E_{ESS}^{level}(t-1) + E_{ESS}^{chage}(t).\eta^{ESS} - E_{ESS}^{Discharged}(t)/\eta^{ESS} \tag{7}$$

whereas η^{ESS} is the efficiency of battery

The ESS's energy should range from EL_{min} and EL_{max}.

$$EL_{min} \leq E_{ESS}^{level} \leq EL_{max} \tag{8}$$

$$E_{ESS}^{level}(at\ t = 1) = EL_o \tag{9}$$

EL_o is the ESS's starting energy level and η^{ESS} is the ESS efficiency.

5.1.3 Home Appliances

In this study, 11 smart home appliances are included and 5 of them are scheduled. These appliances are supposed to be employed with smart sensors [16] so that they can communicate with the HEMS and work according to the given instructions. The appliances are categorized as:

M is shiftable appliances.

M = $(a_1 a_2 \ldots\ldots\ldots a_m)$.

N is the non-shiftable appliances.

N = $(b_1\ b_2 \ldots\ldots\ldots\ldots b_n)$.

Whereas $E_{total}^{appliances}$(t) represents total energy consumed by shiftable and non-shiftable appliances

$$E_{total}^{appliances}(t) = E_N(t) + E_M(t) \tag{10}$$

$$E_M(t) = \sum_{x=1}^{M} Power_{rate}(b_x) \times c(b_x.t) \times \Delta t \tag{11}$$

$$E_N(t) = \sum_{x=1}^{N} Power_{rate}(a_x) \times c(a_x.t) \times \Delta t \tag{12}$$

$$c(b_x.t) = \begin{cases} 1\ Shiftable\ on \\ 0\ Shiftable\ off \end{cases} \tag{13}$$

$$c(a_x.t) = \begin{cases} 1\, Non\, shiftable\, on \\ 0\, Non\, shiftable\, off \end{cases} \tag{14}$$

whereas $\sum_{x=1}^{m} Power_{rate}(b_x)$ & $\sum_{x=1}^{n} Power_{rate}(a_x)$ the power rating of shiftable and non-shiftable appliances.

c (b_x, t) and c (a_x, t) are binary variables that display the state of equipment. energy consumption by all home appliances.

$$\sum_{t=1}^{T} E_{total}^{appliances}(t) = \sum_{t=1}^{T} E_M(t) + \sum_{t=1}^{T} E_N(t) \tag{15}$$

When a shiftable device moves to a low-priced slot, the energy consumption of this system does not change because the shiftable device's operating time is unchanged and its operation is not disrupted.

$$E_{total}^{appliances}(t) = E_{PV}^{load}(t) + E_{ESS}^{load}(t) + E_{MG}^{load}(t) \tag{16}$$

$$E_{MG}^{load}(t) = E_M(t) + E_N(t) - E_{PV}^{load}(t) - E_{ESS}^{load}(t) \tag{17}$$

If $E_{MG}^{Load}(t) \geq 0$, the main grid consistently supplies enough electricity to meet the needs of household appliances.

5.1.4 Load Demand and Cost Function

The energy required from the main grid is denoted as load demand in each time slot.

$$E_{LD}(t) = E_{MG}^{load}(t) + E_{MG}^{charge}(t) \tag{18}$$

$$E_{LD}(t) = E_M(t) + E_N(t) + E_{MG}^{charge}(t) - E_{PV}^{load}(t) - E_{ESS}^{load}(t) \tag{19}$$

$$EC(t) = E_{LD}(t) \times P_{MG}(t) - E_{ESS}^{selling}(t) \times P_{sell}(t) \tag{20}$$

where P_{MG}(t) is the main grid's energy tariff and P_{sell}(t) is the amount of power sold EC(t) is the cost of energy of the main grid in cents.

6 Objective Function

Normally, energy buying price from the main grid and selling price from ESS both are equal so the assumption is made that

$$P_{sell}(t) = \alpha \times P_{MG}(t) \; and \; \alpha = 1 \tag{21}$$

$$Min(\sum_{t=1}^{T}(E_N(t) + \sum_{t=1}^{M} Power_{rate}(b_x) \times c(b_x.t) \times \Delta t + E_{MG}^{charge}(t) - E_{PV}^{load}(t) - E_{ESS}^{load}(t) - \alpha \times E_{ESS}^{selling}(t)) \times P_{MG}(t)) \tag{22}$$

6.1 Peak-To-Average Ratio (PAR)

The PAR measures the ratio of peak demand and 24-h averaged electricity demand. The PAR explains the energy behavior of the system and is closely related to how the primary grid operates. The goal of all electricity distribution companies is to maintain PAR low.

$$\mathrm{PAR} = \frac{\mathrm{Peak\ load\ demand(max}E_{LD}\mathrm{(t))}}{\mathrm{Average\ of\ total\ load\ demand}} \tag{23}$$

7 Results

Now, to illustrate the success of the suggested algorithm which is a hybrid of BPSO and GSA for lowering electricity bills through optimal appliance scheduling. 24 h have been chosen as the length of the scheduling horizon, which is further divided in 1-h slot to create the best start/stop scheduling for each of them.

Figure 4 shows the hourly prices according to day-ahead real time pricing in cents. Figure 5 shows the hourly RES energy generated by the PV system from 6 A.M. to 8 P.M.

Figure 6 shows the hourly energy of the main grid stored in ESS from 12 A.M to 7 A.M due to low price of electricity during these hours. The charging rate of ESS from the main grid is 0.3 kW, and the capacity of a battery is 3 kW. From 7 AM to 8 PM, the ESS energy is used for home load, selling to main grid or it may charge from PV system if extra PV energy is available in any time slot.

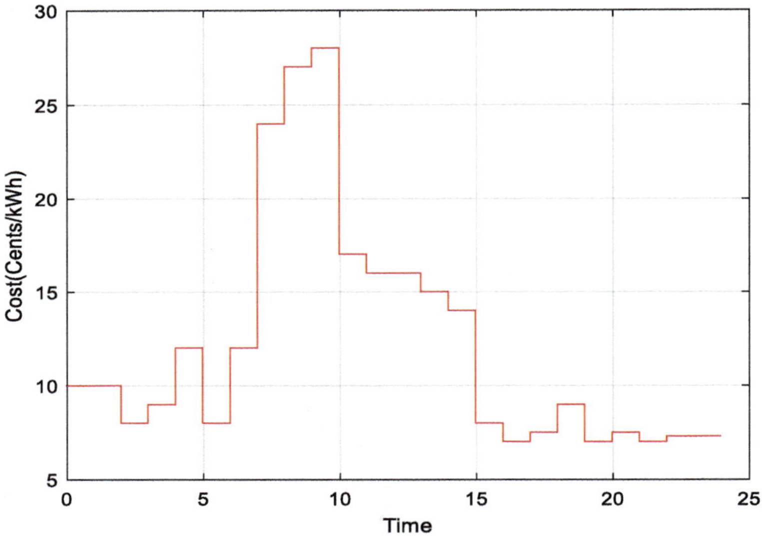

Fig. 4 Hourly prices according to DAP signal

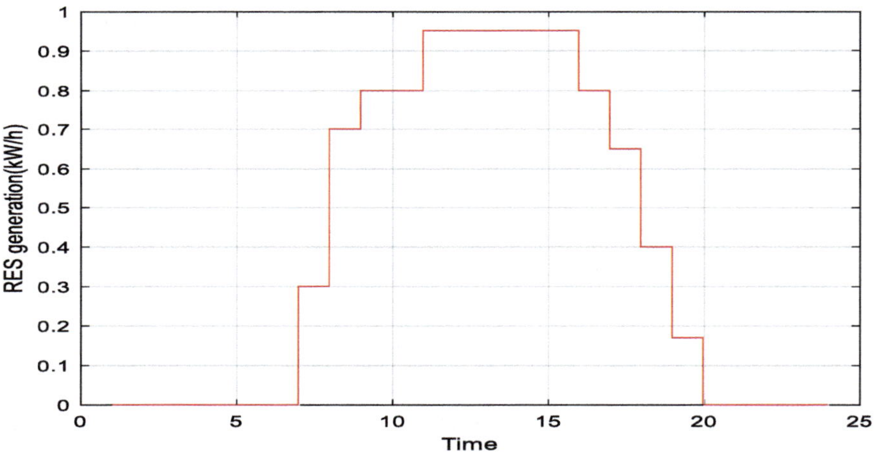

Fig. 5 Hourly RES energy generated by the PV system

Figure 7 shows the dissipation of energy in kW from the ESS to load from 6 AM to 2 PM and from 6 to 7 PM. The rest of the time is employed for selling or for charging the ESS from the main grid and PV system depending on price of electricity from main grid and on availability of extra PV energy.

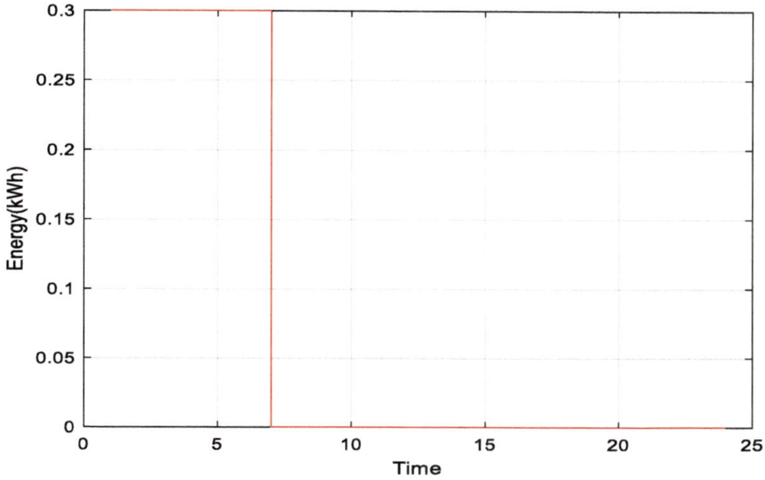

Fig. 6 Hourly energy of main grid stored in ESS

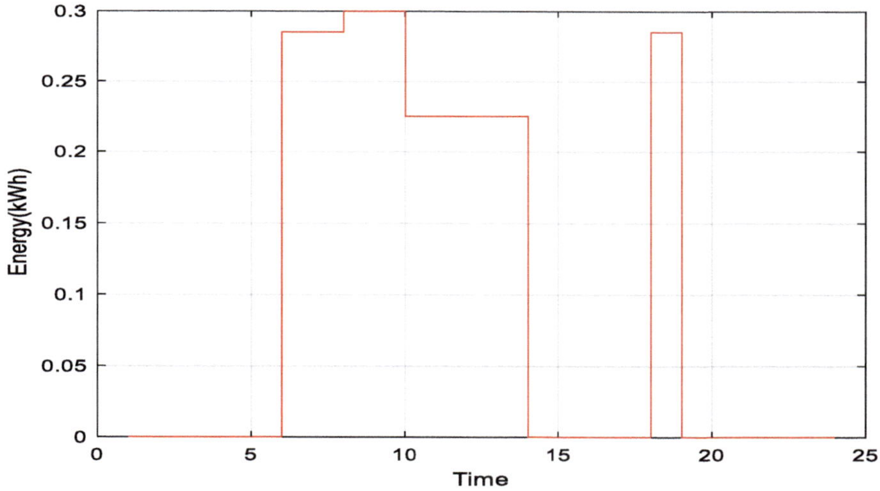

Fig. 7 Hourly ESS energy used for home load

Figure 8 represents the energy sold for a time duration from 7 to 8 AM and 10 AM to 3 PM. Figure 9 shows the hourly energy consumed by the total load from main grid as well as the PV system.

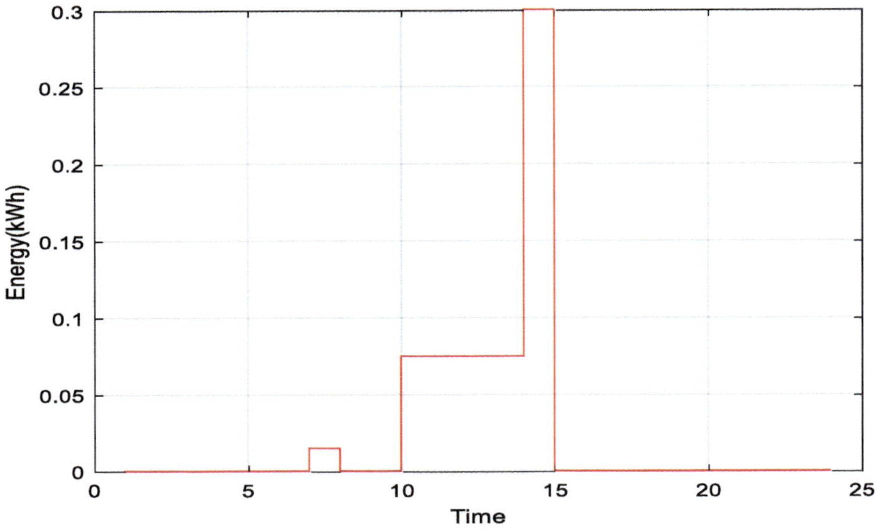

Fig. 8 Hourly ESS energy used for selling

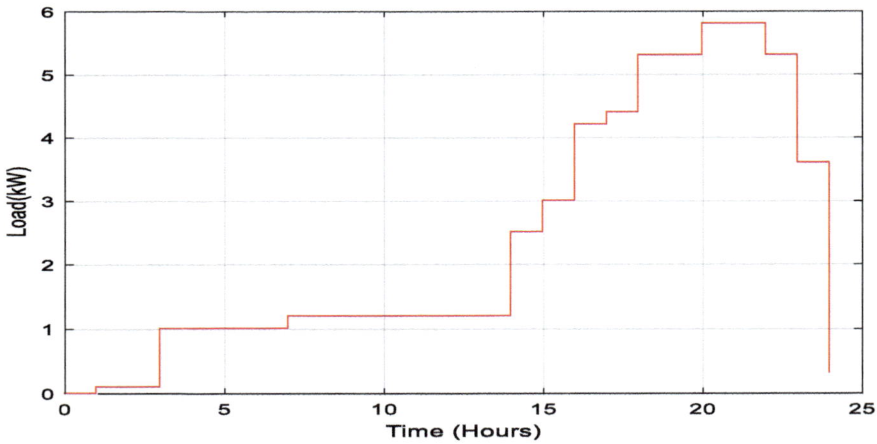

Fig. 9 Hourly energy consumed by the total load

Figure 10 shows the total load demand and power supplied by the main grid, and Fig. 11 shows the energy consumption bill paid by the user. The negative increment in the curve shows the energy selling to the main grid during the particular slots (11–14).

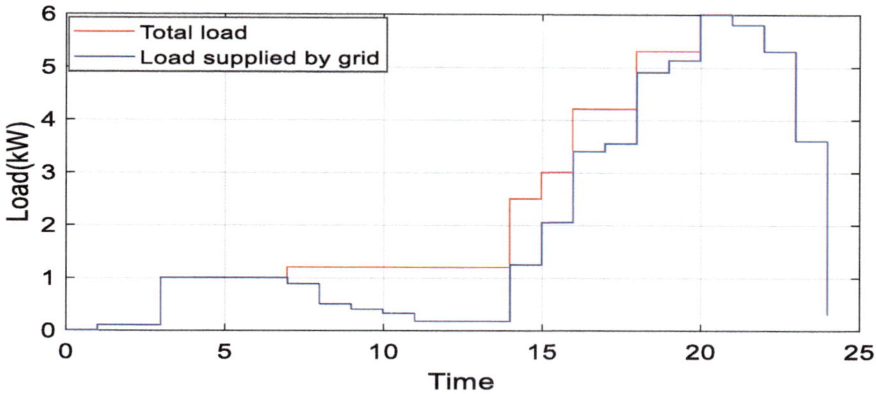

Fig. 10 Total load and load demand from the main grid

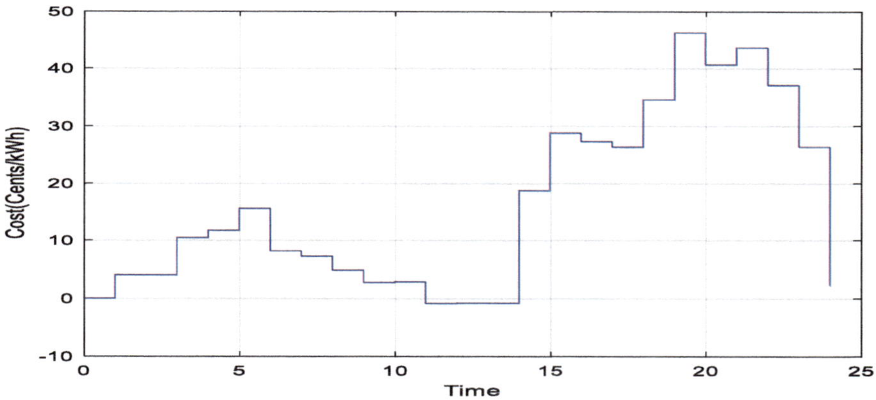

Fig. 11 Hourly energy consumption bill

Figure 12 shows the comparison of peak to average ratio for both BPSO and BPSO-GSA algorithms. It can be clearly seen that using BPSO-GSA, the PAR is reduced by 2.88% as compared to BPSO. Figure 13 shows the total cost of used electricity for BPSO and BPSOGSA algorithms. It shows that the proposed algorithm BPSOGSA has shown a relatively low cost by 10.22% compared to that of the BPSO algorithm. The total cost of electricity achieved using the BPSO-GSA is 400.223 cents, which is 45.557 cents lesser compared to 445.78 cents of the BPSO algorithm.

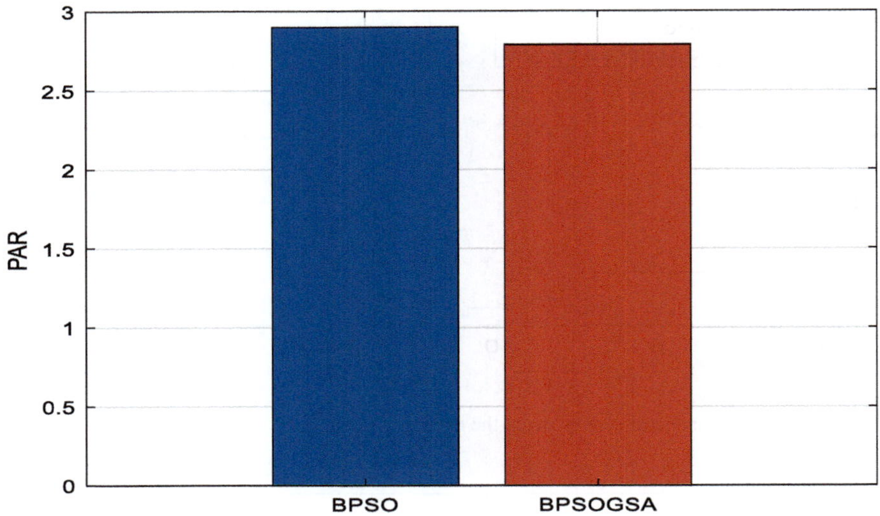

Fig. 12 Comparison of PAR for both algorithms

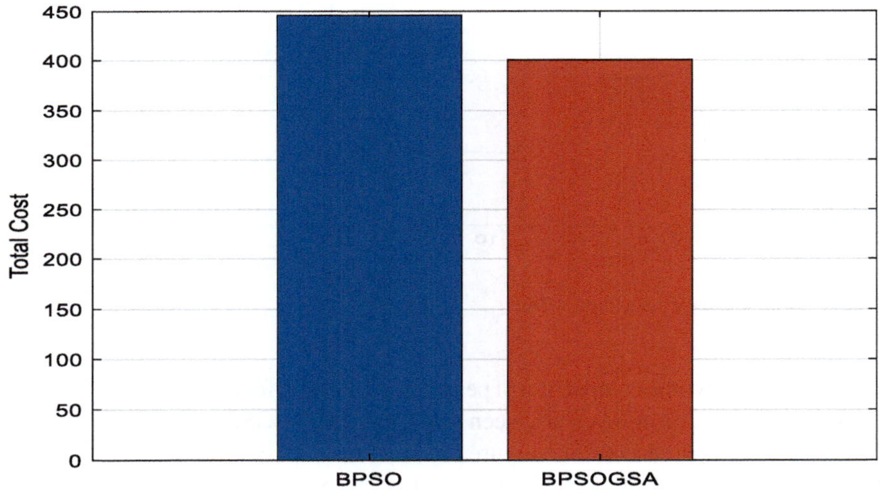

Fig. 13 Comparison of total cost for both algorithm

8 Conclusion

In this paper, BPSO-GSA has been used for the optimal energy scheduling of appliances in an HEMS. The achieved simulation results, using day ahead real-time pricing-based demand response programs, conclude that the electricity bills as well as the peak demand have been significantly decreased. The outcomes of the work as discussed in above section have clearly shown that the suggested method can be identified as a better solution than BPSO. The outcomes have also confirmed that the suggested approach, which includes improved HEMS as well as unified DR services offered by the utility or an aggregator, was successful in reducing peak load via carefully implementation of real-time pricing-based DR model. Overall, it can be concluded that after scheduling home appliances using BPSO-GSA algorithm, total energy cost is reduced by 10.22% and PAR by 2.88% as compared to BPSO.

References

1. Guo Y, Pan M, Fang Y (2012) Optimal power management of residential customers in the smart grid. IEEE Trans Parallel Distrib Syst 23(9):1593–1606
2. Sousa T, Lezama F, Soares J, Ramos S, Vale Z (2018) A flexibility home energy management system to support agreggator requests in smart grids. In: 2018 IEEE symposium series on computational intelligence (SSCI). IEEE, pp 1830–1836
3. Agnetis A, De Pascale G, Detti P, Vicino A (2013) Load scheduling for household energy consumption optimization. IEEE Trans Smart Grid 4(4):2364–2373
4. Kakran S, Chanana S (2018) Smart operations of smart grids integrated with distributed generation: a review. Renew Sustain Energy Rev 81:524–535
5. Juyal V, Kakran S (2023) Optimized cost of energy by a home energy management system employing dynamic power import limit strategy: a case study approach. J Oper Autom Power Eng 11(4):285–294. https://doi.org/10.22098/joape.2022.10254.1728
6. Qdr QJUDE (2006) Benefits of demand response in electricity markets and recommendations for achieving them. US Dept. Energy, Washington, DC, USA, Technical Report
7. Shakeri M, Shayestegan M, Abunima H, Reza SS, Akhtaruzzaman M, Alamoud ARM, Sopian K, Amin N (2017) An intelligent system architecture in home energy management systems (HEMS) for efficient demand response in smart grid. Energy Build 138:154–164
8. Zhou B, Li W, Chan KW, Cao Y, Kuang Y, Liu X, Wang X (2016) Smart home energy management systems: Concept, configurations, and scheduling strategies. Renew Sustain Energy Rev 61:30–40
9. Zhao Z, Lee WC, Shin Y, Song K (2013) B: An optimal power scheduling method for demand response in home energy management system. IEEE Trans Smart Grid 4(3):1391–1400
10. Basit A, Sidhu GAS, Mahmood A, Gao F (2015) Efficient and autonomous energy management techniques for the future smart homes. IEEE Trans Smart Grid 8(2):917–926
11. Dinh H, Kim D (2021) An optimal energy-saving home energy management supporting user comfort and electricity selling with different prices. IEEE Access 1–1. https://doi.org/10.1109/ACCESS.2021.3050757
12. Dinh HT, Yun J, Kim DM, Lee KH, Kim D (2020) A home energy management system with renewable energy and energy storage utilizing main grid and electricity selling. IEEE Access 8:49436–49450
13. Singh A, Deep K (2017) Novel hybridized variants of gravitational search algorithm for constraint optimization. Int J Swarm Intell 3(1):1–22
14. Li J, Dong N (2017) Gravitational search algorithm with a new technique. In: 13th international conference on computational intelligence and security (CIS). IEEE, pp. 516–519. (Dec, 2017)
15. Kazmi S, Javaid N, Mughal MJ, Akbar M, Ahmed SH, Alrajeh N (2017) Towards optimization of metaheuristic algorithms for IoT enabled smart homes targeting balanced demand and supply of energy. IEEE Access 7:24267–24281
16. Ahmad A, Khan A, Javaid N, Hussain HM, Abdul W, Almogren A, Alamri A, Azim Niaz I (2017) An optimized home energy management system with integrated renewable energy and storage resources. Energies 10(4):549

DG Placement of an Islanded Microgrid

Mudasir Farooq, Sachin Mishra, and Javed Dhillon

Abstract As renewable energy is progressively incorporated into the power system, microgrids operating in the islanding mode may experience system frequency deviation, voltage profile deviation, and increase in losses from the required level due to the unpredictable and inconsistent production from renewable energy. Distributed generation (DG) services are required in the power system due to environmental concerns and the rising demand for electrical energy. The placement and sizing of DGs have been employed in this study to achieve the aims of lowering losses and enhancing the microgrid's voltage profile. From a variety of artificial intelligence techniques, the proposed Genetic Algorithm (GA) is described as a solution to the issue. On the network of IEEE 33 buses, the technique is put into practice. This study is presented to show the impact of the selection and placement of DGs in order to cut losses, enhance the voltage profile, and improve frequency profile of islanded.

Keywords Newton Raphson method · Islanded microgrid · Genetic algorithm

1 Introduction

Natural disasters including storms, tornadoes, earthquakes and others have the capacity to seriously harm electricity grid networks. Depending on the extent of the damage, this might cause the network to experience extensive power outages that last for hours, days, or even weeks. Utility-provided power is necessary for the continued functioning of essential services for the general public. By providing emergency power during such grid outages, microgrids, also known as localized low voltage grids, can assist utilities in enhancing the resilience of crucial public assets (i.e. key public resources). Distributed generation sources, both conventional and non-conventional, power storage devices, loads or supply nodes, an interconnection

M. Farooq · S. Mishra · J. Dhillon (✉)
School of Electronics and Electrical Engineering, Lovely Professional University Phagwara, Phagwara, India
e-mail: javeddah@gmail.com

J. Kumar et al. (eds.), *Recent Advances in Power Systems*, Lecture Notes in Electrical Engineering 1192, https://doi.org/10.1007/978-981-97-2788-9_15

switch, a controller, and an energy management system are all components of microgrids [1]. The ability of island-based microgrids to function independently of the main grid during natural disasters, known as islanded mode operation, makes them important resources for utility corporations.

An islanded microgrid often uses wind or solar/photovoltaic-based renewable DGs. Due to the need for land space to build the wind turbines, wind-based DGs are more frequently encountered in the rural regions. This contrasts with solar/ photovoltaic (PV) DGs, which can be quickly installed on rooftops in both rural and urban settings, but the most important thing is where we place the DG in the microgrid because it is most important to decrease the losses in the system, enhance voltage profile and enhance frequency profile of the islanded microgrid. In our case study, we are trying to find the optimal place for the DG in the microgrid for that we are using optimisation technique, i.e. genetic algorithm that helps us in finding optimal place for our DG, modified Newton Raphson load flow technique is used for calculating the losses in the islanded microgrid and also analysed the frequency and voltage variation across the system.

2 Literature Review

In [2], the author discussed positioning several distributed generators in a primary distribution network to decrease losses using an improved analytical technique; however, that research did not consider voltage profile improvement. The positioning and sizing of DGs were also optimized by Gangwar et al. [3] using a multi-objective PSO to minimize power losses and enhance system reliability.

In [4], to economically increase system reliability, a mechanism is proposed for assigning dispatchable distributed generating units in distribution systems. As a result, the primary goal of this research project is to identify the best possible configuration of distributed generation units to be deployed and loads to be shed throughout all potential situations. Using a probabilistic load model, the ideal size of allocated DG units is determined. Singh et al. [5] uses an evolutionary algorithm for optimization, and load levels such as low load, average load, and high load are taken into consideration during scheduling, allowing the utilization of the outcomes for all network conditions. In reference [7], the placement and ideal capacity of DGs are chosen using the plant growth simulator algorithm (PGSA) method to lower losses and enhance the voltage profile of the grid.

The most latest DG integration designs, which have been evaluated by Prakash and Khatod [6], Ehsan and Yang [7], and Adetunji et al. [8], optimize DG integration (for instance, exact location), with the goals of trying to minimize power losses, operational cost, demand satisfied (system reliability), and system resilience, which is typically attained by improving voltage level. To the best of our knowledge, studies that can inform us of the location of the DG in the islanded microgrid are nonexistent because we are aware that it has both voltage and frequency regulation. So, in our case study, we are trying to find the optimal place for DG in the islanded microgrid so that we can minimize the losses and provide supply where it is needed. As a result, the overall objective function accounts for the reliability, resiliency, expenses, and possible power losses of the microgrid in order to reduce the overall cost to a utility. This enables utility providers to assess the investment viability, dependability, and resilience of islanded microgrids when tasked with supplying emergency power to vital public resources in the event that the main grid is compromised.

3 Problem Formulation

In this section, objective function is presented. As already discussed above that the objective of DG placement at proper place is to minimize losses improve frequency and improve voltage in the islanded microgrid. Mathematical formulation for different terms of the objective function is presented as follow:-

Objective Function: min (VDI + FDI – ELI).

Voltage deviation index (VDI): As everyone is aware, maintaining bus voltage within limitations is the most crucial factor, particularly in an island microgrid. A small change in voltage has an impact on the entire system and may cause a block out condition. The Voltage Deviation Index indicates the amount of voltage that deviates from the voltage rated.

$$\mathrm{VDI} = \sum_{i}^{\mathrm{NL}} \frac{|\mathrm{V}_{\mathrm{rated}} - \mathrm{V}_{\mathrm{i}}|}{\mathrm{V}_{\mathrm{rated}}} \tag{1}$$

where Vrated, nominal voltage of the system

Vi Is the Voltage at the ith bus.

NB is the no. of buses.

Frequency deviation index (FDI): Maintaining of frequency within the limits is the most important criteria especially in an islanded microgrid. A slight variation in the frequency affects the whole system and may result in block out condition. The amount of frequency that deviates from the rated frequency is given by the frequency deviation index:

$$\mathrm{FDI} = \frac{|\mathrm{F}_{\mathrm{rated}} - \mathrm{F}_{i}|}{\mathrm{F}_{\mathrm{rated}}} \tag{2}$$

where $\mathrm{F}_{\mathrm{rated}}$, nominal frequency of the system.

Fi Is the frequency at the ith bus.

Energy loss index: In order to make our system more efficient we have to minimize the losses as much as possible when the power losses increase the system it affects the rated voltage, so it affects the performance of the system. In order to use the maximum energy we have to minimize the losses. The change in energy from the base energy due to losses is given by the energy deviation index.

$$\mathrm{ELI} = \sum_{ij}^{Nl} \frac{\mathrm{Losses}_{\text{in case A}} - \mathrm{losses}_{\text{in case B}}}{\mathrm{Losses}_{\text{in case A}}} \tag{3}$$

In case "A" DG is connected on bus 1 and in case "B" DGs are connected at optimal places in an islanded microgrid.

4 Modified Newton Rapson Method

The load flow problem for islanded microgrids is solved by modified Newton Raphson method (MNR) utilizing a novel strategy that combines the traditional Newton Raphson (NR) method with the DG's droop control.

4.1 Describing the System's Buses

The first step in resolving any system's power flow is to determine the different types of buses that are there. The pre-specified quantities determine the bus's type. When tackling the power flow challenge using the traditional model, the NR technique employs three well-defined buses: the PV bus, the load bus, and the slack bus. The premise that any DG will operate as a slack bus in a various source isolated microgrid is invalid since no single DG can maintain a steady local bus voltage and systems frequency. The bus kinds for islanded microgrids are categorized in this research as follows:

- The real and imaginary power of PQ bus is known.
- PV bus: The real power and voltage magnitude are known.
- VF dependent bus: The real and imaginary power of the bus rely on the system frequency as well as the bus voltage. For such voltage ranges with in system, a swing bus is chosen to act as a reference. A generator bus, load bus, or voltage frequency-dependent bus can serve as a swing bus.

4.2 Modified Newton Raphson (MNR)

The real and imaginary power of all the droop buses must be determined t0 be able to determine angle of the voltage and intensity of a droop bus at the (I + 1) iteration (to be added to the mismatch matrix). The real and imaginary powers of the droop bus k can be calculated under the assumption that the output impedance of Distributed Generator is inductive and by setting the references Po & Qo to zero.

$$Q_{GK^{i+1}} = \frac{1}{n_{pk}}\left(|V_0| - \left|V_K^i\right|\right) \tag{4}$$

$$P_{GK^{i+1}} = \frac{1}{m_{pk}}\left(\omega_0 - \omega^i\right) \tag{5}$$

In the event that the droop bus's active and reactive powers go beyond the limit, the relevant power is adjusted to its restricted value. Like what is shown in (1), the frequency of the system affects the active power of the droop bus. All of the droop control buses communicate using the system frequency to be able to distribute the system's loads. In other words, the system angular frequency, which is shared by all droop buses in the microgrid, will be used to deliver active power. The total active powers for d droop buses in the system are expressed as follow:

$$P_{sys} = \sum_{k=1}^{d} P_{GK} = \sum_{k=1}^{d} \frac{1}{m_{pk}}(\omega_0 - \omega) \tag{6}$$

$$Q_{sys} = \sum_{k=1}^{d} Q_{GK} = \sum_{k=1}^{d} (|V_0| - |V_K|) \tag{7}$$

In the case of the modified Newton Raphson load flow approach, starting system frequency of one p.u. is anticipated, as are voltages of the buses at the start of the load flow. Bus 1 voltage magnitude (|V1|) is one more parameter in the updated Newton Raphson approach (cuz of the omission of slack bus). The formula for the differential vector (X′) is

$$X' = \left[|V|^T \omega |V_1|\right]^T \tag{8}$$

$$\Delta' = \left[P^T - P_c^T Q^T - Q_c^T P_{tot} - P_{sys} Q_{tot} - Q_{sys}\right]^T \tag{9}$$

The updated mismatch matrix now contains four new terms. P_{tot} is result of adding the cumulative real power and P_{loss}.. The overall real power requirement in the (P_{load}) syst em & the real power losses of the system can be used to replace the total real powers of the Distribution Generators (P_{tot}). The system's cumulative reactive power (Q_{tot}) is equal to the product of its reactive power loss (Q_{loss}) and total reactive load (Qload).

$$P_{tot} = P_{load} + P_{loss} \tag{10}$$

$$Q_{tot} = Q_{load} + Q_{loss} \tag{11}$$

Losses must therefore be determined in each cycle and are determined using the below equation:

$$P_{loss} = \frac{1}{2}\sum_{k}^{N}\sum_{n}^{N}\{\mathrm{Y_{kn}(V_k * V_n + V_n * V_k)}\} \tag{12}$$

$$Q_{Loss} = \frac{1}{2}\sum_{k}^{N}\sum_{n}^{N}\{Y_{kn}(V_k * V_n + V_n * V_k)\} \tag{13}$$

Next, it is estimated for each bus what their organized real power (P), anticipated imaginary power (Q), computed real power (P_c), & computed imaginary power (Q_C) are. Making the mismatch matrix (Δ/) equal to zero is the goal. The revised Jacobian matrix is acquired by differentiating Eqs. (19) and (20) with regard to, |V |,δ, ω and |V1| to identify the whole modified Jacobian matrix. The updated Jacobian matrix has been provided below

$$\begin{bmatrix} J_{11} & J_{12} & J_{13} & J_{14} \\ J_{21} & J_{22} & J_{23} & J_{24} \\ J_{31} & J_{32} & J_{33} & J_{34} \\ J_{41} & J_{42} & J_{43} & J_{44} \end{bmatrix} \begin{bmatrix} \Delta\delta \\ \Delta V \\ \Delta\omega \\ \Delta V_1 \end{bmatrix} = \begin{bmatrix} \Delta P \\ \Delta Q \\ \Delta P_{sys} \\ \Delta Q_{sys} \end{bmatrix} \tag{14}$$

The updated Jacobian's first sub-matrix can be determined using the standard Newton–Raphson approach because it is the same as the Jacobian of that method. The final 12 sub-matrices, which are updated Jacobian sub-matrices for the updated Newton Raphson approach, are computed utilizing

$$J_{41} = \left[\frac{\partial Q_{sys}}{\partial \delta_2} \cdots \frac{\partial Q_{sys}}{\partial \delta_N}\right] J_{42} = \left[\frac{\partial Q_{sys}}{\partial |V_2|} \cdots \frac{\partial Q_{sys}}{\partial |V_N|}\right] \tag{15}$$

$$J_{31} = \left[\frac{\partial P_{sys}}{\partial \delta_2} \cdots \frac{\partial P_{sys}}{\partial \delta_N}\right] J_{32} = \left[\frac{\partial P_{sys}}{\partial |V_2|} \cdots \frac{\partial P_{sys}}{\partial |V_N|}\right] \tag{16}$$

$$J_{43} = \left[\frac{\partial Q_{sys}}{\partial \omega}\right] J_{23} = \left[\frac{\partial Q_{c2}}{\partial \omega} \cdots \frac{\partial Q_{cN}}{\partial \omega}\right]^T \tag{17}$$

$$J_{44} = \left[\frac{\partial Q_{sys}}{\partial |V_1|}\right] J_{24} = \left[\frac{\partial Q_{c2}}{\partial |V_1|} \cdots \frac{\partial Q_{cN}}{\partial |V_1|}\right]^T \tag{18}$$

$$J_{33} = \left[\frac{\partial P_{sys}}{\partial \omega}\right] J_{13} = \left[\frac{\partial P_{c2}}{\partial \omega} \cdots \frac{\partial P_{cN}}{\partial \omega}\right]^T \tag{19}$$

$$\mathrm{J_{34}} = \left[\frac{\partial \mathrm{P_{sys}}}{\partial |\mathrm{V_1}|}\right] \mathrm{J_{14}} = \left[\frac{\partial \mathrm{P_{c2}}}{\partial |\mathrm{V_1}|} \cdots \frac{\partial \mathrm{P_{cN}}}{\partial |\mathrm{V_1}|}\right]^{\mathrm{T}} \tag{20}$$

5 Genetic Algorithm

The Darwinian idea of "survival of the fittest" serves as the foundation for the optimization technique known as the genetic algorithm. It is a search method that makes advantage of crossover and mutation, two natural genetic processes. It uses artificial intelligence (AI) to find the best site for each DG in an island-based microgrid. Most practical issues can be solved using a basic genetic algorithm that uses the three operators of reproduction, crossover, and mutation.

The following details/facts show how genetic algorithm method is apart from all other optimization methods:

- Rather of coding the parameters directly, GA employs the coding of a group of parameters.
- GA encourages the use of objective function values rather than relying on additional information.
- GA never looks for a single point; it always searches the population.
- GA never employs deterministic rules; instead, it employs probabilistic rules.

For each DG source, the GA takes into account the production of a random population of a location. Following the load flow program's execution, it determines the losses in the anticipated states for the placement of the DG per bus and then distinguishes between optimal states and non-optimal modes. The parent uses these ideal states to produce the population in the following generation. This procedure is repeated until the algorithm's stopping requirements are met.

The algorithm for optimal DG placement in an islanded microgrid is given below (Fig. 1).

6 Results and Discussions

For solving, the optimal DG placement problem for an islanded microgrid optimization technique is required with a load flow algorithm to calculate the voltage, frequency, and losses. It also helps to find optimal placement of DG's. For calculating voltage frequency and losses, modified Newton Raphson method is used and for optimal location of DG's genetic algorithm is used. The results obtained are from 10 bus system and 38 bus system.

Table 1 show the voltages of all the buses for the 10 bus system when the DGs are placed at optimal locations. It is also seen the frequency is also improved when DGs are at optimal locations. Figure 2 shows the comparison of voltages when all the DGs are placed at optimal location with one of the DGs randomly placed at Bus 1.

From the above graph it can seen that voltage profile of all the 10 buses is improved when DGs are connected at the best location in the islanded microgrid, i.e. the deviation in voltage of all the buses is minimum.

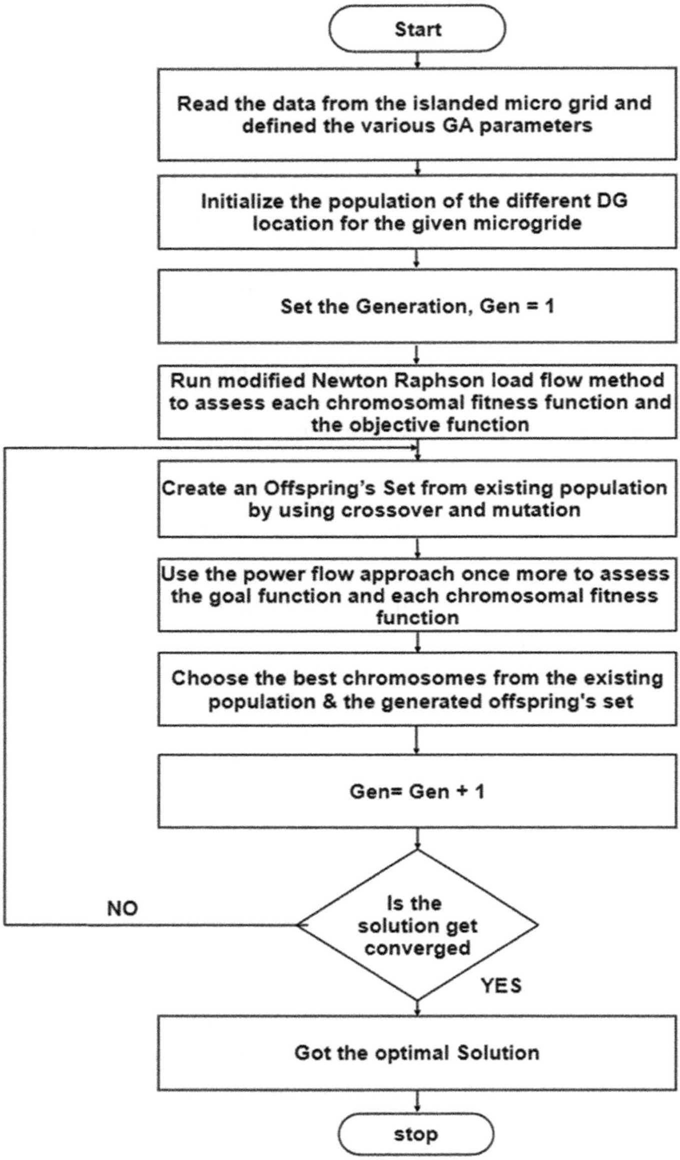

Fig. 1 Genetic algorithm flowchart for DG placement

The results obtained when applying genetic algorithm approach in the 10 bus islanded microgrid grid are given in Table 2. From the below table, it is observed that bus voltages, frequency, and the losses are minimized by placing DG's at the best locations in the islanded microgrid, and the system becomes stable and works effectively.

Table 1 Voltages for 10 bus islanded microgrid

Bus	Voltage	Voltage angle
1	0.999059	0
2	0.999659	0.000155
3	1.001188	0.003364
4	1.003009	0.005802
5	1.004742	0.007848
6	1.001023	0.006794
7	0.998863	0.007517
8	0.999741	0.010302
9	1.001106	0.016416
10	1.00063	0.015726

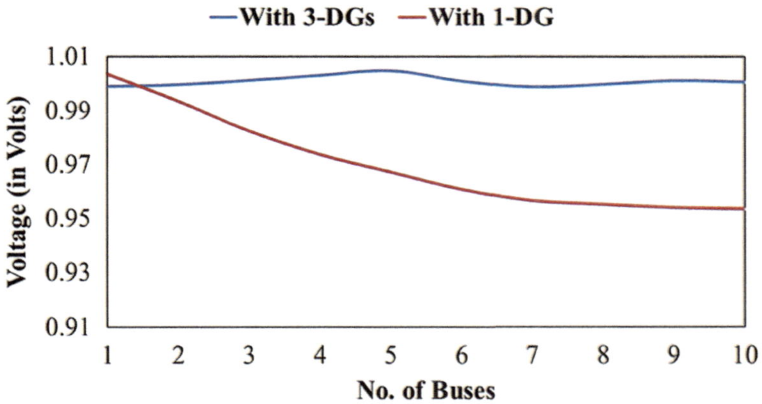

Fig. 2 Voltage profile for 10 bus islanded microgrid

Table 2 Optimal DG placement for 10 bus microgrid

DG parameters	With 1 DG connected on bus 1	With 3 DGs connected at the best locations
Minimum bus voltage	0.9535 V	0.998863 V
Frequency (pu)	0.94821	0.999567
Total active power loss(pu)	0.0136	0.000678
Total reactive power loss(pu)	0.0177 VAR	0.001731 VAR

The genetic algorithm optimization technique is tested on the 38 bus system also. Table 3 shows the voltages of all the buses when DGs placed at optimal location. Table 4 shows the comparison when three DGs are placed at optimal location with single DG placed at random location on Bus 1. Figure 3 shows the voltage profile of all the buses of 38 bus system in an islanded microgrid.

Table 3 Voltages for 38 bus islanded microgrid

Bus	Voltage	Voltage angle	Bus	Voltage	Voltage angle
1	0.990737	0	20	0.986726	−0.00148
2	0.990737	0	21	0.98604	−0.00185
3	0.99248	0.00007	22	0.98542	−0.00223
4	0.990007	−0.00183	23	0.988668	−0.00063
5	0.987809	−0.00388	24	0.981517	−0.0026
6	0.984007	−0.01025	25	0.977552	−0.00374
7	0.988482	−0.01178	26	0.982303	−0.00974
8	0.991092	−0.01743	27	0.980044	−0.009
9	0.991135	−0.0223	28	0.970128	−0.00829
10	0.992787	−0.02652	29	0.963018	−0.00749
11	0.991585	−0.02653	30	0.959906	−0.00604
12	0.989422	−0.02657	31	0.95626	−0.00763
13	0.98213	−0.02943	32	0.955463	−0.00806
14	0.979402	−0.03155	33	0.955221	−0.00819
15	0.977399	−0.03277	34	0.98813	−0.0201
16	0.975256	−0.03377	35	0.988782	−0.02384
17	0.971759	−0.03702	36	0.987207	−0.02821
18	0.97031	−0.03782	37	0.969672	−0.03836
19	0.99022	−0.0002	38	0.977066	−0.00402

Table 4 Optimal DG placement for 38 bus microgrid

DG parameters	With 1 DG connected on bus 1	With 3 DGs connected at the best locations
Minimum bus voltage	0.7331V	0.955221V
Frequency (pu)	0.9829	0.995871
Total active power loss (pu)	0.1725	0.053470
Total reactive power loss (pu)	0.1061VAR	0.042933 VAR

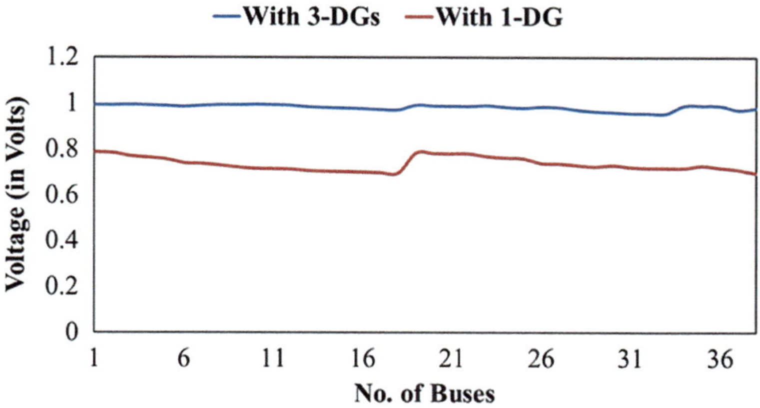

Fig. 3 Voltage profile for 38 bus islanded microgrid

From the above graph, it can be seen that the deviation in the bus voltages is very less when DGs are placed at the best locations in the islanded microgrid.

7 Conclusion

In this research, the genetic algorithm technique is used for optimal placement of DGs in an islanded microgrid to minimize losses, enhance voltage profile and frequency. The research shows that by placing the DGs at optimal locations the voltage profile is enhanced, frequency improved and losses are reduced. Before simulating the load under actual network conditions, decisions cannot be made with certainty. After examining the data, it has been determined that the best DG placement is advantageous since it reduces power loss and significantly raises the voltage profile and maintains the frequency at its rated value. All industrial, residential, and commercial subscribers should be taken into account to provide more precise results. In other words, it is impossible to generalize to all levels of load from the results obtained for the single-level load under nominal conditions.

References

1. Haidar N, Attia M, Senouci SM, Aglzim EH, Kribeche A, Asus ZB (2018) New consumer-dependent energy management system to reduce cost and carbon impact in smart buildings. Sustain Cities Soc
2. Hung DQ, Mithulananthan N (2013) Multiple distributed generator placement in primary distribution networks for loss reduction. IEEE Trans Ind Electron 60(4):1700–1708

3. Gangwar P, Singh S, Chakrabarti S (2019) Multi-objective planning model for multi-phase distribution system under uncertainty considering reconfiguration. IET Renew Power Gener 13:2070–2083
4. Awad A, Salama MMA, Awad ASA, El-Fouly THM (2014) Optimal distributed generation allocation and load shedding for improving distribution. Electr Power Compon Syst 42(6):576–584
5. Singh D, Singh D, Verma KS (2009) Multiobjective optimization for DG planning with load models. IEEE Trans Power Syst 24(1):427–436
6. Prakash P, Khatod DK (2016) Optimal sizing and siting techniques for distributed generation in distribution systems: a review. Renew Sustain Energy Rev 57:111–130
7. Ehsan A, Yang Q (2018) Optimal integration and planning of renewable distributed generation in the power distribution networks: a review of analytical techniques. Appl Energy 210(Oct 2017):44–59
8. Adetunji KE, Hofsajer IW, Abu-Mahfouz AM, Cheng L (2021) A review of metaheuristic techniques for optimal integration of electrical units in distribution networks. IEEE Access 9:5046–5068

PI Based Secondary Control of Microgrids by Considering Communication Delays with Three Sources

Dasari Gopikrishna and Swati Devabhaktuni

Abstract The conventional power system can accommodate the incorporation of distributed generators(DGs) powered by renewable energy sources(RESs), thanks to contemporary power electronic technology. A micro grid (MG), which can function in islanded mode or grid-connected mode, is created by methodically combining DGs with energy storage systems. Significant challenges, including voltage and frequency volatility and the unreliability of RES, are posed by the intermittent character of RES and the fluctuating demand. Non-linear loads and unequal feeder impedances are regarded as current problems in micro grids and micro grid management. Droop control is one of the widely used strategies for sharing active and reactive power in micro grids (MGs). This method's drawback is that in steady state, the MG's frequency deviates from its nominal value and must be corrected using a secondary control system (SCS). In order to rectify the frequency, the signal acquired at the SCS's output is sent through a communication link to the MG's generating sources. The operation of the MG might be jeopardized if the SCS is not designed to account for the fact that communication lines are prone to time delays.

Keywords Primary controller · Secondary controller · Proportional-integral-derivative control · Voltage source converters (VSCs) · Frequency · Active power

1 Introduction

The traditional power system depends on predicted data for electricity production, and the delivery of electricity has long been done passively. However, distributed generation (DG), which is primarily built on renewable energy sources, has shown

D. Gopikrishna (✉) · S. Devabhaktuni
Department of Electrical Engineering, National Institute of Technology, Warangal 506004, India
e-mail: gopikrishnad968@gmail.com

S. Devabhaktuni
e-mail: swatikjm@nitw.ac.in

J. Kumar et al. (eds.), *Recent Advances in Power Systems*, Lecture Notes in Electrical Engineering 1192, https://doi.org/10.1007/978-981-97-2788-9_16

to be a potential new technology for conventional power networks (RESs). The technology combines advanced control, communication, and administration to profile contemporary power networks with DG units based on RES or other sources [1–3]. Micro-sources like photovoltaic (PV) cells, low-power wind generators, and micro turbines are regarded as dispersed generations that are incorporated into the power system. With the integration of distributed generators in the electricity system, there are many benefits for both customers and utilities, especially cost-saving. But the widespread use of DGs could lead to issues with reversible power flow, voltage deviations, and voltage changes in low-voltage distribution networks [4]. However, these problems can be resolved when distributed storage (DS) and DG devices are used together [5, 6]. A micro grid (MG) is created when DG and DS components are arranged in a methodical manner [7]. Comparatively speaking, MG offers greater flexibility and capability than a singular DG that is part of the electricity infrastructure. It needs to have "plug and play" qualities [8] and duplicates the synchronous generators (SGs) linked in parallel in order to keep the standard values of system parameters and to guarantee the reliability of the micro grid during the integration of new DG units. Active and reactive electricity are controlled using frequency and voltage control techniques [9, 10]. As a result, the "plug-and-play" feature does not necessitate changing the applied management system in order to add new DG units or disconnect current DG units. Multiple DG units, DS units, and various kinds of loads connected to MG through various places of connection may make up the fundamental structure of MG. (PCs). In the grid-connected mode, the point of common coupling (PCC), which also offers security and separation from the main grid in case of faults, and establishes the power transfer between the utility and the MG. Electricity converters, such as DC-DC and DC-AC converters, are used to transfer electricity from DGs with DC outputs like PV systems and DS units. However, AC-DC and DC-AC power converters are used to link DG devices with AC outputs. When operating in island mode, DG and DS units only power the MG's directly linked load.

2 Primary Control

The converter output control and power sharing control are provided by the primary control, which is the lowest control. The inside loop and outside loop make up the inverter output control. While the inside loop manages the output current, the outer loop governs the voltage [11]. An appropriate reference frame, such as synchronous (dq), fixed ($\alpha\beta$), or natural (abc), is used by the inverter control [12]. If the control is built on a proportional-integral (PI) controller or only works with DC variables, the synchronous reference frame is used. However, when working with sinusoidal factors or a proportional-resonant (PR) controller, it employs a fixed reference frame. The natural reference frame is used by the hysteresis, slack, PR, and PI controllers [13]. The PI controller is most frequently used to perform inverter output management. It makes use of feed-forward correction to enhance the controller's dynamic efficiency

[12, 14]. Multivariate control techniques have been investigated in order to improve the control's performance in the presence of nonlinear loads and to boost stability [15, 16]. Whether a power sharing control strategy makes use of the concept of droop control determines its classification. The primary controller may be categorized as a secondary controller if it works with power-sharing management.

In MGs, the sharing of active power is typically achieved by changing the phase angle between the DGR voltage outputs. This is further explained using the well-known expression

$$P_{ij} = 3\frac{V_i V_j}{X_{ij}} Sin\left(\delta_{ij}\right) \tag{1}$$

P_{ij} is the active power that is transmitted from source "*I*" to the source "*j*", (V_i, V_j) are the voltage moduli of both sources, δ_{ij} is the phase angle shift between the two vectors, X_{ij} is the equivalent impedance between the two voltage sources. By controlling inverter output current(i) and voltage(V_0) we can control the grid side voltage. The current control loop gives a triggering signal to the pulse generator and this signal will be received as the input of the inverter, so we will call it as the inner current loop and the voltage control loop gives a signal to this inner current control loop. Current control loop is derived from the dynamics of $R_I L_I$ and the outer voltage control loop is derived from the dynamics of the capacitor(C_I) (Fig. 1).

Using the following formulae, we can set the individual operating voltage and operating frequency of the voltage source converter.

$$w_i = w_n - m_{pi} P_i \tag{2}$$

$$V^*_{odi} = V_n - n_{qi} Q_i, V^*_{oqi} = 0 \tag{3}$$

where V_n is the nominal voltage, w_n is the nominal frequency of the system. m_{pi}, n_{qi} are the P-f and Q–V droop control constants. W_i is the operating frequency of the ith inverter and V^*_{odi} is the operating voltage of the ith inverter.

3 PI Based Secondary Control

Reliable and cost-effective running of micro grids is the primary responsibility of the secondary management. As a result, it is known as MG energy management system informally. In this context, the two main approaches are decentralized and centralized. While centralized control frequently relies on the function of the central controller, decentralized control helps communicate with different MG components to facilitate a distributed decision-making process. The implementation of the constantly updated optimization, where all the data are gathered at once, is made possible by the centralized management. It does not, however, provide the required plug-and-play function.

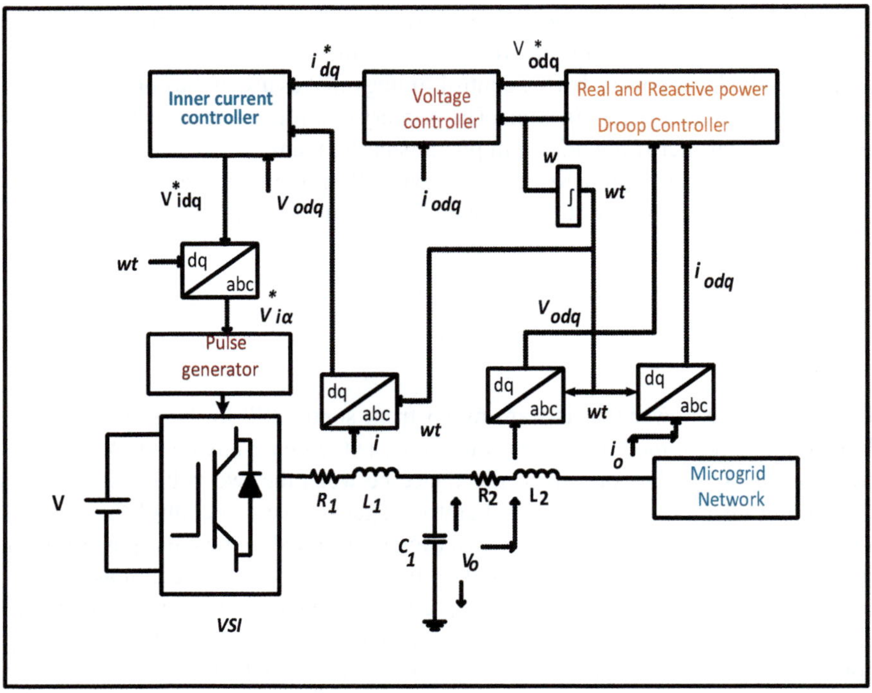

Fig. 1 Voltage source converter block diagram

On the other hand, decentralized control integrates new DG units without making constant changes to controller settings, but there are difficulties in the administration of micro grid operations that call for a high degree of management. A low pass filter is used for decoupling between the droop controller and voltage/current controller (Fig. 2).

The following formulae are used to determine each VSC's operating frequency and voltage set points.

$$w_i = w_n - m_{pi} P_i + \Delta w_{ni} \quad (4)$$

$$V_{odi} = V_n - n_{qi} Q_i + \Delta V_{ni};\ V_{oqi} = 0 \quad (5)$$

Here Δw_{ni} and ΔV_{ni} are coming from the secondary control. Generally, centralized control is suitable for remote MGs with rigid structure and those with critical demand–supply, Conversely, grid-connected MGs with multiple proprietors and a variable number of DG units are best served by the decentralized management (Table 1).

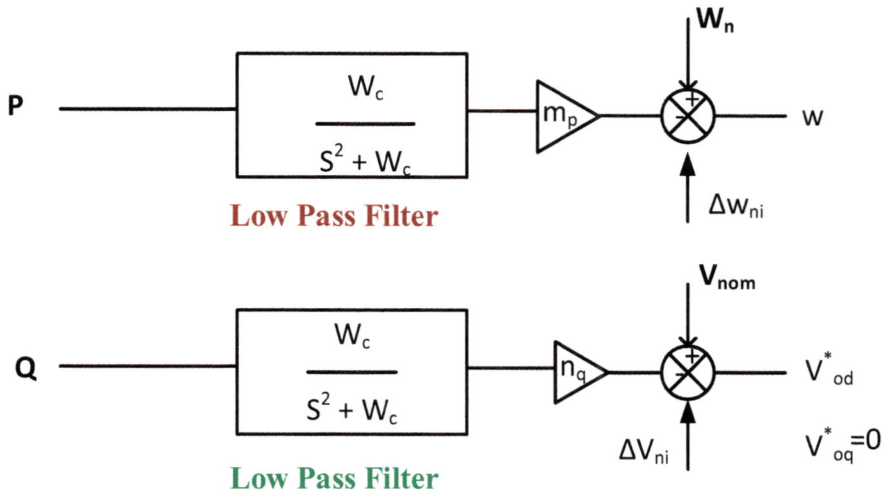

Fig. 2 Voltage and current controller block diagram

Table 1 Micro grid parameters

Parameter	Value	Unit
Filter resistance (R_1)	0.1	[Ω]
Filter inductance (L_1)	3.5	[mH]
Filter capacitance (C_1)	50	[μF]
Line resistance (R_2)	0.05	[Ω]
Line inductance (L_2)	0.15	[mH]
Source voltage (DC)	700	[V]
Inverter O/P current (i)	30	[A]
Inverter rating	20	KVA

4 Results and Discussion

In this paper, we designed three voltage source converters (VSC). Each VSC has an inverter that is connected to a constant DC voltage of 700 V. These voltage sources are connected to three loads of (5.8 + j0.2) KVA, (5.8 + j0.2) KVA, (12.1 + j1.2) KVA. In this instance, it is presumable that the transmission route has a small delay. It is crucial to emphasize that transmission delays can easily reach values in the order of milliseconds.

Or even tens of milliseconds, in contrast to the delays typically used in power electrical systems (which are in the order of μs). From Fig. 3a we observed that three VSCs shared active power in proportion when the load was connected, and the settling time of the active power is approximately 0.7[s] and from (b) when the load was disconnected (after 1[s]) the active power falls to 25% of its initial value. Here time delay is 0.1 [s], it does not make much difference compared to without communication delay. Figure 4a shows that the settling time of frequency is approximately 0.9[s], and the frequency of all three VSCs is settling at the same point. Figure 4b shows the waveforms of the load disconnection case, and when the load was disconnected the frequency was raised and settled at a point which is higher than the previous case.

This is because we knew that frequency and active power are inversely proportional. Here for Figs. 3 and 4 the secondary controller values are $K_p = 0.1$ and $K_i = 10$. Figure 5 shows that the settling time of frequency is approximately 1.1[s], it is higher when compared with plant delay of 0.1[s] case, and more oscillations are also present in this case. Active power waveform is almost same in both plant delay cases (i.e. 0.1[s] and 0.3[s]). From Fig. 6 we observed that settling of frequency is

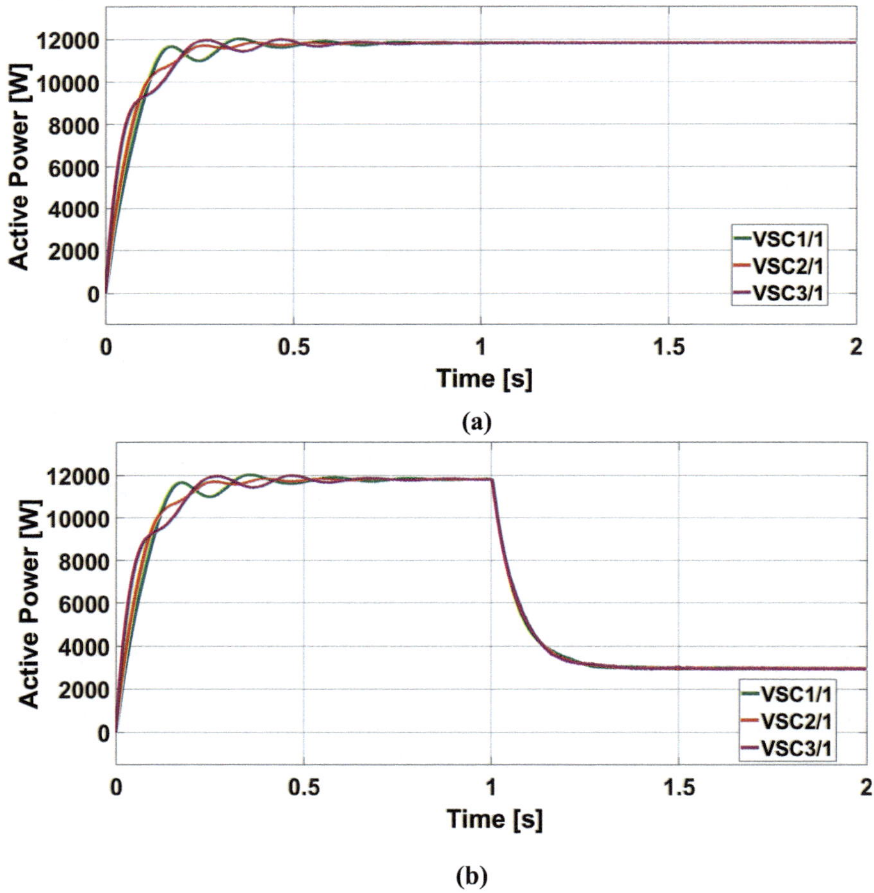

Fig. 3 Active power in inverters **a** load connection and **b** load disconnection after 1[s] (time delay 0.1 [s])

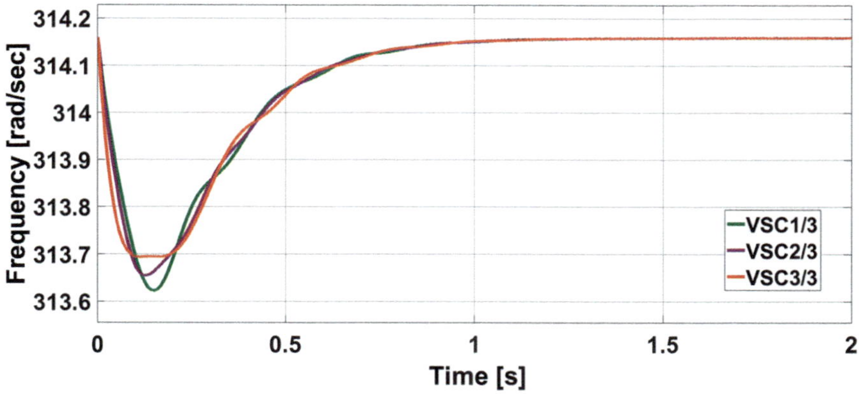

(a)

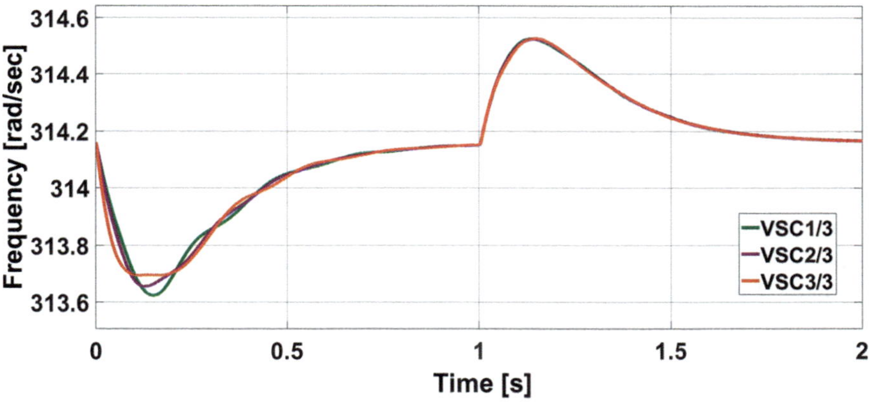

(b)

Fig. 4 Frequency from inverters **a** for a load connection with plant delay 0.1[s], and **b** for a load disconnection with plant delay 0.1[s]

1.8[s] and settling time of active power is more than 2[s]. Here, we observed that the settling times of both frequency and active power rise and active power is also not sharing proportionally among the voltage source converters when the communication delay from the secondary control to the primary control lengthens. Oscillations of frequency and active power are also more when communication delay is 0.5[s] and they may go to an unstable condition as communication delay increases.

In the upcoming years, it is anticipated that solar energy use will rise globally. After local and global disasters, it will be the primary source of energy sustaining the economy, particularly at medium and low altitudes. Solar energy will be essential for the eco-friendly generation of industrial gases, a critical component of effective, eco-friendly, distributed energy storage, in part because of its lower historical environmental impact. Photovoltaic and solar-thermal concentration facilities will

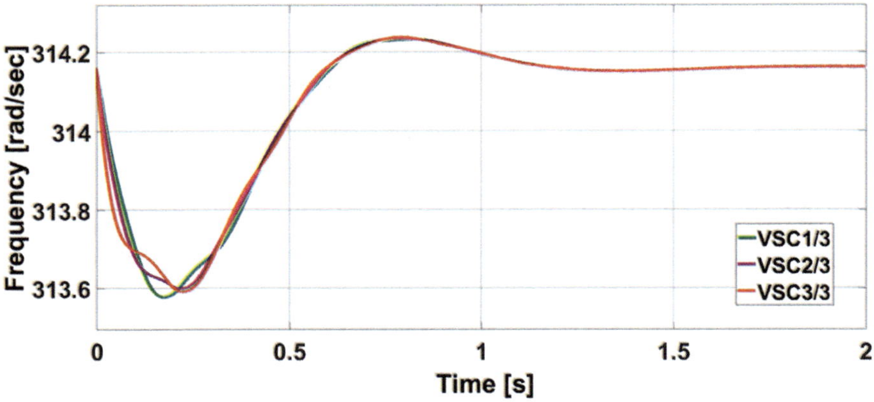

Fig. 5 Frequency of inverters for a load connection with plant delay 0.3[s]

guarantee dispersed electrical energy generation on a wide range of scales, from very large to tiny local generation. The demand for green energy from the solar cell will increase due to the new corporate models and growing popularity of electrical transportation. High caliber research output from academic and commercial R&D divisions is required to drive this transformation. Here we designed a solar cell that is giving approximately 700 V to the inverter of all three voltage source converters by using the Perturb & Observe maximum power transfer technique and boost converter.

Figure 7 indicates that the active power falls to 40% of initial sharing when the load was disconnected. Secondary controller parameters for Figs. 7, 8 and 9 are $K_p = 0.2$, $K_i = 15$.

Figure 8 shows that the settling time of frequency is 1.7[s] when the communication delay is 0.1[s] and the settling time is approximately 2[s] when the communication delay is 0.3[s]. Frequency is more deviated from the rated value and the settling time also more while there is increase in communication delay. Figure 9 shows that when the load was disconnected for some time and again connected (a) active power sharing falls during that period after which it was settled to its rated value, (b) frequency was increased when the load was disconnected then it was settled to its rated value. Here communication delay is 0.1[s], it does not make much changes.

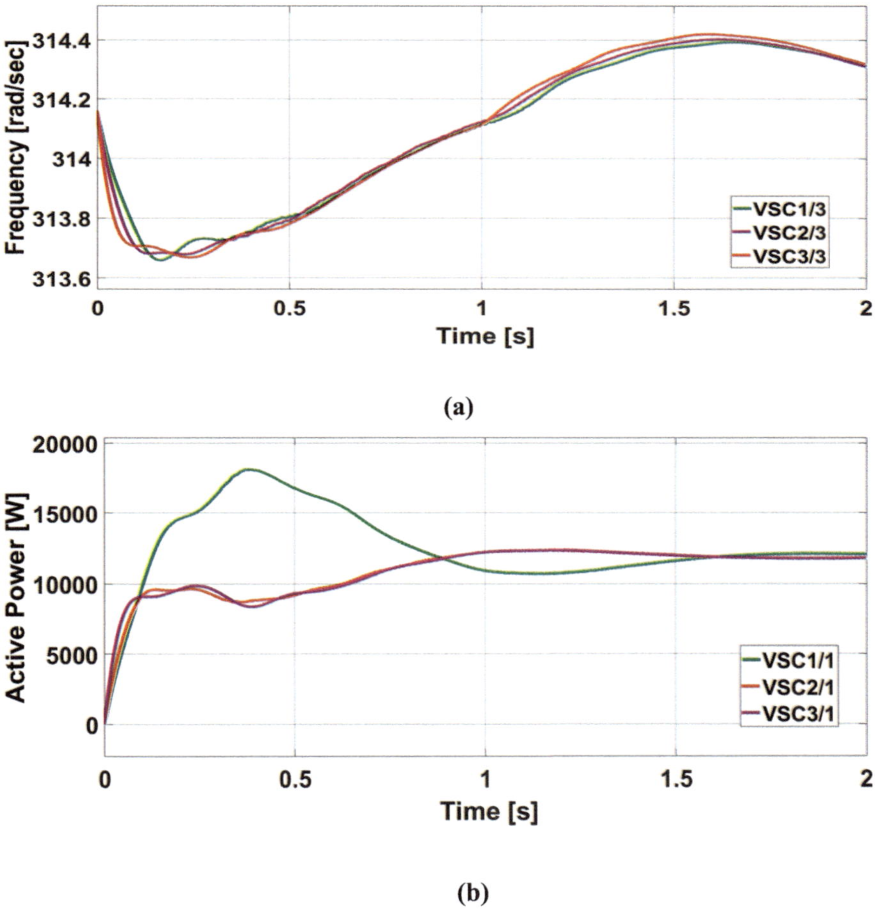

Fig. 6 Frequency from inverters **a** for a load connection with plant delay 0.5[s], and **b** Active power in inverters with plant delay 0.5[s]

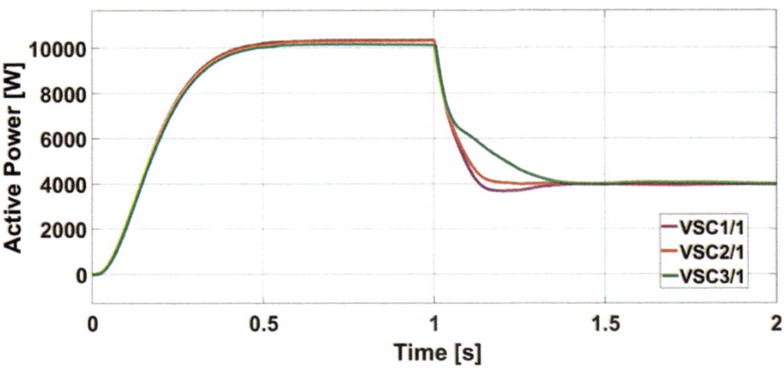

Fig. 7 Active power in inverters load disconnection after 1[s] with plant delay of 0.1[s] (solar)

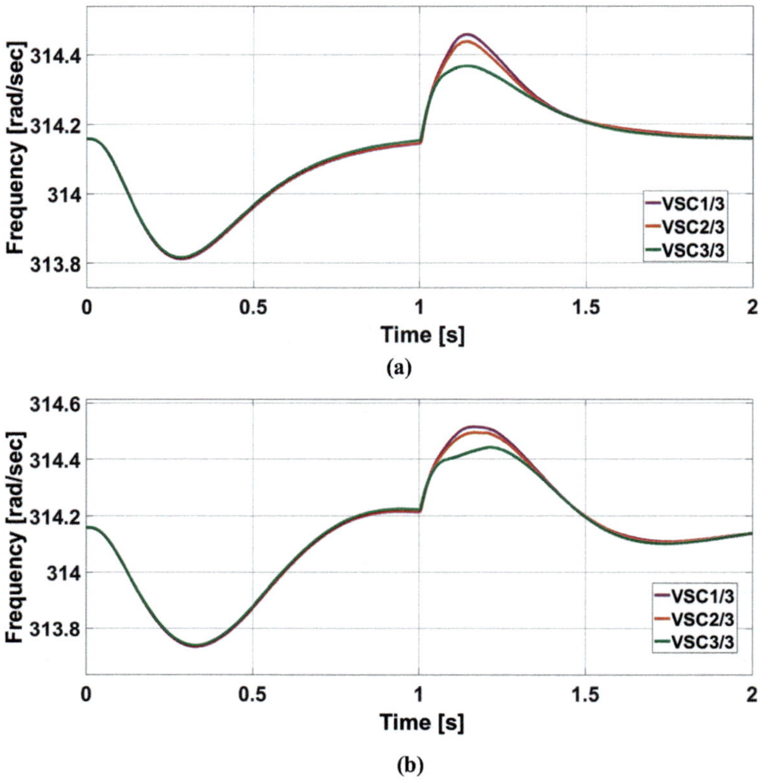

Fig. 8 Frequency from inverters for load disconnection with plant delay **a** 0.1[s], and **b** with plant delay 0.3[s]. (solar connected VSCs)

From Table 2 we observed that as the communication delay from the secondary control to the primary control increases the settling time of both frequency and active power also increases and they may go to an unstable condition for higher order communication delays.

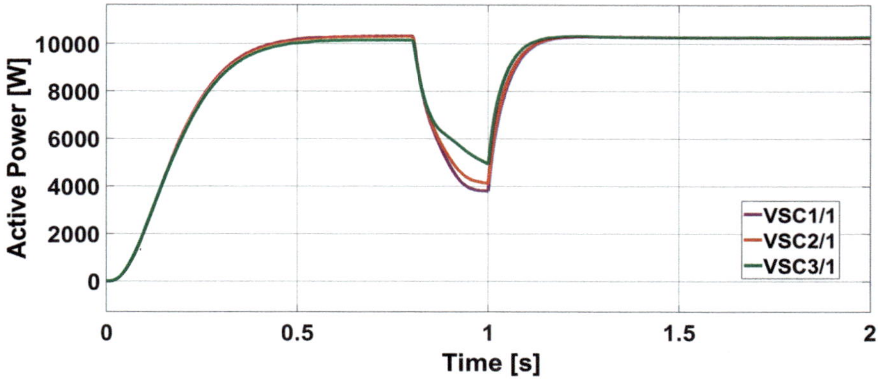

(a)

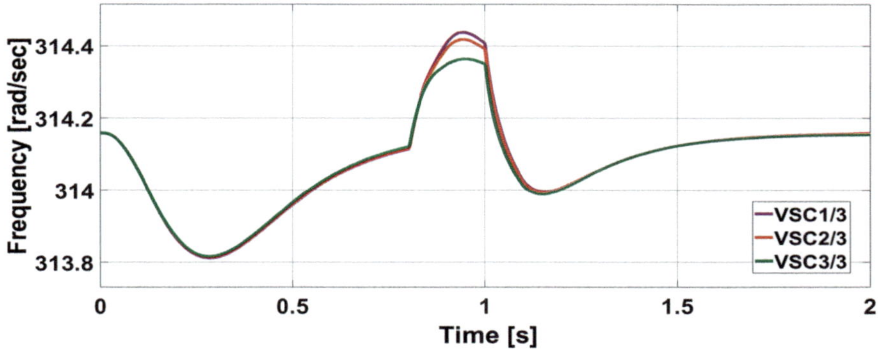

(b)

Fig. 9 **a** Active power in inverters load disconnection from 0.8[s] to 1[s] with plant delay 0.1[s] and **b** Frequency from inverters for a load disconnection from 0.8[s] to 1[s] with plant delay 0.1[s] (solar connected VSCs)

Table 2 Settling time of frequency and active power

Plant delay [s]	Parameter	Settling time[s]
0.1	Frequency	0.7
0.3	Frequency	1.1
0.5	Frequency	1.8
0.1	Active power	0.7
0.5	Active power	More than 2[s]

5 Conclusion

This article presents a reliable control technique for micro grid secondary management. Small signal analysis and simulation studies have been used for analyses and to evaluate secondary control strategies based on PI control. With robust performance, this management strategy has been created to work in micro grids with varying and unknowable transmission delays. Here we observed that with proper tuning of K_p and K_i of the secondary controller we can get robust values of frequency and active power. Changes in communication delays are impacting frequency and active power i.e. increase in communication delays increases settling time, oscillations, frequency and active power that is deviating more from its rated values. We observed that an increase in communication delays from secondary controls to inverters makes frequency and active power unstable.

References

1. Wang C, Yang X, Wu Z et al (2016) A highly integrated and reconfigurable microgrid testbed with hybrid distributed energy sources. IEEE Trans Smart Grid 7(1):451–459
2. Blaabjerg F, Chen Z, Kjaer SB (2004) Power electronics as efficient interface in dispersed power generation systems. IEEE Trans Power Electron 19(5):1184–1194
3. Marzal S, Salas R, González-Medina R et al (2018) Current challenges and future trends in the field of communication architectures for mi- crogrids. Renew Sustain Energy Rev 82:3610–3622
4. Kim Y-S, Kim E-S, Moon S-I (2016) Frequency and voltage control strategy of standalone microgrids with high penetration of intermittent renewable generation systems. IEEE Trans Power Syst 31(1):718–728
5. Kim J-Y, Jeon J-H, Kim S-K et al (2010) Cooperative control strategy of energy storage system and microsources for stabilizing the microgrid during islanded operation. IEEE Trans Power Electron 25(12):3037–3048
6. Tang X, Qi Z (2012) Energy storage control in renewable energy based microgrid. In: Proceedings of IEEE PES general meeting, San Diego, USA, Jul 2012, pp 1–6
7. Lasseter RH (2002) Microgrids. In: Proceedings of IEEE PES winter meeting, New York, USA, Jan 2002, pp 305–308
8. Ashabani SM, Mohamed Y (2014) New family of microgrid control and management strategies in smart distribution grids–analysis, comparison and testing. IEEE Trans Power Syst 29(5):2257–2269
9. Majumder R, Ghosh A, Ledwich G et al (2010) Operation and control of hybrid microgrid with angle droop controller. In: Proceedings of 2010 IEEE region 10 conference, Fukuoka, Japan, Nov 2010, pp 509–515
10. Barklund E, Pogaku N, Prodanovic M et al (2008) Energy management in autonomous microgrid using stability-constrained droop control of inverters. IEEE Trans Power Electron 23(5):2346–2352
11. Chandorkar MC, Divan DM, Adapa R (1993) Control of parallel connected inverters in standalone AC supply systems. IEEE Trans Ind Appl 29(1):136–143
12. Blaabjerg F, Teodorescu R, Liserre M et al (2006) Overview of control and grid synchronization for distributed power generation systems. IEEE Trans Ind Electron 53(5):1398–1409
13. Kawamura A, Chuarayapratip R, Haneyoshi T (1988) Deadbeat control of PWM inverter with modified pulse patterns for uninterruptible power supply. IEEE Trans Ind Electron 35(2):295–300

14. Chung I-Y, Liu W, Cartes DA et al (2010) Control methods of inverter interfaced distributed generators in a microgrid system. IEEE Trans Ind Appl 46(3):1078–1088
15. Karimi H, Davison EJ, Iravani R (2010) Multivariable servomechanism controller for autonomous operation of a distributed generation unit: design and performance evaluation. IEEE Trans Power Syst 25(2):853–865
16. Bahrani B, Saeedifard M, Karimi A et al (2013) A multivariable design methodology for voltage control of a single-DG-unit microgrid. IEEE Trans Ind Inf 9(2):589–599

A DVR-Based Compensation System Under Various Power System Faults in a Multi-microgrid

A. V. Pavan Kumar, Akkenaguntla Karthik, and P. Balamurali Krishna

Abstract In today's contemporary world, electricity is an essential resource. Access to electricity accelerates a country's growth. The majority of power is generated from fossil fuels, which have a negative impact on the environment. This encouraged people to produce power using renewable energy sources such as solar and wind. This expanded the volume of microgrids available at the consumers' end. The power system is a massive and complicated network that includes the generating, transmission, and distribution systems. Numerous power quality issues such as voltage sag, voltage swell, harmonics, reactive power compensation, and voltage distortions, arise as a result of building microgrids at the consumer ends and connecting various types of loads. Among these, the work aims to reduce voltage sag, voltage swell, and harmonics induced by the rise in load and non-linear loading at the consumer ends. Flexible AC Transmission Systems (FACTS) devices are used to address these power quality challenges. A Dynamic Voltage Restorer (DVR) has been considered as a solution to this problem. A multi-microgrid is modelled in MATLAB and simulated with varied load situations that cause power quality concerns in multi-microgrids, and the findings are provided in this paper.

Keywords Multi-microgrid · DVR · Voltage sag · Voltage swell · Harmonics and voltage distortions

A. V. P. Kumar
Department of EEE, Madanapalle Institute of Technology & Science, Madanapalle, AP, India

A. Karthik
PGDM (Energy Management), NTPC School of Business, Noida, UP, India

P. B. Krishna (✉)
School of Electrical Engineering, KIIT University, Patia, Bhubaneshwar, Odisha, India
e-mail: pbmk20031993@gmail.com; balamuralip20@gmail.com

J. Kumar et al. (eds.), *Recent Advances in Power Systems*, Lecture Notes in Electrical Engineering 1192, https://doi.org/10.1007/978-981-97-2788-9_17

1 Introduction

As the days are progressing the demand for electricity is also increasing a lot. Investigators [1] suggest that the electricity consumption will affect the GDP of a country. As the loads at the distribution end are collectively increasing, electricity producers are making a lot of efforts to match the supply and demand. Electricity has become a mandatory resource for humans. It is there in everyone's life. When the world gets to know about the adverse impact of using fossil fuels, more research is made to explore alternate energy resources which are clean and sustainable. In this exploration Photovoltaic (PV), wind, fuel cells technology etc. has been discovered and using these reduce the pollution as well. But mostly PV and wind are extensively used [2] for large production and some are installed at homes thus increasing the use of renewable energy sources [3]. Thus microgrid technology has become an alternative [4]. When the number of microgrids is more than one it is termed as multi-microgrid. Many architectures and control systems are available in literature [5]. A multi-microgrid structure is shown in Fig. 1. The volume of sensitive loads at the distribution ends are growing, and these loads are impacted by voltage anomalies when occurring in the system, which are referred to as power quality concerns. Impulse transients, voltage sag, voltage swell, harmonics and flickers are a few examples of power quality issues. When a huge inductive load is suddenly connected to a line voltage sag occurs, and energizing of high rating capacitors causes voltage swell and connecting a non-linear load to the line causes harmonics.

Various conventional methods like synchronous condensers, capacitor filter banks, phase advancers etc., are used to mitigate these problems, but their range is limited. This made investigators to focus on development of devices like FACTS. Currently, FACTS like Static Synchronous Compensator (STATCOM), Static Synchronous Series Compensator (SSSC), Thyristor-Controlled Reactor (TCR), DVR etc., are used to diminish the power quality challenges. Among these FACTS devices DVR provides an effective way for load compensation irrespective of voltage sag, voltage swell, unbalanced loading, harmonics and faults like phase-phase and ground faults [6, 7]. DVR is a solid state inverter that detects the voltage abnormalities in the network and injects the voltage in series to compensate the load voltage in the power distribution system [8]. A DVR is mainly composed of 1 Voltage source Inverter (VSI), 1 DC storage system, transformer and filters [9]. During the design of DVR

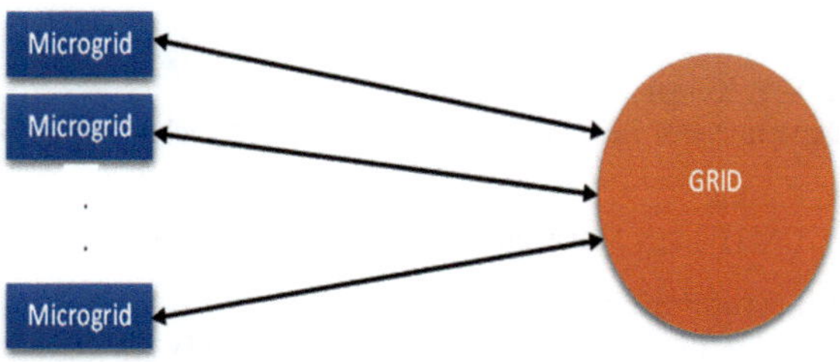

Fig. 1 Schematic of multi-microgrid

proper care should be taken for designing asuitable algorithm for injecting voltage and also suitable converter topology to be chosen. In this paper multi-micro gird is modelled and unified with DVR designed using a PI controller. Each micro grid consists of a PV and wind generation system. The functionality of the DVR when incorporated with multi-microgrid is investigated for mitigating power quality problems like harmonics and voltage sag, voltage swell. The entire system is modelled in MATLAB Simulink and simulated under various loading circumstances. Section 2 provides information about the Dynamic Voltage Restorer. Section 3 talk about the multi-microgrid structure. Section 4 escribes the output results followed by conclusion in Sect. 5.

2 Dynamic Voltage Restorer

In transmission and distribution lines, SSSC, booster transformers, DVR etc., are being employed to keep the load voltage constant. Among these DVR is frequently employed at the distribution end to provide a constant voltage irrespective of loading conditions. DVR is a series connected power device for mitigating voltage sag, voltage swell, harmonics etc., that occur at the distribution ends. A series transformer connects a DVR to the line, and voltage injection is done in series with the line voltage as shown in Figs. 2 and 3 shows the flowchart of DVR. It's DC side is connected to a battery, capacitor or energy source and AC side is connected to the line. If any voltage abnormality occurs V_{DVR} is injected in series to the line voltage thus delivering constant voltage to the load possible. When there is no voltage abnormality of the voltage injected by the DVR is $V_{DVR} = 0$. Due to faults or any other loading conditions the sending end or source voltage will be subjected to sag or swell, and during this time DVR injects voltage ($V_{DVR} > 0$) to maintain constant load voltage. Equation (1) gives the injected voltage of DVR.

$$Voltage\,(injected)_{DVR} = Voltage_{Source} - Voltage_{Load} \tag{1}$$

3 Designing Multi-microgrid Model

The simulation system is composed of two parallelly connected microgrids, each with its own PV and wind generation systems. These 2 microgrids are synchronised with the grid and the total generation system is connected to the load by transmission lines. Thus, the multi-microgrid is modelled.

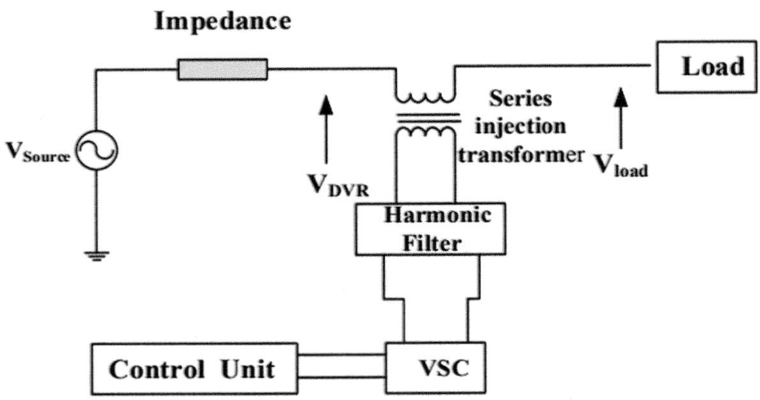

Fig. 2 Representation of DVR arrangement

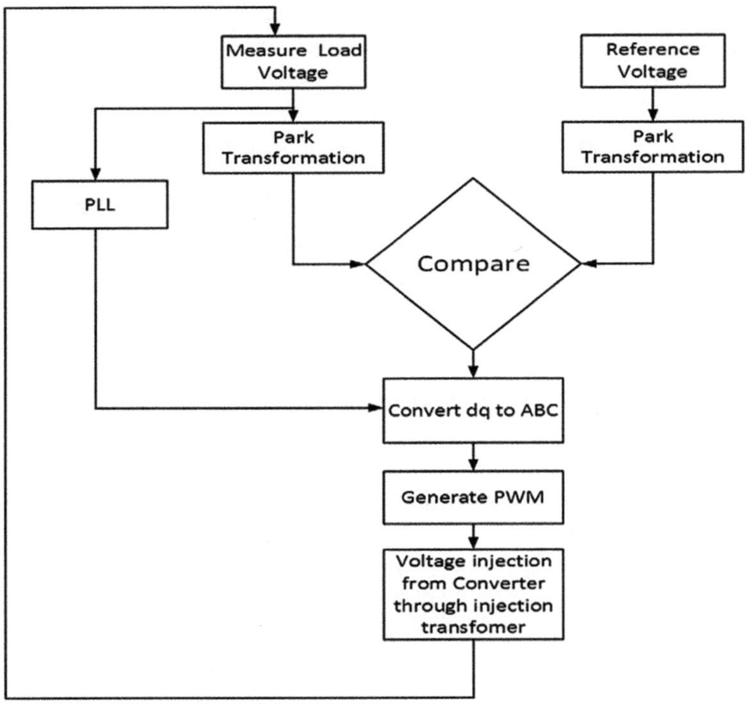

Fig. 3 Flowchart of DVR control scheme

3.1 Multi-microgrid Modelling

In this study, multi-microgrid is made up of 2 microgrids, each with 2 Distributed Energy Resources (DERs), one of which is a Photovoltaic and the other is a wind system. The considered multi-microgrid block diagram is shown in Fig. 4. The modelling of wind and PV systems is presented in further detail.

3.1.1 Photovoltaic Modelling in Multi-microgrid

Mathematical modelling is followed for developing a PV module. Depending upon the number of series cells connected the required characteristics can be obtained. The PV current equation is represented in Eq. (2).

$$I_{pv} = I_{ph} - I_0\left[\exp\left(\frac{V_{PV+}R_S I_{PV}}{\mathrm{n}(\eta K_B \mathrm{T/q})}\right) - 1\right] - \left[\frac{V_{PV+}R_S I_{PV}}{R_{Sh}}\right] \tag{2}$$

where I_0 is Reverse saturation current in amps, I_{ph} is photo current in amps, V_{PV} is PV Voltage in volts, R_S is series resistance, R_{Sh} is shunt resistance, T is temperature in Kelvin, K_B is Boltzmann's constant, n is number of cells in series, q is electric charge and η is ideality factor. The photo current I_{ph} is represented in Eq. (3)

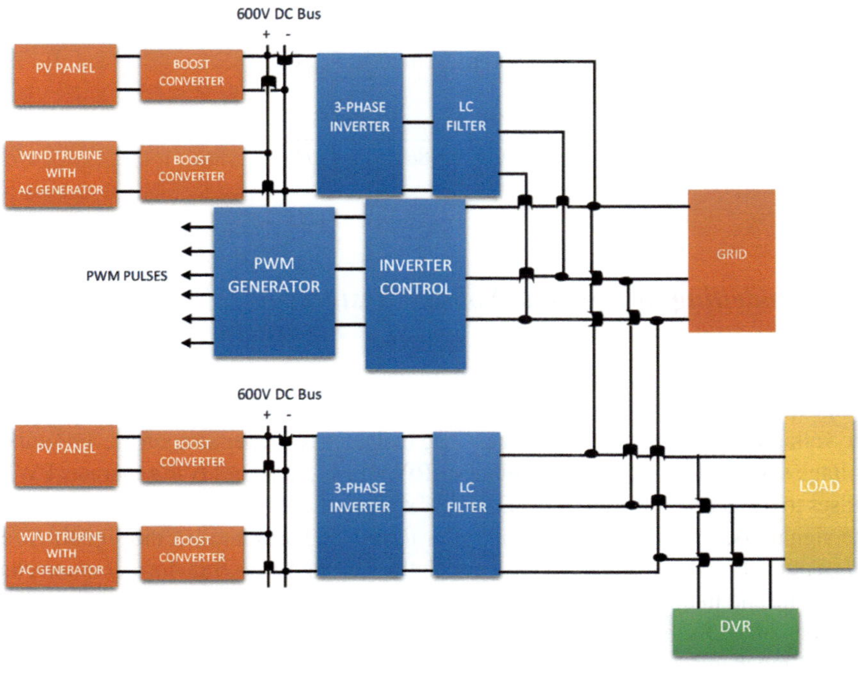

Fig. 4 Block diagram of multi-microgrid with DVR

$$I_{ph} = I_{sco}\left[\frac{G}{G_0}\right](1 + \propto_1(T - T_0)) \tag{3}$$

where G_0 is the standard irradiance at STC, G is the irradiance, T_0 is the temperature at STC, I_{sco} is the short circuit current at STC and $\propto_1$ is the temperature coefficient of the cell.

3.1.2 Wind Generator Modelling in Multi-microgrid

Various research studies provided details regarding wind turbine and PMSG in the literature. The power output equation of the wind turbine is represented in Eq. (4).

$$P_o = \frac{1}{2}\rho A V_{wind}^3 C_p(\lambda, \theta) \tag{4}$$

where P_o is the turbine mechanical power output, λ is the TSR of rotor blades, ρ is the density of air, θ is the pitch angle of the blade, A is the turbine swept area V_{wind} is the wind speed, $C_p\ (\lambda,\theta)$ is the turbine power coefficient which is given by the Eq. (5).

$$C_p(\lambda, \theta) = C_1[C_2\frac{1}{\beta} - C_3\beta\theta - C_4\theta^x - C_5]e^{-C_6\frac{1}{\beta}} \tag{5}$$

where C_1 to C_6 are coefficients, θ is the angle between the rotation plane and cross section chord of the blade and β is represented in Eq. (6).

$$\frac{1}{\beta} = \frac{1}{\lambda + 0.08\theta} - \frac{0.035}{1 + \theta^3} \tag{6}$$

3.2 *Modelling of Dynamic Voltage Restorer*

As discussed in Sect. 2, the DVR is used for detecting voltage sags, swell, harmonics etc., depending on the design of the DVR controller it functions. The DVR injects the voltage depending on the amount of sag that occurs. Depending on the amount of voltage sag, the Pulse Width Modulation (PWM) generator of DVR generates PWM pulses to trigger the inverter of the DVR. As soon as the inverter generates AC output, the signal goes through the filters in order to reduce harmonics in the signal. After this a pure sinusoidal signal passes through the step-up series transformer connected to the line and in this way the abnormalities in the voltage are detected and according to it the voltage is injected in order to maintain a constant load voltage. The system's control mechanism relies on the error signal produced by comparing genuine and benchmark values. Error signal generates PWM pulses to trigger the power electronic switches in the inverter.

The suggested approach is employed for mitigating voltage sag and voltage swell that are caused due to sudden connection & disconnection of loads, and harmonics that are caused when a non-linear load gets connected.

4 Simulation and Results

A multi-microgrid is developed consisting of two microgrids. Each microgrid consists of 2 DERs one is PV and the other is the wind power source. The output voltage is 400 V. The output voltage of the multi-micro grid is shown in Fig. 5. The proposed system is tested with voltage sag, voltage swell, harmonics and voltage distortions.

4.1 Under Voltage Sag and Voltage Swell Condition

Voltage sag and voltage swell are produced by employing a capacitive and inductive load between 0.2–0.4 and 0.6–0.8 s, respectively. When DVR is not connected to the line, the swell and sag of the load voltage can be observed from Fig. 6.

When DVR is coupled to the system, voltage sag and voltage swell are mitigated. The result is illustrated in Fig. 7.

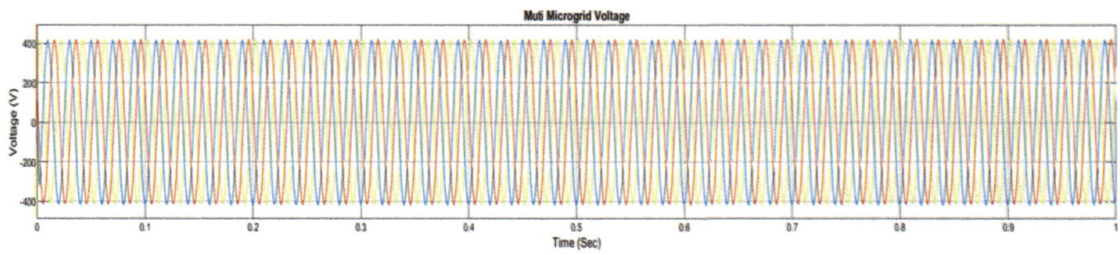

Fig. 5 Multi-microgrid output voltage

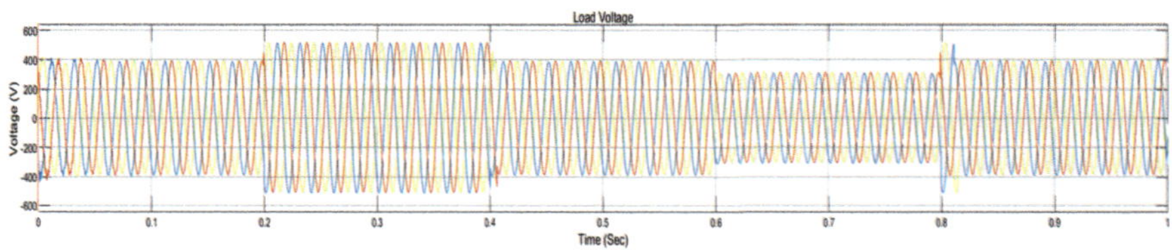

Fig. 6 Voltage sag and voltage swell in load voltage

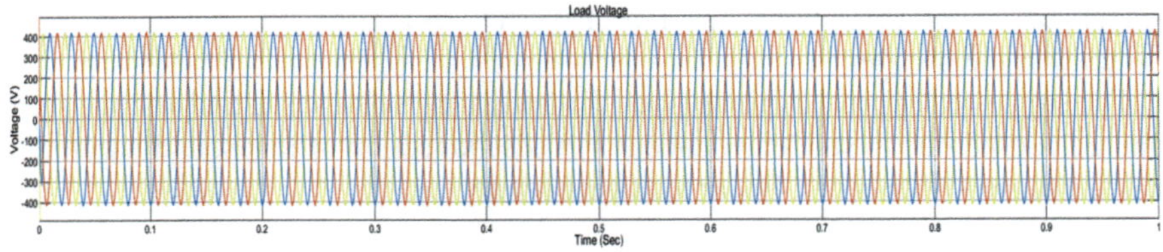

Fig. 7 Load voltage sag and voltage swell compensation in multi-microgrid employing DVR

Therefore, load voltage is maintained constant even though voltage sag, voltage swell occurred at various times. In the next section a non-linear load is considered and the effect of DVR on Total Harmonic Distortion is shown.

4.2 When Non-linear Load is Connected

Diode rectifier is considered to create a non-linear load in the multi-microgrid systems. Results are taken when the DVR is connected to the multi-microgrid and when not connected to the multi-microgrid FFT analysis is carried out. When DVR is not connected to the system, Total Harmonic Distortion (THD) is 32.24% and Fig. 8 shows the Fast Fourier Transform (FFT) analysis. Number of cycles considered for FFT analysis is 1 cycle with start time of 0.2 s and with maximum frequency of 5000 Hz (Fig. 8).

When DVR is connected to the system, THD is 1.32% and the FFT analysis is shown in Fig. 9. This shows that by connecting DVR the THD is reduced by 30.92%.

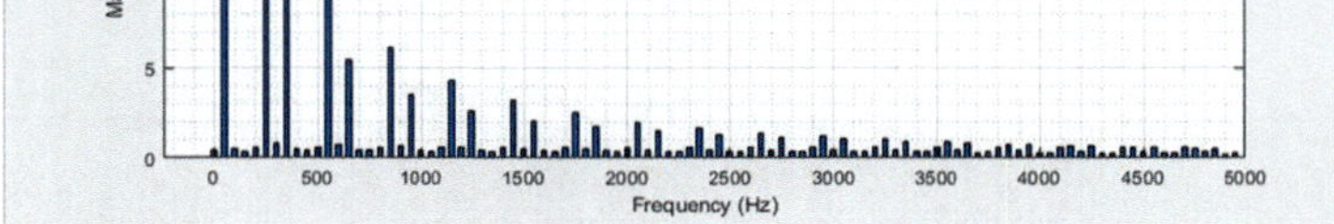

Fig. 8 THD without DVR

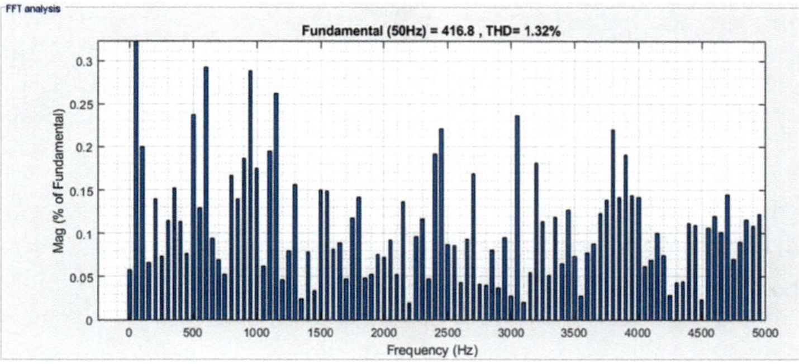

Fig. 9 THD with DVR

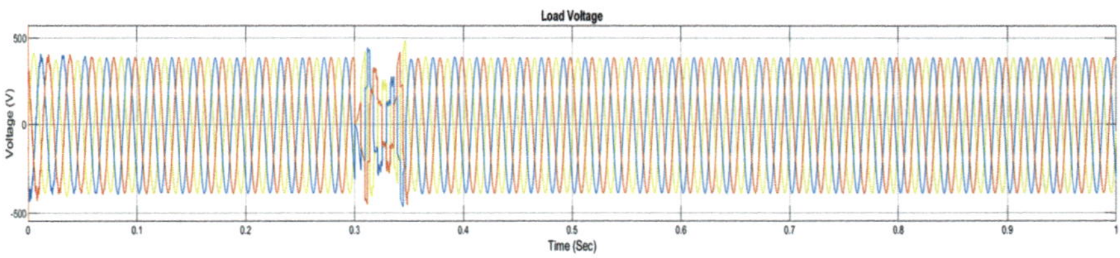

Fig. 10 Load voltage without DVR

4.3 When Induction Motor is Connected

An induction motor is connected to the multi-microgrid. During a total simulation time of 1 s the induction motor is switched on at 0.3 s. When DVR is not connected, abnormalities in the load voltage can be seen as shown in Fig. 10. With DVR, at 0.3 s no abnormalities occurred in the load voltage and the results are illustrated in Fig. 10.

As shown in Fig. 10, when the induction motor is switched on at 0.3 s distortions in the load voltage can be observed up to 0.35 s when DVR is connected. As shown in Fig. 11 when the induction motor is switched on at 0.3 s when DVR is connected, no distortions are observed in the load voltage. Load voltage is constant irrespective of loading conditions.

From these results it is evident that DVR plays a vital role in supplying a constant voltage to load, irrespective of the loading conditions. Post employing DVR voltage sag, voltage swell, harmonics and load voltage distortions are mitigated.

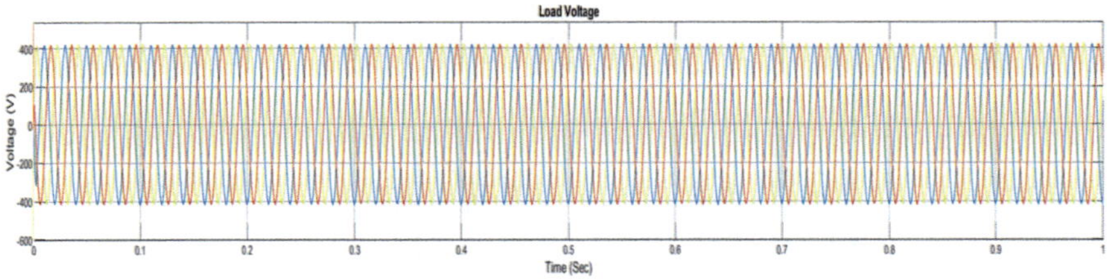

Fig. 11 Load voltage with DVR

5 Conclusion

In this research, a multi-micro grid comprises of 2 micro grids with two DERS each considered to check the performance of the DVR. The DVR's aim is to counteract on issues like voltage sag, voltage swell, harmonics and load voltage distortions that occur in the distribution end. DVR reduced harmonics and voltage sag; voltage swell occurs when an induction load is attached. The control system of the DVR is a PI controller, which compares the reference voltage to the actual load voltage and generates PWM pulses based on the error signal. The DVR is also capable of handling the multi-imbalanced microgrid's loading circumstances.

References

1. Mahadevan R, Asafu-Adjaye J (2007) Energy consumption, economic growth and prices: a reassessment using panel VECM for developed and developing countries. Energy Policy 35:2481–2490. https://doi.org/10.1016/j.enpol.2006.08.019
2. Ahshan R, Nasiri N, Al-Badi AH, Hosseinzadeh N (2020) Control and dynamic analysis of an isolated microgrid system: a case study. In: 2020 6th IEEE international energy conference (ENERGYCon), Gammarth, Tunisia, pp 254–259. https://doi.org/10.1109/ENERGYCon48941.2020.9236500
3. Magdy G, Ali H, Xu D (2022) Effective control of smart hybrid power systems: cooperation of robust LFC and virtual inertia control systems. CSEE J Power Energy Syst 8(6):1583–1593. https://doi.org/10.17775/CSEEJPES.2020.05230
4. León LF, Martinez M, Ontiveros LJ, Mercado PE (2022) Devices and control strategies for voltage regulation under influence of photovoltaic distributed generation. A review. IEEE Lat Am Trans 20(5):731–745. https://doi.org/10.1109/TLA.2022.9693557
5. Zhao Z et al (2022) Distributed robust model predictive control-based energy management strategy for islanded multi-microgrids considering uncertainty. IEEE Trans Smart Grid 13(3):2107–2120. https://doi.org/10.1109/TSG.2022.3147370
6. Lu J, Zhang X, Zhang B, Hou X, Wang P (2022) Distributed dynamic event-triggered control for voltage restoration and current sharing in DC microgrids. IEEE Trans Sustain Energy 13(1):619–628. https://doi.org/10.1109/TSTE.2021.3123372
7. AppalaNaidu T (2016) The role of dynamic voltage restorer (DVR) in improving power quality. In: 2016 2nd international conference on advances in electrical, electronics, information, communication and bio-informatics (AEEICB), Chennai, India, pp 136–141. https://doi.org/10.1109/AEEICB.2016.7538259.

8. Hossain MI, Rahaman I, Rahman MN, Hasan MF, Hasan MM, Sarkar RC (2020) Voltage sag compensation in distribution system using dynamic voltage restorer. In: 2020 2nd international conference on advanced information and communication technology (ICAICT), Dhaka, Bangladesh, pp 492–497. https://doi.org/10.1109/ICAICT51780.2020.9333502
9. Sasitharan S, Mishra MK, Kalyan Kumar B, Jayashankar V (2010) Rating and design issues of dvr injection transformer. Int J Power Electron 2(2):143–163

Techno Economic Analysis of a Mobile Automatic Teller Machine Powered by Off Grid PV-DG-Battery Hybrid System

Rajib Lochan Dash, Banaja Mohanty, and Prakash Kumar Hota

Abstract The study presents the technical and economical evaluation of mixed hybrid renewable energy(RE) system to run truck mounted mobile automatic teller machines (ATM).Considering the dynamic nature of mobile ATM load, optimized combination of energy sources is found out using HOMER (Hybrid optimization model for electric renewable) Pro software. The combination having lowest cost of energy (COE) will provide reliable and cheap operation of mobile ATM. It will enhance the banking services in the Puri, India which is a busiest pilgrimage destination in India. The simulation results show that 34.9 kW PV, 6.60 kW DG and 50 kWh Battery is the optimum design for the location. The pollutant emission also decreases considerably by the PV/DG/Battery combination with increased renewable energy penetration as the diesel consumption is reduced considerably. Thus making the environment more clean and green leading to better air quality index of the area and providing reliable banking service to millions of pilgrimage across Puri, India.

Keywords ATM · SPV · DG · RE · COE

1 Introduction

Renewable energy is a modern day substitute to the high cost fossil fuel based fuel and modern environmental problem like global warming, acid rain, ozone gas depletion.There are many sources such as solar, wind, geothermal, biomass, tidal energy etc. which are renewable in nature [1]. Fossil fuel are more used in all cases due to popularity and high end technology of IC engines and readily availability of thermal power plant. Thermal power plant based on coal causes a lot of environmental issue while generating electricity [2]. Increase pollutant emission make the world to heat further due to the presence of green house gases (GHG) and causing air quality index to drop [3]. As renewable energy is exhaustible in nature and cleaner and greener in

R. L. Dash (✉) · B. Mohanty · P. K. Hota
VSSUT, Burla, India
e-mail: rajibatbpd@gmail.com

J. Kumar et al. (eds.), *Recent Advances in Power Systems*, Lecture Notes in Electrical Engineering 1192, https://doi.org/10.1007/978-981-97-2788-9_18

nature it will reduce the environmental pollution. Because they were never depleted, it is a dependable supply of energy. Technology advances make RES more available, economical, and effective [4]. Even though solar and wind systems have many benefits. Due to intermittent nature of solar generation, a single energy source system cannot reliably supply electricity to loads. They have a very tough time keeping up with the load demand of the customers since the load pattern changes often. As a result, new storage systems are needed, but they are quite expensive [5]. This ensures that the load receives power continuously. Off grid system often is expensive due to the cost of battery and low life of the storage system [6].

Since automated teller machines (ATMs) were used in India and their distribution across the country has increased quickly. But for inaccessible areas like sea beach, Yatra, it is impossible to provide mobile ATM service to a largely present population. Banks are spending a lot of money deploying these mobile ATMs to reach the crowd [7]. Due to the expansion of official banking hours, the quality of service provided to consumers has significantly increased as a result of investment in mobile ATMs [8]. More mobile ATMs have been installed by various banks around the country as a result of Govt. of India's recent introduction of banking for all initiative to guarantee that bank clients have access to their monies whenever they need it, regardless of banking hours [9]. Truck mounted mobile ATMs require power to run the mobile ATM due to mobile nature of duty. Because of this, the banking industry has been forced to deploy fossil fuel-powered generators, The operation of DG is characterized by no of financial, logistical, and pollution related issues, making their usage a source of power insufficient [10]. Renewable energy technologies now hold great promise as sustainable energy sources because of the rising need for it [11]. The most popular type of renewable energy is solar [12]. Solar energy is widely distributed and cost-free. They produce no greenhouse emissions and are clean energy sources. However, compared to traditional (diesel) power producing systems, it requires high investment initially for solar panel installation and commissioning [13]. Furthermore, because of their reliance on weather, renewable energy sources are intermittent. Renewable energy sources' erratic nature might provide an unstable power supply. Therefore, an energy storage system like a battery is employed to balance out unexpected power fluctuations and stabilize voltage and frequency [14].

The mixed hybrid energy system mixes different sources to optimize the COE and Net present cost NPC. A mixed energy system, when correctly built is dependable and less costly than a solo energy source. However, potential investors should be informed of the best design requirements, capacity predictions, and the techno-economic consequences of mixed systems, including various renewable technologies, since they will help investors choose the best technology or combination [15].

2 HOMER-Pro

HOMER-Pro is an optimization software developed by the National renewable energy laboratory. It can optimize both grid connected and off grid systems with various modules like wind, PV, Biomass, hydro, hydrogen etc. Based on the load estimation, it optimizes the best possible combination with lowest COE and NPC. Techno economic evaluation of a project before the commissioning makes the decision maker to take decision more accurately.

3 Hybrid System Description

3.1 Load Profile

A major problem is frequently the power supply for these mobile ATMs and associated ancillary devices. A typical mobile ATM uses roughly 3 kW of electricity (0.7 kW for the mobile ATM itself and 3.82 kW for various auxiliary equipment like the air conditioner, lights, etc.). Depending on the kind, quantity, and style of installation of the installation devices, the power consumption pattern varies depending on the mobile ATM site. Table 1 displays the power requirements for a typical mobile ATM installation (including auxiliary equipment) at an off-site location in India.

3.2 System Components

3.2.1 DG

Diesel generator in a power system, HOMER-Pro can design up to three generators at once. Each generator can be alternating current or direct current, and they use different types of fuel. The maximum and minimum output is an important parameter of DG. The fuel curve indicates the amount of fuel burned in relation to the amount of electrical power produced. The diesel generator's capital cost is 370$, while replacement and operating maintenance expenses are 300$ and 0.150$, respectively. The

Table 1 The power consumption of a typical off-site mobile ATM installation in Puri, India

Equipment	Quantity	Power (kW)	Load (kW)
ATM	3	0.70	2.10
AC	3	1	3
Lightning	6	0.09	0.54
CCTV	1	0.28	0.28

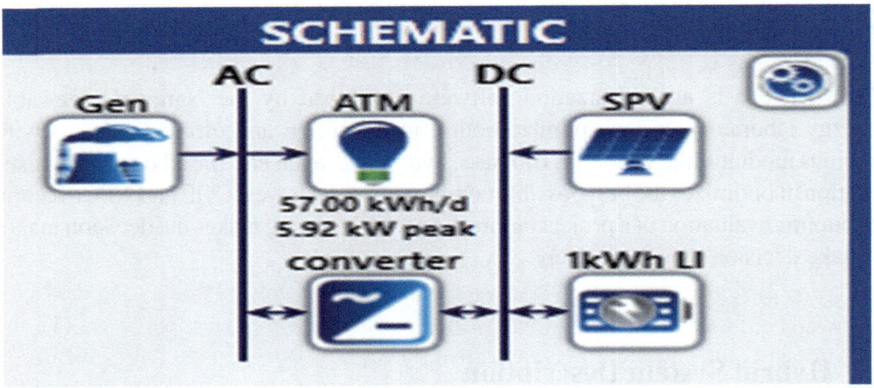

Fig. 1 Mixed energy system

minimum load ratio is considered to be 25%, and the maximum life time of DG is taken to be 15,000 h.

$$\nu_c = \alpha_{DG} Y_{gen} + \beta_{DG} P_{gen} \quad (1)$$

where, α_{DG}: fuel consumption curve coefficient in L/h/kW

β_{DG}: Slope of fuel curve L/h/kW.

$Y_{gen:}$Genset rated capacity (kW).

$P_{gen;:}$Genset power output (kW).

3.2.2 Solar Photovolatic

Solar PV panel produces electrical energy when exposed to solar radiation due to difference in voltage. When diesel prices are high and wind speeds are low, solar voltaic is the best option. The initial cost of a 1 kW SPV panel is around 500$; the replacement cost is 450$ and the operating cost is around 1$ respectively. The PV system has a 25-year lifespan.

The power generated from the PV is determined using the equation

$$P_{SPV} = C_{SPV} P_{SPV} \frac{I_{Tr}}{I_S} \quad (2)$$

where,

CS_{PV}: derating factor of SPV cell,

PS_{PV}: Rated capacity (kW).

I_{Tr}: Quantity of solar radiation that falls on solar panel (kW/m2).

I_S: Standardized radiation 1 kw/m^2.

3.2.3 Battery

The battery under consideration is a standard 6 V lithium-ion battery with a capacity of 1kwh. The initial cost of a 1kwh Li-ion battery is around 500$,whereas the replacement cost and O&M cost are respectively 455$ and 0 $.The battery provides back up power supply when sufficient generation is not achieved through renewable sources. As solar energy is not prevalent through night time, power is supplied from the battery at night time. However as the battery is expensive it increases the cost of energy (COE) and NPC considerably. It is generally assumed that the battery specifications remain consistent during its whole lifespan and it is independent of temperature variation. Lithium ion battery is preferred as comparison to other battery as it requires less size for same AH of rating.

3.2.4 Converter

A converter is an electronics device which either acts as an inverter or a rectifier. It converts ac to dc based on thyristor based technology. The rectification efficiency is considered as constant for HOMER Pro. The initial cost of a 1 kW inverter is 154$, with replacement and the maintenance cost is 154$ and 15.40$, respectively. It has a life expectancy of roughly 15 years and an efficiency of 95%.Convention source of energy has low capital cost whereas renewable energy set up requires high capital cost. But the O&M cost of the conventional is much higher as compared to renewable.

The equation representing COE and NPC is represented as follows

$$COE = \frac{TAC}{E_{\text{Generator}}} \tag{3}$$

where

TAC = Total annualized cost.

$E_{\text{generator}}$: annual energy production.

Net present cost (NPC) of the system is calculated by the following equations. NPC is a more suitable economic parameter as compared to COE with deep mathematical calculation.

$$NPC = \frac{C_{\text{total}}}{CRF} \tag{4}$$

where *NPC: Net Present Cost*

$C_{total,}$: Systems annual cost.

CRF: Capital recovery factor.

$$i = \frac{i - f}{1 + f} \quad (5)$$

The CRF is represented by the following Eq. 6

$$CRF = \frac{i(1 + i)^{N}}{(1 + i)^{N} - 1} \quad (6)$$

where,

i: Nominal interest rate.

f = Annual inflammation rate.

N: Number of years.

4 Results and Disusssions

The study region is located at 19°38'N latitude and 26°32'E longitude. Figure 2 depicts the average value of each day's solar radiation data as well as the clearness index values. Solar radiation ranges from 4 kWh/m2/day to 6.49 kWh/m2/day, with an annual average of 4.97 kWh/m2/day.

Table 2 displays the overall optimal sizing findings of the investigated hybrid energy system designs. As demonstrated, the photovoltaic-DG-battery hybrid energy system has the lowest prices. The energy cost of photovoltaic-DG-battery hybrid systems is US$ 0.297 per kWh, whereas the initial capital required and net present cost are US$ 51,245 and US$ 80,019, respectively. The ideal configuration includes

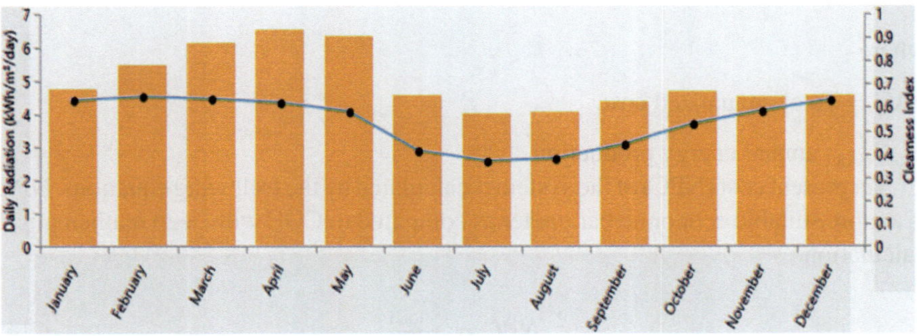

Fig. 2 Monthly variation of solar radiation

34.9 kW PV arrays, 6.60 kW diesel generators, a 40 kW converter, and a 50kWh battery bank.

Different optimized systems were given in Table 3 whereas the comparison of the optimal system with the reference system is given in Table 4. The NPC and initial cost of the optimal system is much more less as compared to thr base system.

DG fuel consumption happens when the output power from the photovoltaic is very less and the battery is drained out which is represented in Fig. 3.

The SOC of the batteryvaries throughout the year as it can seen from Fig. 4.

Table 5 gives the data of the generic flat PV parameter with grid connection. This shows that the PV penetration is 98.6% in 384 h of operation. The auto size genset only produces 1.36% of the total load.

Table 2 Homer optimized result parameters

Cost summary (Net Present)	System ($)	DG ($)	PV ($)	Generic LI-ion ($)	Converter ($)
Capital	51,244.75	2442	17,442.75	25,000	200
Replacement	13,000.92	0	0	10,387.40	0
O&M	13,393.50	4914.52	450.98	0	64.64
Fuel	4008.49	4008.49	0	0	0
Salvage	1628.82	170.76	966.17	966.17	0
Total	80,018.84	11,194.26	17,893.73	34,421.22	264.64

Table 3 Different optimized systems

Architecture						Cost		
⚠	SPV (kW)	Gen (kW)	1kWh LI	converter (kW)	Dispatch	COE ($)	NPC ($)	Operating cost ($/yr)
	34.9	6.60	50	40.0	LF	$0.298	$80,019	$2,226
	63.9		48	40.0	LF	$0.311	$83,157	$1,611
⚠	61.7	6.60		40.0	LF	$0.766	$205,948	$12,864
		6.60			LF	$0.869	$233,784	$17,880
		6.60	2	40.0	LF	$0.928	$249,696	$18,557

Table 4 Comparision of optimized system with base system

Architecture					Cost	
⚠	SPV (kW)	Gen (kW)	1kWh LI	converter (kW)	NPC ($)	Initial capital ($)
		6.60			$233,784	$2,642
	34.9	6.60	50	40.0	$80,019	$51,245

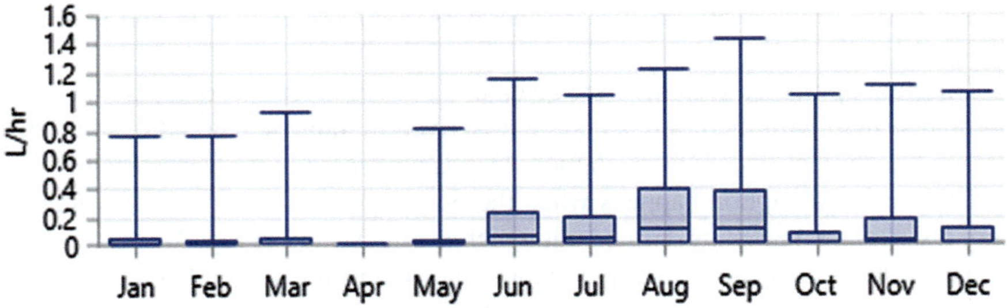

Fig. 3 Diesel consumption throughout the year

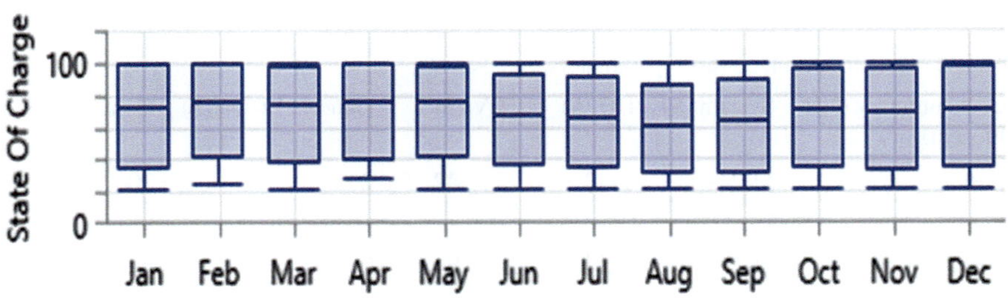

Fig. 4 Battery state of charge throughout the year

Table 5 Energy produces the different components

Production	Kwh/yr	Percentage
Generic flat plate PV	53,837	98.6
Genset	744	1.36
Total	54,582	100

The monthly electricity production is presented in Fig. 5. It is shown that throughout the year, SPV generation remains dominant only in month of August as it requires DG largely.

The parameters of the different components are given in Table 6. The LCOE is 0.297 $ /kWh whereas the capacity factor is 1.29%. Electrical efficiency is 23.4%. The fuel consumption is 384 h/year. The mean output is 1.94kW and the minimum output is 1.65kW.

The emission parameters are described in Table 7. The carbon dioxide emission is 845 kg/yr and carbon monoxide is 5.33 kg/year. The values of the emission are very less as compared to the stand alone DG system. The carbon dioxide emission is 22,102 kg/year and the carbon monoxide emission is 139 kg/year.

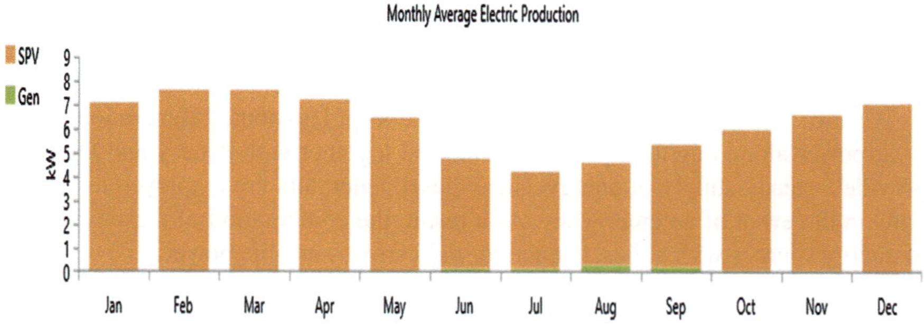

Fig. 5 Monthly average electricity production

Table 6 Different parameters of optimized system

Parameters	Value	Units	Quantity	Value	Units
Rated capacity	6.60	kW	Electrical efficiency	23.4	%
Mean output	1.94	kW	Fuel consumption	323	kL
Minimum output	1.65	kW	PV penetration	98.6	%
Capacity factor	1.29	%	Hours of operation	384	hrs/yr
Total production	744	kWh/yr	Levelized cost	0.2975	$/kWh

Table 7 Emission parameters

Quantity	Value (kg/yr)
Carbon dioxide	845
Carbon monoxide	5.33
Unburned hydrocarbons	0.233
Particulate matter	0.0323
Sulfur dioxide	2.07
Nitrogen oxides	5.01

5 Conclusion

The optimal size of a hybrid energy system for dependable and cost-effective operation of a distant load under Indian climatic circumstances is described in this study. A hybrid system with two major energy sources (solar, diesel), battery storage, and a power converter was investigated. Energy balance calculations based on HOMER software are performed over a one-year period, utilizing long-term weather data and the load profile of a typical mobile AT installation. The purpose of this study is to provide a concept for designing, simulating, and analyzing stand-alone hybrids in mobile ATM machines incorporating renewable energy resources. The primary goal of this study is to provide the best solution to the world's energy challenge. In the current context, nonrenewable fuels are abundant enough to meet our expanding need taking into account the preceding data.

We have made an attempt to analyze the benefits of Distributed Generation. While solar power alone might vary, when connected together with battery and DG, they provide a steady supply of energy, since diesel generators are employed as backup units in the event of an emergency. As a result, the combination of these two types of energy sources with a diesel generator delivers dependable power.

The optimized photovoltaic-battery hybrid system, consisting of 34.9 kW PV arrays, 6.60 kW, 40 kW converter, and 50 kWh battery bank, has the lowest cost of US$ 0.297 per kWh of energy consumed when compared to the diesel-alone system having COE 0.869$..Furthermore, the yearly fuel consumption of the mobile ATM installation location might be lowered. According to the analysis, not the quality of the financial services will be enhanced.

According to the analysis, not only would the quality of banking services be improved by making mobile ATM operations more reliable and less expensive, but CO_2 and other harmful gas emissions from mobile ATM installation sites in India would be reduced, making the environment much more friendly and safe.

Appendix: Battery Specifications

Nominal voltage = 6v, Nominal capacity = 1 kwh, Nominal capacity = 167 Ah, Round trip efficiency = 90%, Maximum charge current = 167 amp, Maximum dischage current = 500.

References

1. Pradhan K, Mohanty MK, Kar SK (2017) Techno-economic evaluation of stand-alone hybrid renewable energy system for remote village using HOMER-Pro software. Int J Appl Power Eng (IJAPE) 6. https://doi.org/10.11591/ijape.v6.i2.pp74-89
2. Suresh V, Mahankali M, Kiranmayi R (2020) Modelling and optimization of an off-grid hybrid renewable energy system for electrification in a rural areas. Energy Rep 6:594–604. https://doi.org/10.1016/j.egyr.2020.01.013
3. Bazdar E, Shirzadi N (2022):Economic analysis and simulation of solar PV, Wind Turbine Hybrid Energy System Using HOMER Pro. J Solar Energy Res 2(3):53–58. https://jser.ut.ac.ir/article_63944.html. Accessed 24 Jul 2022
4. Mokheimer EM, Al-Sharafi A, Habib MA, Alzaharnah I (2015):A new study for hybrid PV/wind off-grid power generation systems with the comparison of results from homer. Int J Green Energy 12(5):526–542. https://doi.org/10.1080/1543507.2013.833929
5. Babatunde OM, Munda JL, Hamam Y (2022) Off-grid hybrid photovoltaic–micro wind turbine renewable energy system with hydrogen and battery storage: effects of sun tracking technologies. Energy Convers Manag 255:115335. https://doi.org/10.1016/j.enconman.2022.115335

6. Saheli MA, Lari K, Salehi G, Azad T (2022) Techno-economic assessment of a hybrid on grid PV-wave system: a case study in Caspian Sea. Renew Energy 186:596–608. https://doi.org/10.1016/j.renene.2021.12.146
7. Okundamiya MS, Akpaida VO, Omatahunde BE (2014) Optimization of a hybrid energy system for reliable operation of automated teller machines. J Emerg Trends Eng Appl Sci 5(8):153–158
8. Babaei R, Ting DSK, Carriveau R (2022) Feasibility and optimal sizing analysis of stand-alone hybrid energy systems coupled with various battery technologies: a case study of Pelee Island. Energy Rep 8:747–4762. https://doi.org/10.2139/ssrn.3945337
9. Ramesh M, Saini RP (2020) Effect of different batteries and diesel generator on the performance of a stand-alone hybrid renewable energy system. In: Energy sources, part A: recovery, utilization, and environmental effects, pp 1–23. https://doi.org/10.1080/15567036.2020.1763520
10. Li J, Liu P, Li Z (2020) Optimal design and techno-economic analysis of a solar-wind-biomass off-grid hybrid power system for remote rural electrification: a case study of west China. Energy 208:11838. https://doi.org/10.1016/j.energy.2020.118387
11. Azimoh CL, Klintenberg P, Wallin F, Karlsson B, Mbohwa C (2016) Electricity for development: mini-grid solution for rural electrification in South Africa. Energy Convers Manag 110:268–277.https://doi.org/10.1016/j.enconman.2015.12.015
12. Baseer MA, Alqahtani A, Rehman S (2019) Techno-economic design and evaluation of hybrid energy systems for residential communities: case study of Jubail industrial city. J Clean Prod 237:117806. https://doi.org/10.1016/j.jclepro.2019.117806
13. IMwakitalima IJ, Rizwan M, Kumar N (2021) Standalone solar photovoltaic electricity supply to rural household in Tanzania. IETE J Res 1–16. https://doi.org/10.1080/03772063.2021.1920854
14. Chen L, Wang S, Yousefi N (2022) An optimal arrangement for photovoltaic/diesel/battery management system applying Crow Search Algorithm: a case of Namib Desert. Int J Ambient Energy 43(1):4977–4989. https://doi.org/10.1080/01430750.2021.1909130
15. Pradhan AK, Mohanty MK, Kar SK (2017) Techno-economic evaluation of stand-alone hybrid renewable energy system for remote village using HOMER-pro software. Int J Appl 6(2):73–88. https://doi.org/10.11591/ijape.v6.i2.pp74-89

Optimal Scheduling of Vehicle-To-Grid Power Exchange Using Hybrid PSO and GSA Approach

Vikas Kumar Badhan and Shelly Vadhera

Abstract Electric vehicles (EVs) are set to revolutionize the transportation industry due to their benefits and incentives. They possess the unique characteristic of being a distributed energy storage device due to their onboard batteries. This function allows them to provide valuable ancillary services through vehicle-to-grid (V2G) operation, including load variance minimization on the grid. However, the issues of uncertain EV availability and ever-changing SOC present obstacles to practical V2G implementation. This paper proposes a scheduling method for V2G power exchange that addresses these challenges by examining the unpredictable nature of EV grid connectivity. In this paper, we suggest a novel technique for optimizing the scheduling of V2G power exchange utilizing particle swarm optimization (PSO) with gravitational search algorithm (GSA). This hybrid optimization algorithm of PSO with GSA has a faster convergence rate compared to using either PSO or GSA alone, making it a more effective method for optimizing the scheduling of V2G power exchange. The suggested V2G scheduling technique significantly flatter the load profile, according to the statistical evaluation of the information. Comparisons were made between the production of the suggested technique best-case phase and the accurate performance of approach suggested in earlier work on a related topic in order to concern the outcomes and validate the suggested approach.

Keywords Electric vehicle · Vehicle-to-grid · Particle swarm optimization · Gravitational search algorithm

---

V. K. Badhan (✉) · S. Vadhera
Department of Electrical Engineering, NIT Kurukshetra, Kurukshetra, India
e-mail: vikas_32114323@nitkkr.ac.in; vikasbadhan2021@gmail.com

S. Vadhera
e-mail: shelly_vadhera@nitkkr.ac.in

J. Kumar et al. (eds.), *Recent Advances in Power Systems*, Lecture Notes in Electrical Engineering 1192, https://doi.org/10.1007/978-981-97-2788-9_19

1 Introduction

On comparing electric cars (EVs) to traditional gasoline-powered automobiles, environmental benefits are introduced. By lowering oil imports, they improve energy stability while lowering air pollution and greenhouse gas emissions. Future predictions indicate that EV market penetration will increase dramatically. The restricted on-board energy storage and electric driving range of plug-in EV are a major worry. A vehicle movement optimization technique is needed in addition to battery technological improvements for the car battery to use its energy stored as efficiently as possible.

The concentrate is now on efficiency as conventional gasoline car innovation has almost reached saturation. Several optimization algorithms for driving speed on multiple pathways have been investigated. For EVs, many methods have been created to regulate the energy flow among the gasoline source & batteries optimally [1]. This begs the question of whether plug-in electric vehicles could have a similar ideal driving style taking into account how motion of the vehicle affects the batteries' SOC. SOC is outlined as the proportion of battery charge left over versus a full charge [2]. To establish optimal control, one requirement is to have a precise SOC value. The SOC control method for EVs has been studied in few works [3], despite the fact that due to its nonlinearity it is challenging to construct an appropriate computational formula.

The driving characteristics throughout a trip, including battery SOC, and also vehicle speed, power usage of key elements, drivers' conduct, etc. have a significant impact on a plug-in EV's range [4]. The driving distance is further influenced by outside variables such the state of the roads, their slope, the posted speed, rolling resistance, or aerodynamics. Systems to anticipate, sustain, or increase the driving scale are presented in paper [5] that describes an optimization approach to find the optimum solution to certain optimal solutions that account for battery autonomy, driving comfort indices, and journey time. The driving situation has been assessed in [6]. Although some methods keep an eye on the operations in real time, others rely on previous knowledge or predictions of the future driving state provides a description of a few common or specially developed operating modes for different purposes [7–9] (Fig. 1).

This paper is structured as obeys: Sect. 2 shows the hybrid PSO-GSA. Section 3 shows the objective function and Sect. 4 shows the input data. The illustration and summary of modeling results for various situations are found in Sect. 5. Section 6 provides the conclusion in the end.

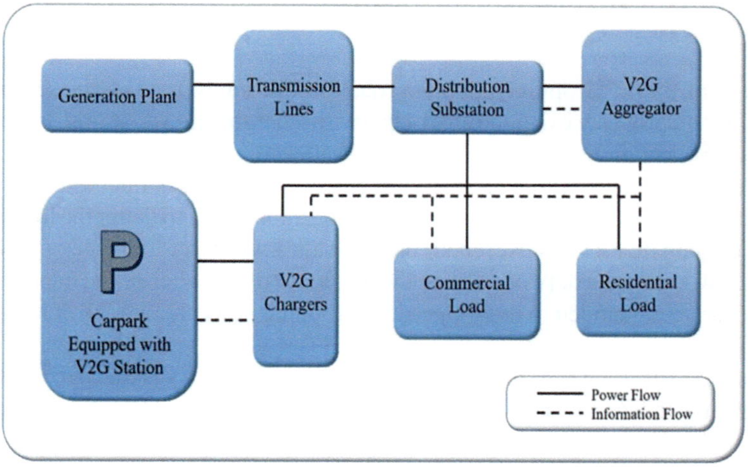

Fig. 1 V2G power exchange scheme

2 Hybrid PSO-GSA

To optimize the advantages of PSO social coordination & the exploratory capabilities of hybrid PSO-GSA were introduced [10]. The basic idea of PSO is to model a swarm of particles, where each particle represents a potential solution to the optimization problem. The particles move in the search space, and their movements are influenced by their own experience (personal best) and the experience of the swarm (global best). The personal best represents the best solution that the particle has found so far, while the global best represents the best solution found by any particle in the swarm's, on the other hand, is based on the law of gravity and the motion of celestial objects in the universe. In GSA, the optimization process is modeled as the motion of particles under the influence of gravitational and other forces. The algorithm starts with an initial population of particles, and each particle's position is updated based on the attractive and repulsive forces it experiences from other particles and a global best particle. However, combining the two algorithms may lead to better optimization performance. For instance, PSO can be used to update particle positions, while GSA can be used to calculate the gravitational forces between the particles. The combined PSO-GSA algorithms can help to overcome the limitations of each algorithm individually, such as premature convergence and slow convergence speed. The hybrid algorithm can also enhance the worldwide expansion and regional extraction skills of the swarm, resulting in a better convergence rate and more optimal solutions.

Overall, the PSO-GSA algorithm can be a better optimization technique for V2G power exchange, as it can efficiently handle the complex and dynamic nature of the optimization problem and provide more accurate and robust solutions.

3 Objective Function

The goal of the V2G optimization approach in this study is to decrease grid load variability by properly using the charging and discharging power of grid-linked EVs. This may be accomplished by completing the valley filling (when going to charge) and peak demand shaving (while discharging) procedures. To carry out these operations, a targeted load (Pt) is set, around which the algorithm will try to flatter the load profile. When the targeted load is exceeded, the algorithm will direct all accessible EVs to deliver power into the network. When the true load (pl) falls below the intended load, the EVs are commanded to draw electricity from the power network by charging the internal batteries.

When the true load matches the targeted load, no electricity is transferred between the EV and the power network. This goal function is stated mathematically by equation number (1)

$$\min\ \Delta p = pt(t) - pl(\mathrm{t}) - \mathrm{PEV(t)} \tag{1}$$

$$\mathrm{PEV(t)} = \sum_{i=1}^{N} pev^{i}(t) \tag{2}$$

In Eq. (1), Δp is the change in grid load relative to the goal load, and PEV is the power shared by all (N = 2200) EVs, as shown in Eq. (2).whereas the output contributed by the *i*th vehicle is referred to as $\mathrm{pev}^{\mathrm{i}}$.

By dissolving Eq. (1) on the basis of just one vehicle, we obtain

$$\min\ \Delta p = \frac{pt(t)}{K} - \frac{pl(t)}{K} - pev^{i}(t) \tag{3}$$

K is a term in Eq. (3) that simply represents the total population of EVs that are grid-tied while meeting the SOC limit. [11, 12].

4 Input Data

4.1 Load Curve

For modeling purposes, a regional grid's 24-h load curve, as shown in Fig. 2 was created. In the actual world, the V2G power exchange could be scheduled in real time or a day in advance using load curve forecasting [13].

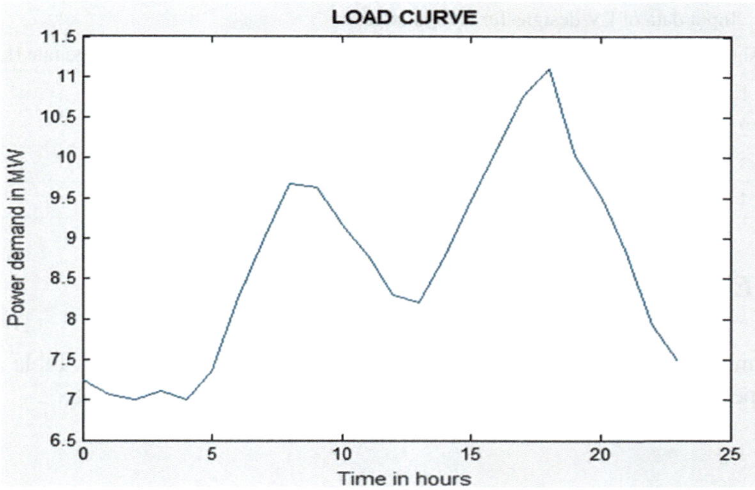

Fig. 2 Load curve

4.2 EV Availability

Only when EVs are linked to the grid, the V2G power exchange is feasible. Because of their mobility and stochastic nature, each EV's accessibility, however, cannot be guaranteed.

However, it is possible to build a preliminary collective driving design of EVs for the computation analysis of the suggested V2G approach, which is depicted in Fig. 3.

Assume a typical town with roughly 2200 EVs and availability to chargers in parking lots. The parking lots are separated into two groups: residential parking lots & commercial parking lots. In residential parking lots, the large percentage of EVs are accessible at night and until six in the morning. Following that, as individuals depart for work, school, and other destinations, the amount of EVs in residential parking lots starts to decline. Contrary to the popular belief, commercial parking lots have the most electric vehicles (EVs) present from 9.00 am to 6:00 pm. However, commercial parking lots on the opposite side are infrequently filled at night [11].

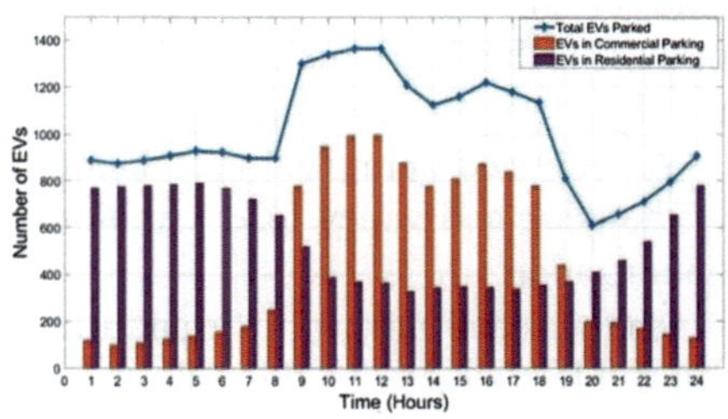

Fig. 3 EV Grid connection probability

Table 1 Input data of EV designs for computation

EV design	Battery capacity (kWh)	Power exchange rate (kW)
Model 1 (Hyundai Kona Electric)	64	100
Model 2 (Ford Mustang Mach-E)	88	115
Model 3 (Audi e-tron)	95	150
Model 4 (Nissan Leaf)	62	50

4.3 EV Specification

The simulation analysis in this paper uses four different EV models. Table 1 lists their specifications, including battery capacity and power exchange rate.

4.4 EV SOC Limit

To avoid battery pack damage and take into account overcharging difficulties, the limit of maximum SOC is set at 90%, while the minimum SOC is set at 50%.

Enable discharging only if SOC > SOCmin

Enable charging only if SOC < SOCmax

5 Results and Discussion

The suggested V2G algorithm-achieved improved load curves are shown beside the actual load curve. The chart in the illustration probably depicts two different curves, one of which represents the actual load curve and the other the optimal load curve attained by implementing the hybrid PSO-GSA.

The target load in the best situation depicted in Fig. 4 was set at 8 MW, and the V2G method is effective in maintaining this goal load all day. This indicates that the method was successful in controlling the energy usage and redistribution because the optimized load curve closely resembled the goal load curve.

All things considered, figures shows how well the V2G algorithms optimized the load curve to reach a certain target load. It draws attention to the potential advantages of these technologies in improving energy efficiency and sustainability and stresses the significance of the ongoing research and development in this field.

A suggested optimization technique is shown in Fig 5 along with a comparison of the peak and valley load variations percentage improvements before and after optimization.

The peak and valley prior to and after optimization, accordingly, are the parameters utilized to characterize this enhancement and Pi is the comparison factor. A graph or chart that displays the improvement in peak and valley load differences as a

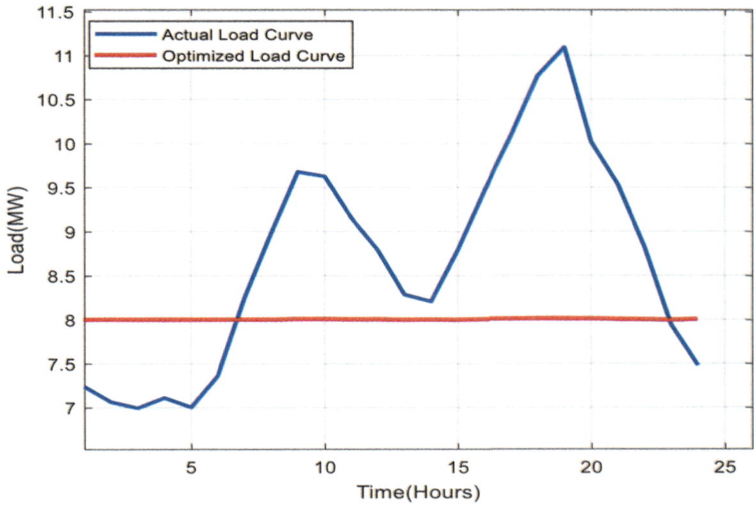

Fig. 4 Optimized load curve with targeted load 8 MW

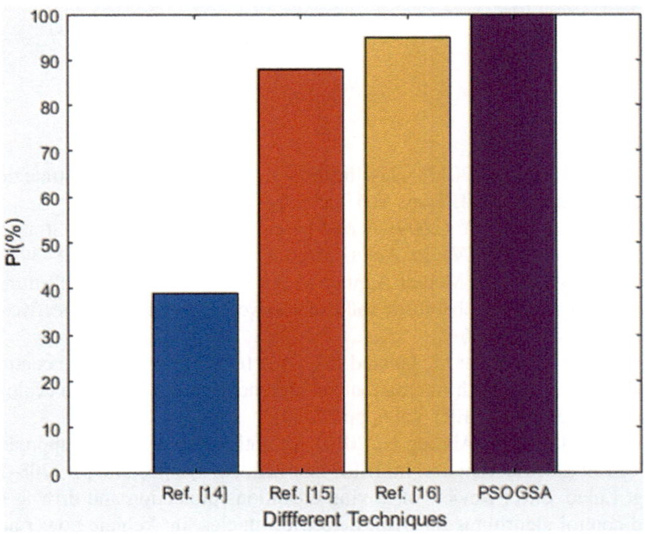

Fig. 5 Comparison of the suggested V2G algorithm

percentage over a given time period is most usually included in Fig. 5. Other details may also be included, such as the precise algorithm that was utilized, the desired load or energy consumption objective, as well as any pertinent data points or annotations. Comparisons were made between the production of the suggested technique best-case phase and the accurate performance of approach suggested in earlier work on a related topic in order to concern the outcomes and verified the suggested approach.

Finally, Fig. 5 illustrates how well the suggested optimization approach works to lessen the disparity between peak and valley loads. This measure can save costs and advance sustainability, hence it is a major indicator of energy efficiency.

6 Conclusion

An optimal scheduling method for V2G action is suggested in this study to reduce load fluctuations in the power grid around a certain targeted load. The accessibility of EVs for V2G function, the regularly shifting SOC of EV batteries and other issues make the efficient deployment of V2G power exchange difficult. Through examining the unpredictable nature of EV grid connection and continuously shifting the variables at certain intervals, the proposed optimization technique addresses these issues. Analyzing the outcomes of all cases, it is shown that the suggested technique may considerably and successfully flatten the load curve.

Comparisons were made between the production of the suggested technique best-case phase and the accurate performance of approach suggested in earlier work on a related topic in order to concern the outcomes and validate the suggested approach.

References

1. Wirasingha SG, Emadi A (2011) Classification and review of control strategies for plug-in hybrid electric vehicles. IEEE Trans Veh Technol 60:111–122
2. Banvait H, Sohel S, Chen Y (2009) A rule-based energy management strategy for plug-in hybrid electric vehicle (PHEV). In: American control conference, pp 3938–3943
3. Dehghanian P, Hosseini SH, Moeini-Aghtaie M, Arabali S (2013) Optimal sitting of DG units in power systems form a probabilistic multi-objective optimization perspective. Int J Electric Power Energy Syst 51:14–26
4. Caillard P, Gillon F, Hecquet M, Janiaud N (2013) Influence of machine control strategy on electric vehicle range. In: 8th International conference and exhibition on ecological vehicles and renewable energies (EVER), vol 6, pp 27–30
5. Grossard M, Kachroudi S, Abroug N (2012) An optimal energy-based approach for driving guidance of full Electric Vehicles. In: IEEE international symposium, pp 1708–1713
6. Rui Wang, Lukic, S.M.: Review of driving conditions prediction and driving style recognition based control algorithms for hybrid electric vehicles. In: Vehicle power and propulsion conference (VPPC) (2011)
7. Higgs B, Abbas M (2015) Segmentation and clustering of car-following behavior: recognition of driving patterns. IEEE Trans Intell Transp Syst
8. Wang B, Hu Y, Xiao Y et al (2018) An EV charging scheduling mechanism based on price negotiation. Futur Internet
9. Liu BC, Ieee M, Chau KT et al (2013) Opportunities and challenges of vehicle-to-home, vehicle to-grid technologies. Proc IEEE
10. Mirjalili S, Wang GG, Coelho LS (2014) Binary optimization using hybrid particle swarm optimization and gravitational search algorithm. Neural Comput Appl
11. Mulla A, Jadhav HT (2022) Optimal scheduling of vehicle-to-Grid power exchange using particle swarm optimization technique. Int J Comput Appl 44(7):687–704

12. He Y, Venkatesh B, Guan L (2012) Optimal scheduling for charging and discharging of electric vehicles. IEEE Trans Smart Grid
13. Cho H, Goude Y, Brossat X et al (2013) Modeling and forecasting daily electricity load curves: a hybrid approach. J Am Stat Assoc
14. Alam MJE, Muttaqi KM, Sutanto D (2014) A controllable local peak-shaving strategy for effective utilization of PEV battery capacity for distribution network support. IEEE Trans Ind Appl
15. Tan KM, Ramachandaramurthy VK, Yong JY, Padmanaban S, Mihet-Popa L, Blaabjerg F (2017) Minimization of load variance in power grids—Investigation on optimal vehicle-to-grid scheduling
16. Jian L, Zheng Y, Xiao X, Chan CC (2015) Optimal scheduling for vehicle-to-grid operation with stochastic connection of plug-in electric vehicles to smart grid

Design and Experimental Testing of Solar-Powered Desalination Using HDH Process

I. William Christopher, V. Giri Ranganathan, M. S. Niveda, and S. Robin Rosario

Abstract The goal of the paper is to examine and contrast various methods for extracting pure water from seawater in order to determine which method produces the most output while utilizing the least amount of input. Although there are various desalination techniques that use fossil fuels like electricity or gasoline, this project focuses on using solar energy, one of the main renewable resources. It uses a variety of equipment, including a black surface, aluminium fins, solar heaters, aluminium reflectors, and phase-change materials, to boost the heat within. With the aid of a PCM and a solar heater, this technology helps transform seawater into pure water even at night, so water continues to flow out consistently throughout the day. The effectiveness of various combinations is tested, and the outcomes are compared. The HDH procedure of solar-powered desalination is discovered to be optimal.

Keywords Humidification and de-humidification (HDH) · Photovoltaic (PV) system

1 Introduction

One of the major problems the world is facing now is the water crisis. As the population increases daily, the need for water also keeps increasing. Water being one of the basic needs for a human to live on this planet the availability and value must

I. W. Christopher (✉)
Department of Electrical and Electronics Engineering, Loyola-ICAM College of Engineering and Technology, Chennai, India
e-mail: iwchristop@gmail.com

V. G. Ranganathan
Bank of Singapore, Asia's Global Private Bank, Singapore, Singapore

M. S. Niveda
ITBA Department, Philip Capital, Singapore, Singapore

S. R. Rosario
PRE Machines (India) Pvt. Ltd, Chennai, India

J. Kumar et al. (eds.), *Recent Advances in Power Systems*, Lecture Notes in Electrical Engineering 1192, https://doi.org/10.1007/978-981-97-2788-9_20

be very high. The ironic information is that 71% of the Earth is covered by water but still, there is a huge water demand. The future is based on solar energy and we figured out a way to use solar energy to purify and desalinate the water so that it can be used for our day-to-day activities. We have targeted to increase the performance in a cost-efficient way so that poor people near the shore can afford it and might help them solve their water problem.

There are many different kinds of desalinating methods like multi-stage flash distillation (MSF), Ion exchange Membrane processes, Electro dialysis reversal (EDR), Freezing desalination, Geothermal desalination, Methane hydrate crystallization, and High-grade water recycling. The common factor in all these types is that they use a non-renewable form of resources that once used can never be regained back and there are a lot of processes involved in getting the power for the process and lots of money required to desalinate 1 L of water, whereas in Solar desalination the source is free of cost and can always be regained once used up i.e. renewable form of energy.

A useful PCM for energy storage in a latent heat storage system is paraffin wax. With a latent heat of 210 kJ/kg and a good transition temperature range of 58–60 °C, it has high latent heat. Moreover, there is no subcooling to be seen. With suitable charging and discharging periods as well as heat release rates, an easy-to-use tube-in-tube heat exchanger system [1] can be employed for energy storage.

A variety of thermal and membrane water desalination techniques created in recent decades have been reviewed [2]. Each desalination technology's benefits and drawbacks, including economic ones, have also been discussed. The design of the membrane module, energy recovery, and pre-treatment techniques have all undergone significant advancements in recent years, making membrane desalination systems cost-competitive with thermal ones.

The performance of a tar solar water heater is evaluated experimentally and theoretically by the authors [3] and contrasted with that of a typical type collector. Under the same circumstances, both collectors' performance is evaluated. Both collectors are glazed and have the same surface area. The water tubes are welded to the absorber plate in the conventional type, but in the tar type, the tar serves as an absorber plate and covers the water tubes.

This paper [4] presents some of the most popular solar collector types. Flat-plate, compound parabolic, evacuated tube, parabolic trough, Fresnel lens, parabolic dish, and Heliostat field collector are only a few of the different types of collectors mentioned (HFC).

Here [5] is a thorough assessment of the most recent research on several solar-powered desalination methods. Together with an economic study and cost comparison of various solar energy-based technologies with conventional desalination methods, this research also emphasizes the environmental effects of desalination technology. This analysis is a component of research into how solar thermal desalination can be integrated into the Australian grid system.

Solar energy was employed for the closed-air cycle humidification–dehumidification process to desalinate water [6]. Hot water from an electrical heater or a flat plate solar collector was used to heat and humidify the air that was circulated by natural or forced convection. The condenser recovered the latent heat of condensation to warm the saline feed water. Two units of various sizes were built from various components. It was discovered that the production of these units was significantly higher than that of the single-basin stills.

With a focus on technologies and economics, an attempt has been made to provide a concise summary of the work of the highlights that have been accomplished in recent years globally and the state-of-the-art for the most significant initiatives in the field of desalination by renewable energies [7]. Water supplies, demand, the availability of drinkable water, and purification techniques are also reviewed. Up until this point, the classification of distillation units has been done via a literature review. Desalination systems driven by various renewable energy technologies have been compared in terms of economics.

A comparison of a small solar pond-assisted solar still's theoretical and experimental study is provided [8]. Experiments with various salinities were carried out in a miniature solar pond. The micro solar pond's ideal salinity level was discovered to be 80 g/kg of water. The effects of adding sponge cubes to the still, integrating a small solar pond with it, and combining the two are addressed.

An innovative method for merging DCMD and HDH processes, as driven by an FPC system, was proposed by the authors [9]. Instead of using thermal energy to increase the feed temperature, the FPC system was configured with Gizan sun irradiation data. The heated saltwater in this system is first fed into the DCMD and then into the HDH system.

Experimental research is done to determine how well a humidification-dehumidification (HDH) desalination system works with a heat pump unit [10]. A humidifier, a heat pump, a plate heat exchanger, a first-stage dehumidifier, and a second-stage dehumidifier make up the newly created HDH system. The suggested method makes use of a heat pump unit, which simultaneously absorbs heat from the air through a second-stage dehumidifier and transfers heat to seawater via a plate heat exchanger.

The design and working principle of a four-stage cross-flow HDH desalination system with freshwater absorption dehumidification have been established [11]. The experiments were carried out in both a transient and a steady state. On the system's performance, the impacts of the air volume flow rate, feed seawater mass flow rate, and temperature are investigated.

Experimental research on the HSHDH for desalinating very saline water was done [12]. When using a reflector, the HSHDH can create fresh water even when the feed temperature is below 60 °C, saving 16.5% of the daily power usage. When compared to brackish water, the amount of electrical energy consumed dropped by 65% while using extremely saline water. When using saline water that is 85 °C in temperature, the HSHDH can produce 72 kg per day.

In this work [13], a closed-air cycle-based solar HDH water desalination system is developed, and its outcomes are compared to calculated values. Investigated are the effects of various parameters on gain output ratio, productivity, humidifier efficiency, and dehumidifier efficiency.

There are two proposed and demonstrated solar-aided variable pressure HDH desalination systems [14]. Sub-atmospheric pressures enable more effective operation of the proposed HDH system. The suggested HDH desalination method relies just on a vacuum pump; a compressor is not used.

Experimental research was done to determine how well a desalination machine using humidification–dehumidification technology performed using new corrugated packing aluminium sheets in the humidifier [15]. From the current investigation, certain crucial conclusions can be drawn. By raising the inlet temperature, the mass flow rate of water entering the humidifier, and the rate of cooling water in the dehumidifier, the generation of freshwater utilizing the HDH system is increased.

The solar still and humidification dehumidification desalination systems were recently modified utilizing renewable energy sources to increase the rate at which fresh water is produced, according to this article [16]. The impact of integrating fins, using energy-storing materials and wick materials, mixing nanoparticles, agitation effect, transparent cover cooling, integrating a thermoelectric cooler, multi-effects of solar still, preheating with a water heater and a photovoltaic thermal collector, improving the condensing cover, operating with a heat pump and refrigeration, and integrating with waste heat recovery are all discussed in relation to solar still.

To increase freshwater productivity and lower the price of fresh water, a novel solar-assisted heat pump with an internal heat recovery desalination system is proposed [17] in this study. The design and construction of an experimental setup has been completed. The system performance has also been investigated under various operating situations based on the experimental findings.

A desiccant wheel or heat exchanger-based humidification–dehumidification desalination unit's performance has been investigated [18]. The total distillate rate, GOR, STE, and RR have all been examined in relation to the inlet air and water flow rates, temperature, inlet air humidity, and regeneration temperature.

Mathematical modelling is used to explore [19] the performance study of the heat pump-aided HDH system with direct contact dehumidifier. We can infer that the system's equivalent SEEC is higher when the system is operating in a manner that results in a higher water production rate. With system working parameters of T_{sw}, t = 70 °C, T_{fw}, t = 20 °C, and m_{rsw}/fw = 1.05, the optimum RR of around 5.8% and SEEC of 335.4 kWh/m^3 can be accomplished.

An HDH desalination system based on a direct absorption sun humidifier is being researched [20]. The SH-HDH desalination system operates at its optimum water flow rate. The proposed performance curve compares GOR to normalized gain based on humidity. In terms of performance, the SH-HDH system outperformed the C-HDH system by 18.4%.

Several HDH setups' thermos-economic performance is compared [21]. A comparison and analysis of single, dual, and triple humidifiers is made. The efficacy of humidifiers and dehumidifiers affects the amount of freshwater produced. The HDH cycle of the water-heated multi-humidifier produces more fresh water.

A novel solar system for humidification and dehumidification is described [22]. Experimental testing reveals that new packing materials can enhance the humidifier's evaporation process. It is estimated that the system can produce up to 22 L/m^2 of solar collector area.

To increase freshwater production, a solar-assisted MS-VHDH desalination system was created [23]. A three-stage humidifier with liquid ring vacuum pumps for reducing humidifier pressure and solar heaters for heating the air and water was constructed to study its performance. While this study was carried out experimentally at sub-atmospheric pressure humidification, earlier MS-HDH processes were examined theoretically and at atmospheric pressure.

A new SNVD system was suggested [24] in order to make efficient use of the system's waste heat. It was created to make use of the brine and distilled water waste heat. To transmit the heat to the system feed water, concentrator heat exchangers were offered. Additionally, the evaporation process was heated using a flat-plate solar collector while operating under natural vacuum pressure.

Investigated [25] is a unique NVSD system with an inbuilt condenser. To remove NCG and restore Hoover, the system uses a procedure of loading and emptying water. Because the inbuilt condenser allows for the recovery of latent heat of vapour during desalination, the NVSD system exhibits a high water yield rate. The performance of the system is examined as an experimental prototype is constructed and developed.

Humidification and dehumidification are the fundamental processes driving this solar desalination. A similar process is used within the desalination box as it does throughout the water cycle, where water evaporates, is stored in clouds, and then condenses to create rainwater, the purest form of water. Because salt has a higher density than water and doesn't evaporate, a miniature water cycle is produced inside the box. Water that has been separated from salt hits the top surface of the glass, slides downward, and is gathered using an outflow pipe. According to the literature review, evaporation can be boosted in three different ways: by raising the temperature, by producing a vacuum inside the box, and by stirring the air on the surface of the liquid.

To put into practice, the concept is by raising the temperature within the box using various tools and testing techniques. Different types of apparatus include the black surface, which absorbs heat and releases less of it, the mirror-like aluminium reflectors, which focus solar radiation onto the glass container to heat it more, and the aluminium fins, which help transfer heat from the base to water over a large surface area. A solar-powered heater that uses PV cells to convert solar energy into electric energy and a battery to store the electricity from the panel and use it throughout the night so that the process never stops. Figure 1 shows the block representation of the solar-powered desalination using the HDH process.

Last but not least, PCM material is a phase-changing material that transforms into a different phase when heated to its melting point. Paraffin wax, which has a melting point of just 37 °C, and zinc oxide, which is being manufactured in a laboratory, both change phases and were our two materials of choice. The PCM's function is to alter its state in the morning and release heat at night when the ambient and internal temperatures drop.

This article is structured so that Sect. 1 is an introduction explaining the goal and block diagram, Sect. 2 is details about the desalination using the HDH process, Sect. 3 is explanations of solar-powered desalination methods, Sect. 4 is discussions of

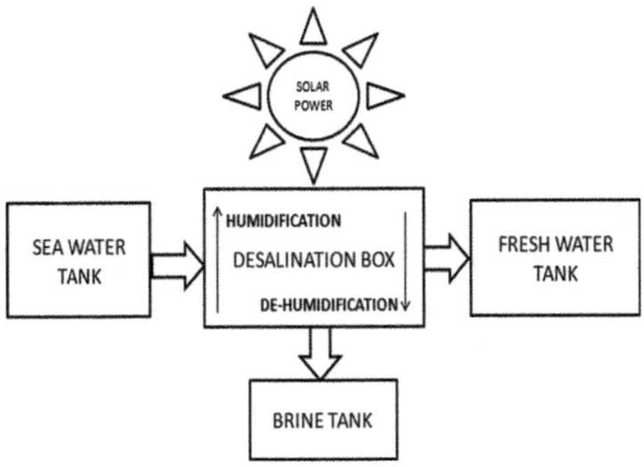

Fig. 1 Block diagram of HDH process

design and fabrication, Sect. 5 discusses the experimental setup, and Sect. 6 discusses experimental results followed by a conclusion.

2 Desalination Using HDH Process

Desalination techniques remove dissolved minerals and salts from water. Seawater, brackish water, wells, surface water (rivers and streams), wastewater, and industrial feed and process waters are all potential sources of feed water. To successfully overcome natural osmotic pressures and force water through membrane processes, membrane separation requires driving factors such as pressure (applied and vapour), electric potential, and concentration. Because of this, technology is energy-intensive, and research is constantly being done to increase efficiency and lower energy use.

2.1 Types of Desalination Process

In the desalination process, the following kinds are frequently employed:

(1) Multi-Stage Flash Distillation (MSD)
(2) Multiple-Effect Distillation (MED)
(3) Vapour-Compression Evaporation (VCE)
(4) Cogeneration
(5) Water desalination using solar energy

A technique referred to as the Solar Humidification-Dehumidification (HDH) process, also known as the multiple-effect humidification-dehumidification process, Solar Multistage Condensation Evaporation Cycle (SMCEC), or Multiple-Effect Humidification (MEH), imitates the natural water cycle on a shorter time scale by evaporating and condensing water to separate it from other substances. Thermal solar energy is the driving force behind this process, which creates water vapour that is then condensed into a different chamber. By capturing the heat from the condensing water vapour and preheating the incoming water source, waste heat is reduced in complex systems. Due to the relatively low cost of solar thermal collectors, this technology is efficient for small- to mid-scale desalination plants in isolated regions.

3 Solar-Powered Desalination Using HDH Process

This paper focuses on the utilization of solar energy, one of the main renewable resources. There are many different types of desalination processes that employ non-renewable sources like electricity or fuels. Although the efficiency is relatively low, direct solar energy alone is sufficient to heat the water and turn it into vapour. Therefore, the various testing techniques employing apparatus such as black surfaces, aluminium fins, solar heaters, aluminium reflectors, and phase change materials in order to boost the heat within.

3.1 Working Principle

The Humidification Dehumidification (HDH) process resembles the hydrological process that occurs naturally when the ocean is heated by the sun and the ocean water subsequently evaporates to generate clouds. When the clouds finally condense, water is obtained in the form of rainfall. This natural hydrological cycle tends to be replicated by the HDH process.

In the HDH process, evaporation occurs when hot water and dry air are combined in the humidification chamber. This mixture is then transferred to the dehumidification chamber, where a condenser is utilized to produce fresh water. The HDH process is thought to be the most promising desalination method that is compatible with renewable energy sources, especially solar energy, which is one of its benefits. In the HDH system, the humidified air circulates between the humidification and dehumidification chambers in a circuit that is driven by natural or forced convection.

After being spread on a wetted surface, the air travels in either a co-current or a counter-current flow through the humidification chamber to the hot feed (seawater), where it gets saturated. Water vapour from the humid air condenses on a heat exchanger in the dehumidification chamber. The distillate drips into a basin after flowing along the heat exchanger. The cold seawater feed passing through the heat exchanger receives the majority of the heat from condensation. As a result, the feed temperature in the condenser increases. This feed can be discarded or heated further using waste heat or a solar collector. Figure 2 depicts the process flow for solar-powered desalination utilizing HDH.

Similar principles are used within the desalination box to produce the purest form of water, rain, just as they do in the water cycle when water evaporates, condenses in clouds, and then falls to the ground as rain. Salt does not vaporize because of its larger density than water, creating a miniature water cycle inside the box. When

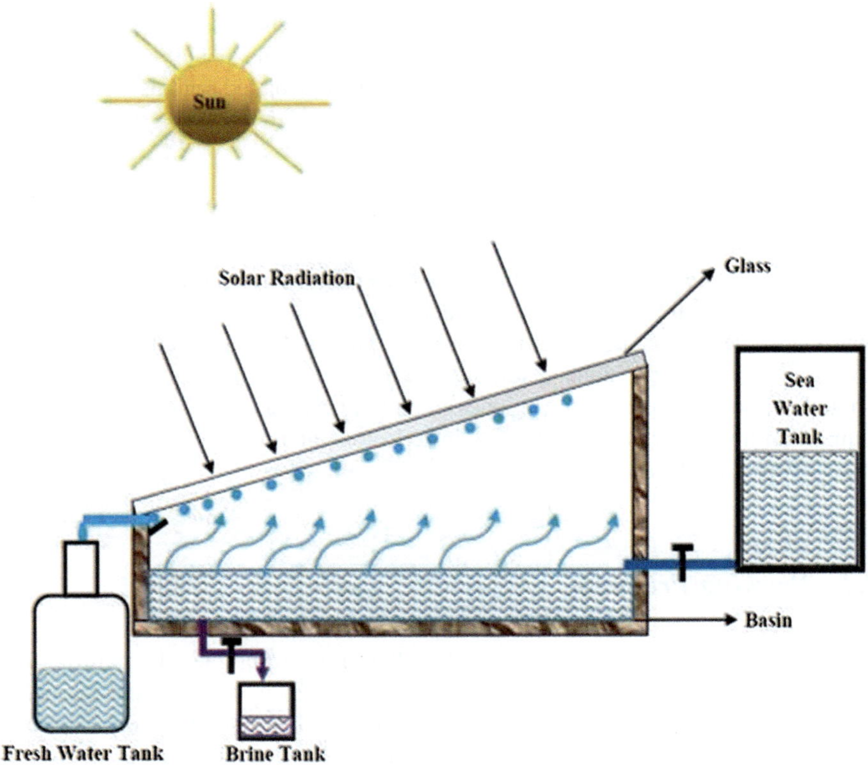

Fig. 2 Process diagram of the solar-powered desalination using HDH

water that has been separated from salt strikes the top glass surface, it falls downward and is gathered using an outflow pipe. According to the literature review, evaporation can be boosted in three different ways: by raising the temperature, by producing a vacuum inside the container, and by stirring the air on the water's surface.

4 Design and Fabrication

4.1 Tank Design

The tank is made of annealed glass sheets of six-millimeter thickness. Annealed glass material is chosen because of its low cost and durability. The tank is designed in such a way that it can hold a hundred liters of water (stagnant water). Before designing the tank, it is very important to know the maximum possible temperature that will be developed inside the tank. Only based on such data the material is chosen. Six-millimeter thickness is more than enough for the maximum possible temperature of

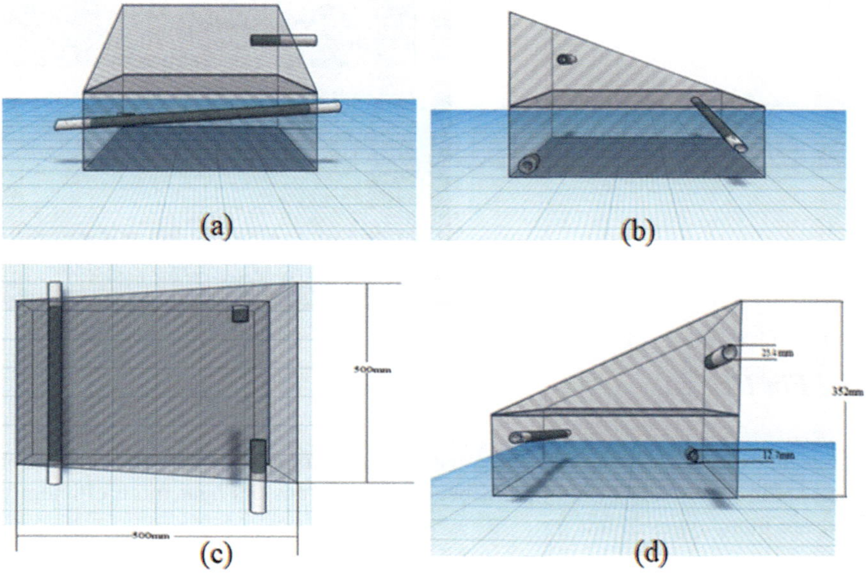

Fig. 3 Tank design using solid work design. **a** Front view, **b** left side view, **c** right side view, **d** top view

100 °C (boiling point of water). By solid work design, the tank has been designed and it is illustrated in Fig. 3.

The roof of the tank is designed in such a way that the water which evaporates due to the heat developed inside the tank, gets in contact with the roof, gets condensed, and is collected. There are many roof styles available but the most commonly used roof style is of two types

(1) Single direction sloped roof
(2) Two-direction sloped roof

Single direction sloped roof is preferable because water can be collected from one side and the rate of condensation is more so this increases productivity. The angle of the slope is an important factor that influences the velocity of the sliding water or the sliding velocity of the condensed water. The slope angle should be neither too high nor too low. The slope angle is decided based on certain calculations and is taken as 22.5°.

Fig. 4 Fin design

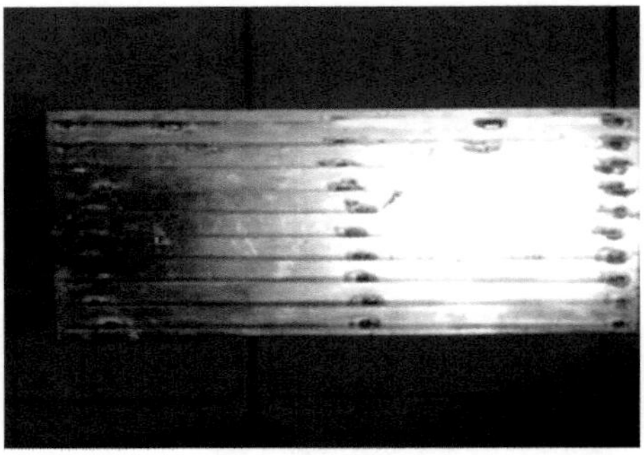

4.2 Fin Design

A fin is an extended surface of a body that is used to transfer heat by the convection process. Fins are made of aluminum because they transfers heat more efficiently. There are many types of fins

(1) Constant area straight fin
(2) Variable area straight fin
(3) Pin fin
(4) Annular fin

Constant area straight fins or rectangular fins are more widely used in IC Engines and microprocessors because of their higher efficiency. Fins are designed based on the requirements and based on certain calculations. A plate of thickness three millimeters is chosen and the fin thickness is 1.5 mm and the fin design is shown in Fig. 4. The ratio for welding fins is 1:2. By assuming the efficiency as hundred percent the length, width, and height of the fins are decided. The total number of fins is decided based on the requirements. But it is impossible to attain an efficiency of 100%.

4.3 Design Calculations

These are the designs of various components that are to be used in the experiments to find out which gives the highest productivity and study the ways to improve it.

Fin Design

Efficiency expected ~ 100%
$\eta_{f} = \tanh(mL)/mL$

Assuming,

L = 2 cm for a constant input of 4 liters
m = Sqrt (hP/kA)

h = 49.95 w/m^2 K
k = 205 W/m K
L = 500 mm, B = 1.5 mm

Solar Heater

Immersion heater power = 25 W
Solar panel power = 100 W
Voltage = 12.8 V
Battery Capacity = 13 V and 1000 Ah

Reflector

$$\text{Reflector angle} = 90^{\circ} + \sin^{-1}\left(\begin{array}{l}(-\text{reflector length}/4 * \text{target length}) \\ +0.25 * \left(\text{Sqrt}\left(\text{reflector length}^2/\text{target length}^2\right) + 8\right)\end{array}\right)$$
$$= 123.3^{\circ}$$

Solid Desalination Box

The solid desalination box is designed for the following dimensions

Total height, $Tl = 352$ mm.
Length of the cuboid = 500 mm.
Breadth of the cuboid, $Bc = 500$ mm.
Height of the cuboid, $Hc = 150$ mm.
Length of the slope, $Ls = x$.
Height of the top triangular box,
$Ht = Tl - Hc = 352 - 150 = 202$ mm.

By Pythagoras' theorem,

$$\begin{aligned} x^2 &= Ht^2 + Bc^2 \\ &= 202^2 + 500^2 + 40804 + 250000 \\ &= 539.2\text{–}540 \text{ mm} \end{aligned}$$

Slope angle, $s = \theta$,

$$\begin{aligned} \cos\theta &= (Bc/x) \\ \theta &= \cos -1(Bc/x) \\ &= \cos -1(500/540) \\ \theta &= 21.9\text{–}22^{\circ} \end{aligned}$$

5 Experimental Setup

5.1 Components Required for Experimental Setup

Annealed Glass. Most architectural glass is made by melting silica sand, soda ash, dolomite, limestone, and salt cake to about 2700 °F, and then allowed to cool slowly. Again it passes through an annealing process, a furnace that controls the cooling process. It is produced without internal stresses imparted by heat treatment. The specifications of Annealed Glass are listed in Table 1.

Aluminium Reflectors. Aluminum is a good reflector of visible light as well as heat, and that together with its low weight makes it an ideal material for reflectors. The specification of Aluminium Reflectors is listed in Table 2.

Aluminium Sheet. Aluminum Sheet (Plate) is available in various alloys. It has medium strength, very good corrosion resistance, and very good chemical resistance. The specifications of Aluminium Sheet are listed in Table 3.

Solar Panel. Light is converted into electricity by the photovoltaic effect. When light is incident to the cell, the absorbed energy excites bound electrons. This allows them to jump their atomic bonds and become free. The free electrons travel through the material, and the resulting current is harnessed when conductors are attached to either side of the cell. The specifications of the Solar Panel are listed in Table 4.

Phase Change Material. A Phase Change Material (PCM) is a substance with a fusion of heat which, melting and solidifying at a certain temperature, is capable of storing and releasing large amounts of energy. Heat is absorbed or released when the

Table 1 Specifications of annealed glass

S. no.	Parameters	Dimension (in mm)
1	Thickness	6
2	Base	500 * 500
3	Front	180 * 500
4	Rear	425 * 500

Table 2 Specifications of aluminium reflectors

S. no.	Quantity	Dimension (in mm)
1	4 Reflectors sheets	600 * 600/sheet

Table 3 Specifications of aluminium sheet

S. no.	Parameters	Dimension (in mm)
1	Thickness	3
2	Size	510 * 400

Table 4 Specifications of a solar panel (100 W)

S. no.	Parameters	Specifications
1	Maximum peak voltage, V_{MP}	16.9 V
2	Maximum peak current, I_{MP}	5.19 A
3	Open circuit voltage, V_{OC}	21.2 V
4	Short circuit current, I_{SC}	5.75 A

Table 5 Specifications of a PCM

S. no.	Parameters	Specifications
1	Latent heat of fusion	184.48 kJ/kg
2	Volume	42.92×10^{-6} m^3
3	Density	54.32C
4	Melting temperature	833.6 kg/m^3

material changes from solid to liquid and vice versa. The specifications of a PCM are listed in the Table 5.

5.2 *Auxiliary Materials*

Immersion Heater. An immersion heater has an electrical resistance heating element encased in a tube and directly placed in the water (or other fluid) to be heated.

Thermostat. A thermostat is a component that senses the temperature of a physical system and performs actions so that the system's temperature is maintained near a desired set point. The thermostat is used to find the temperature of the tank.

PWM Charge Controller. The PWM solar controller is used to stabilize the dc output from the solar panel to a constant voltage.

Black Paint. Black paint is used for painting the base of the tank.

Inverter. The inverter is used to convert stabilized dc into ac output.

Pipelines. Since water is the major source pipelines act as a guide way to guide the purified water into the outlet tank and saltwater into the tank from the inlet tank.

The components and the materials required for the proposed Solar-Powered Desalination using HDH process are illustrated in Fig. 5.

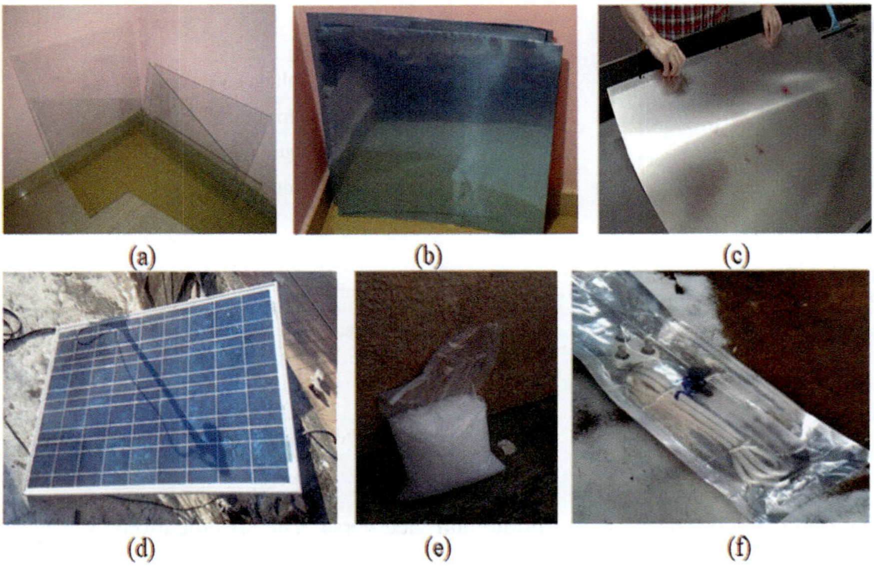

Fig. 5 Components and materials required for the proposed system. **a** Annealed glass, **b** aluminium reflectors, **c** aluminium sheet, **d** solar panel, **e** phase change material, **f** immersion heater

5.3 Experimental Setup

The completed experimental setup for the solar-powered desalination using the HDH process is illustrated in the Fig. 6.

Fig. 6 Experimental setup for the proposed system

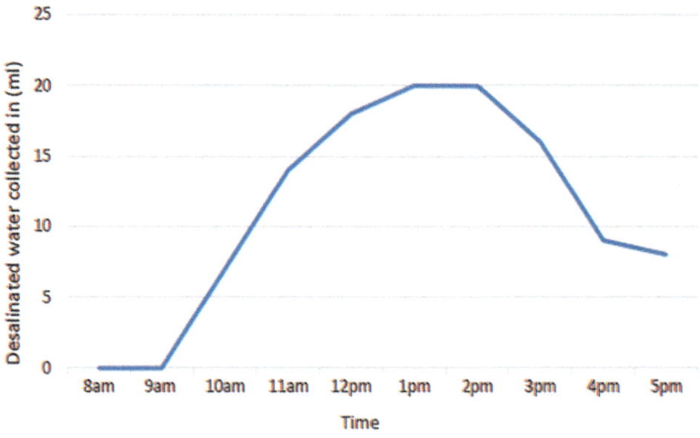

Fig. 7 Plain glass box test observation

6 Experimental Results and Discussions

To determine the maximum temperature developed inside the tank and to estimate the maximum quantity of pure water produced from four liters of seawater several tests were conducted. The duration of each test is nine hours. This section discusses the different experimental test results along with the temperature graph and cumulative graph.

6.1 Plain Glass Box Test

The test started around 8:00 am. The ambient temperature was recorded as 29.5 °C. After every 60 min, the temperature inside the tank was recorded. Sea water (4 L) was the input. The inference obtained from the plain glass box test is that the maximum temperature of 49.9 °C was recorded at 1:00 pm and 118 mL of the total quantity of pure water was produced. The observation can be understood by the following graph and it is illustrated in Fig. 7.

6.2 Black Surface Test

In this test, the bottom of the glass tank was painted black. The test started around 8:00 am. The ambient temperature was recorded as 29.5 °C. After every 60 min, the temperature inside the tank was recorded. Sea water (4 L) was the input. The inference obtained from the black surface test is, the maximum temperature of 53.2 °C was recorded at 1:00 pm and 222 mL of the total quantity of pure water was produced.

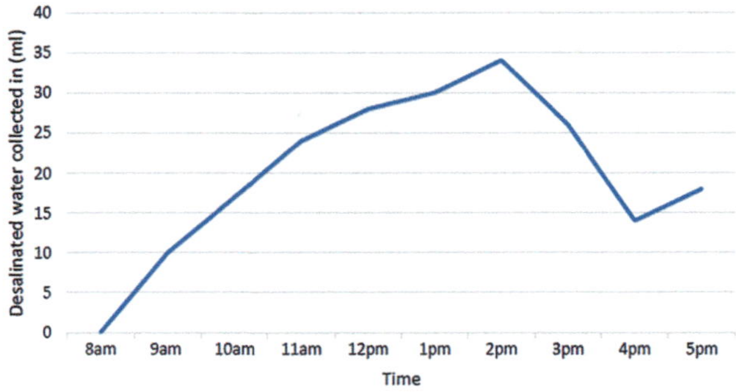

Fig. 8 Black surface test observation

The observation can be understood by the following graph and it is illustrated in Fig. 8.

6.3 *Aluminium Fin Test*

The aluminium fins were placed inside the tank. This test was carried out after painting the bottom of the tank with black paint. The test started around 8:00 am. The ambient temperature was recorded as 29.5 °C. After every 60 min, the temperature inside the tank was recorded. Sea water (4 L) was the input. The inference obtained from the aluminium fin test is that the maximum temperature of 58.4 °C was recorded at 12:00 pm and 319 mL of the total quantity of pure water was produced. The observation can be understood by the following graph and it is illustrated in Fig. 9.

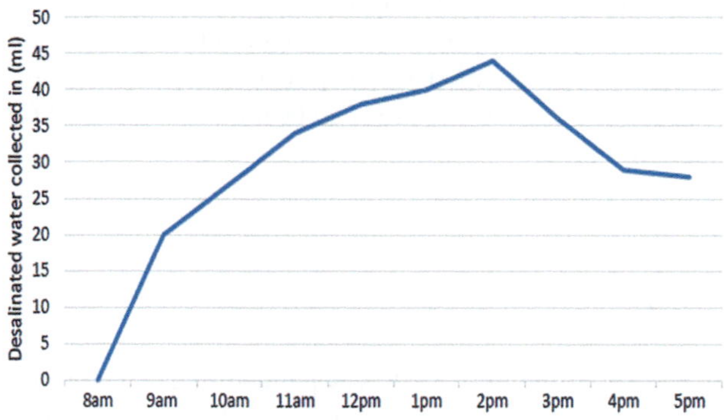

Fig. 9 Aluminium fin test observation

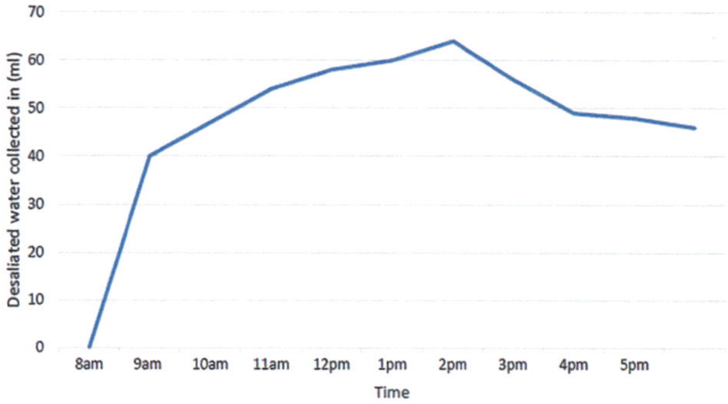

Fig. 10 Solar-powered immersion water heater test observation

6.4 *Solar-Powered Immersion Water Heater Test*

A 25 W solar-powered immersion water heater was used to carry out this test. A solar panel with a capacity of 80 W was used as the power source. A 13 V, 100 Ah battery was used to store the energy drawn by the solar panel. A dc to ac converter was also used in this test. The test started around 8:00 am. The ambient temperature was recorded as 29.5 °C. After every 60 min, the temperature inside the tank was recorded. Sea water (4 L) was the input. The inference obtained from the solar-powered immersion water heater test is that the maximum temperature of 61.4 °C was recorded at 12:00 pm and 562 mL of the total quantity of pure water was produced. The observation can be understood by the following graph and it is illustrated in Fig. 10.

6.5 *Aluminium Reflector Test*

The aluminium reflector sheets were fixed on all sides of the glass tank near the roof. They were inclined at an angle of 123°. The tank was placed in the north–south direction. The test started around 8:00 am. The ambient temperature was recorded as 29.5 °C. After every 60 min, the temperature inside the tank was recorded. Sea water (4 L) was the input. The inference obtained from the aluminium reflector test is that the maximum temperature of 68.7 °C was recorded at 12:00 pm and 750 mL of the total quantity of pure water was produced. The observation can be understood by the following graph and it is illustrated in Fig. 11.

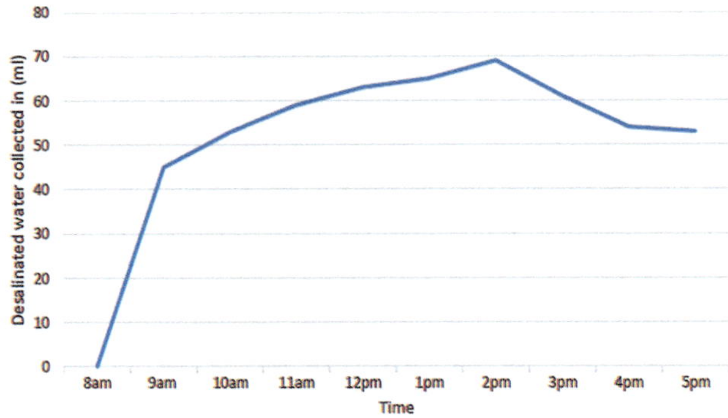

Fig. 11 Aluminium reflector test observation

6.6 Phase Changing Material (PCM) Test

Phase-changing materials such as paraffin wax and zinc oxide were used in this test. Zinc oxide in colloidal form (80 mL) was sealed inside a copper tube. Paraffin wax in crystal form was also sealed in a copper tube. The copper tubes were placed inside the glass tank during day time so that the heat developed inside the tank was absorbed by these copper tubes. The test was carried out at the night. The test started at 6:00 pm. The ambient temperature was recorded as 31.6 °C. After every 60 min, the temperature inside the tank was recorded. Sea water (4 L) was the input. The inference obtained from the phase-changing material test is that 250 mL of the total quantity of pure water was produced.

6.7 All Apparatus Test

In this test, all the components such as aluminium fins, aluminium reflectors and solar powered immersion water heater were used at the same time. The test started around 8:00 am. The ambient temperature was recorded as 29.5 °C. After every 60 min, the temperature inside the tank was recorded. Sea water (4 L) was the input. The inference obtained from the aluminium reflector test is that the maximum temperature of 70.2 °C was recorded at 12:00 pm and 1050 mL of the total quantity of pure water was produced. The observation can be understood by the following graph and it is illustrated in Fig. 12.

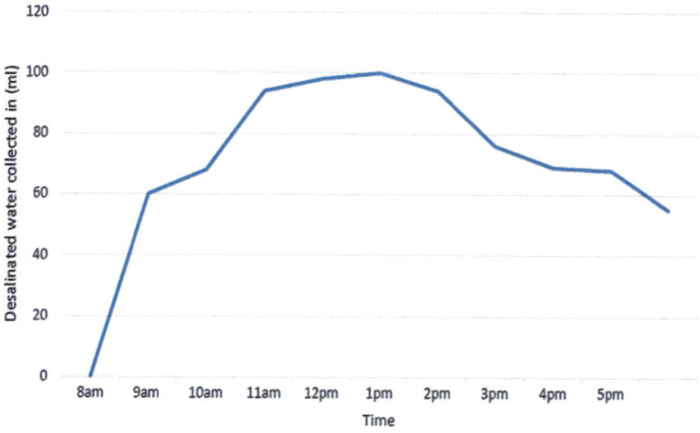

Fig. 12 All apparatus test observation

6.8 Overall Result Analysis

From the above-mentioned tests, the performance of the evaporation process improvement can be seen and the overall test consists of all the apparatus giving the maximum output with maximum temperature, and the process runs all day and night so water gets purified 24/7. For a maximum of 4000 mL of seawater input a purified water output of 1050 mL is obtained which is approximately 1:4 ratio. The overall cumulative observation of the various tests is listed in Table 6.

Table 6 The overall cumulative observation of various tests

Amount of purified water collected from different tests (in mL)						
Time	Plain glass test	Black surface test	Aluminium fin test	Solar-powered immersion water heater test	Aluminium reflectors test	All apparatus test
8 am	0	0	0	0	0	0
9 am	0	10	20	40	68	83
10 am	7	17	27	47	79	89
11 am	14	24	34	54	87	108
12 pm	18	28	38	58	93	135
1 pm	20	30	40	60	86	119
2 pm	20	34	44	64	79	118
3 pm	16	26	36	56	71	110
4 pm	9	19	29	49	67	102
5 pm	8	18	28	48	63	98
6 pm	6	16	26	46	57	88
Total (mL)	**118**	**222**	**319**	**522**	**750**	**1050**

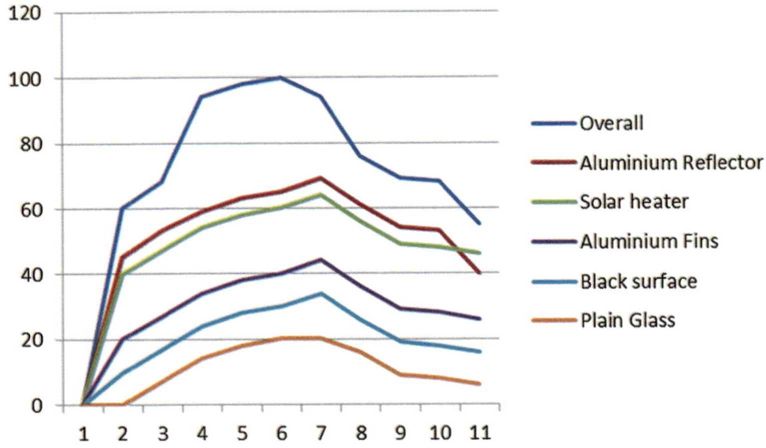

Fig. 13 Overall cumulative observation of various tests

Table 6 shows the readings taken during the month of February 2018 in Chennai from morning 8 am to evening 6 pm under different apparatuses that are used to increase the temperature inside the box and the corresponding variation has been observed. The observation of the various tests can be understood by the following graph and which is illustrated in Fig. 13. This graph shows the variation of temperature from the 1st hour to the last hour in different tests. There is an improvement in every test.

6.9 Desalinated Water Report

The desalinated water which is obtained as the purified output from different experimental tests is given for lab testing to check the quality of the water. The tested desalinated water report is shown in Fig. 14. The following lab report was obtained from Swasti Enviro Solutions Private Limited, Chennai) shows that the desalinated water collected from the solar-powered desalination using the HDH process meets the requirements of IS 10500 R.2012 for portability with respect to the parameters tested.

7 Conclusions

The results of the subsequent test show how the evaporation process is performing better, and the whole test, which includes all the apparatus, produces the most with the highest temperature and runs round-the-clock to purify water. The maximum

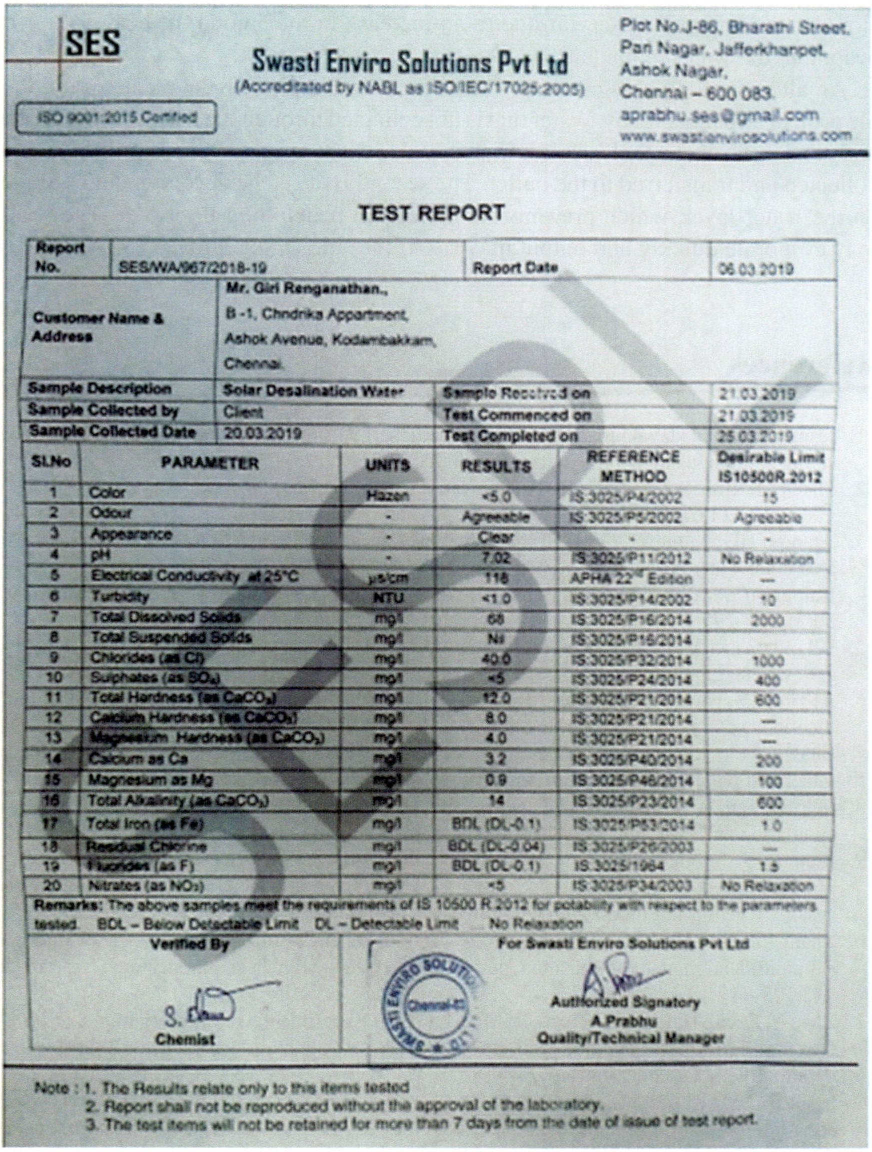

SES
ISO 9001:2015 Certified

Swasti Enviro Solutions Pvt Ltd
(Accreditated by NABL as ISO/IEC/17025:2005)

Plot No.J-86, Bharathi Street,
Pari Nagar, Jafferkhanpet,
Ashok Nagar,
Chennai – 600 083.
aprabhu.ses@gmail.com
www.swastienvirosolutions.com

TEST REPORT

Report No.	SES/WA/967/2018-19	**Report Date**	06.03.2019
Customer Name & Address	**Mr. Giri Renganathan.,** B -1, Chndrika Appartment, Ashok Avenue, Kodambakkam, Chennai.		
Sample Description	**Solar Desalination Water**	**Sample Received on**	21.03.2019
Sample Collected by	Client	**Test Commenced on**	21.03.2019
Sample Collected Date	20.03.2019	**Test Completed on**	25.03.2019

Sl.No	PARAMETER	UNITS	RESULTS	REFERENCE METHOD	Desirable Limit IS10500R.2012
1	Color	Hazen	<5.0	IS:3025/P4/2002	15
2	Odour	-	Agreeable	IS:3025/P5/2002	Agreeable
3	Appearance	-	Clear	-	-
4	pH	-	7.02	IS:3025/P11/2012	No Relaxation
5	Electrical Conductivity at 25°C	µs/cm	118	APHA 22nd Edition	—
6	Turbidity	NTU	<1.0	IS:3025/P14/2002	10
7	Total Dissolved Solids	mg/l	68	IS:3025/P16/2014	2000
8	Total Suspended Solids	mg/l	Nil	IS:3025/P16/2014	
9	Chlorides (as Cl)	mg/l	40.0	IS:3025/P32/2014	1000
10	Sulphates (as SO_4)	mg/l	<5	IS:3025/P24/2014	400
11	Total Hardness (as $CaCO_3$)	mg/l	12.0	IS:3025/P21/2014	600
12	Calcium Hardness (as $CaCO_3$)	mg/l	8.0	IS:3025/P21/2014	—
13	Magnesium Hardness (as $CaCO_3$)	mg/l	4.0	IS:3025/P21/2014	—
14	Calcium as Ca	mg/l	3.2	IS:3025/P40/2014	200
15	Magnesium as Mg	mg/l	0.9	IS:3025/P46/2014	100
16	Total Alkalinity (as $CaCO_3$)	mg/l	14	IS:3025/P23/2014	600
17	Total Iron (as Fe)	mg/l	BDL (DL-0.1)	IS:3025/P53/2014	1.0
18	Residual Chlorine	mg/l	BDL (DL-0.04)	IS:3025/P26/2003	—
19	Fluorides (as F)	mg/l	BDL (DL-0.1)	IS:3025/1984	1.5
20	Nitrates (as NO_3)	mg/l	<5	IS:3025/P34/2003	No Relaxation

Remarks: The above samples meet the requirements of IS 10500 R.2012 for potability with respect to the parameters tested. BDL – Below Detectable Limit DL – Detectable Limit No Relaxation

Verified By

Chemist

For Swasti Enviro Solutions Pvt Ltd

Authorized Signatory
A.Prabhu
Quality/Technical Manager

Note : 1. The Results relate only to this items tested
2. Report shall not be reproduced without the approval of the laboratory.
3. The test items will not be retained for more than 7 days from the date of issue of test report.

Fig. 14 Desalinated water report

volume for a 4 L water input is 1050 mL, or around a 1:4 ratio. Water is made pure by letting sunlight shine on static water in a tank that is enclosed by a reflector, which in turn heats the water to cause it to evaporate. The salt is left behind as a residue after the evaporated droplets are collected over the top of the enclosed glass and then stored in a temporary reservoir due to an increase in the amount of heat needed to evaporate the water in the tank.

An alternative process can improve performance even more. We discovered that the project is losing money. Water that is not collected through the exit pipe evaporates and condenses on the tank's sides. Hence, productivity might increase if that water is collected and transferred to the outlet. The second issue is the development of Skims on the water layer, which prevents sunlight from penetrating the water. The water may evaporate quickly and output may rise if the skims are removed.

References

1. Jesumathy SP, Udayakumar M, Suresh S (2012) Heat transfer characteristics in latent heat storage system using paraffin wax. J Mech Sci Technol 26(3):959–965
2. Shatat M, Riffat S, Gan G (2016) An innovative psychometric solar-powered water desalination system. Int J Low-Carbon Technol 11(2):254–265
3. Ammari HD, Nimir YL (2003) Experimental and theoretical evaluation of the performance of a tar solar water heater. Energy Convers Manag 44(19):3037–3055
4. Kalogirou SA (2004) Solar thermal collectors and applications. Prog Energy Combust Sci 30(3):231–295
5. Ullah I, Rasul MG (2019) Recent developments in solar thermal desalination technologies: a review. Energies 12(1):119
6. Al-Hallaj S, Farid MM, Tamimi AR (1998) Solar desalination with a humidification–dehumidification cycle: performance of the unit. Desalination 120(3):273–280
7. Eltawil MA, Zhengming Z, Yuan L (2009) A review of renewable energy technologies integrated with desalination systems. Renew Sustain Energy Rev 13(9):2245–2262
8. Velmurugan V, Srithar K (2007) Solar stills integrated with a mini solar pond—analytical simulation and experimental validation. Desalination 216(1–3):232–241
9. Memon S, Soomro SH, Kim W-S, Kim Y-D (2022) Performance investigation of solar-powered direct-contact membrane distillation and humidification–dehumidification hybrid system. Chem Eng Process - Process Intensif 174:108889
10. Zhang Y, Zhu C, Zhang H, Zheng W, You S, Zhen Y (2018) Experimental study of a humidification–dehumidification desalination system with heat pump unit. Desalination 442:108–117
11. Zhao Y, Zheng H, Liang S, Zhang N, Ma XL (2019) Experimental research on four-stage cross flow humidification dehumidification (HDH) solar desalination system with direct contact dehumidifiers. Desalination 467:147–157
12. Shalaby SM, Kabeel AE, Moharram BM, Fleafl AH (2021) Experimental study of hybrid solar humidification dehumidification system for extremely saline water desalination. Energy Convers Manag 235:114021
13. Mohamed ASA, Shahdy AG, Ahmed MS (2021) Investigation on solar humidification dehumidification water desalination system using a closed-air cycle. Appl Therm Eng 188:116621
14. Rahimi-Ahar Z, Hatamipour MS, Ghalavand Y (2018) Solar assisted modified variable pressure humidification–dehumidification desalination system. Energy Convers Manag 162:321–330
15. Ahmed HA, Ismail IM, Saleh WF, Ahmed M (2017) Experimental investigation of humidification–dehumidification desalination system with corrugated packing in the humidifier. Desalination 410:19–29
16. Srithar K, Rajaseenivasan T (2018) Recent fresh water augmentation techniques in solar still and HDH desalination—a review. Renew Sustain Energy Rev 82(Part 1):629–644

17. Xu H, Zhao Y, Dai YJ (2019) Experimental study on a solar assisted heat pump desalination unit with internal heat recovery based on humidification–dehumidification process. Desalination 452:247–257
18. El-Agouz SA, Sathyamurthy R, Manokar AM (2018) Improvement of humidification–dehumidification desalination unit using a desiccant wheel. Chem Eng Res Des 131:104–116
19. Dehghani S, Date A, Akbarzadeh A (2018) Performance analysis of a heat pump driven humidification–dehumidification. Desalination 445:95–104
20. Dave T, Ahuja V, Krishnan S (2021) Economic analysis and experimental investigation of a direct absorption solar humidification–dehumidification system for decentralized water production. Sustain Energy Technol Assess 46:101306
21. Khass TM, Mohammed RH, Qasem NAA, Zubair SM (2022) Different configurations of humidification–dehumidification desalination systems: thermal and economic assessment. Energy Convers Manag 258:115470
22. Abdel Dayem AM, AlZahrani A (2022) Psychometric study and performance investigation of an efficient evaporative solar HDH water desalination system. Sustain Energy Technol Assess 52(Part A):102030
23. Rahimi-Ahar Z, Hatamipour MS (2021) Performance evaluation of a solar and vacuum assisted multi-stage humidification-dehumidification desalination system. Process Saf Environ Prot 148:1304–1314
24. Dayem AMA (2023) Innovative solar natural vacuum desalination system. Case Stud Therm Eng 44:102881
25. Wang L, Zheng H, Zhao Y, Ma X (2021) Solar-driven natural vacuum desalination system with inner condenser. Appl Therm Eng 196:117320

Performance Analysis of a Li-Ion Battery Using Generic Electro-thermal Model Under Different Temperature and Charging/Discharging Environment

Vikas Verma and Bhavnesh Kumar

Abstract Globally, in the automotive industry, internal combustion engines (ICE) based driven vehicles are the major source of environmental degradation because of their poisonous emissions in the atmosphere. Electric vehicles are the most appropriate alternative for this change. The rapid growth of electric vehicles is the need of today's census. The growth of this lithium-ion battery provides an influential factor in powering the rise of electric vehicles. The effective operation of the battery is supreme, so to achieve it the critical factors that affect its performance are needed to be analyzed for effective management. The battery performance is greatly varied with the rise in temperature. This paper presents an electro-thermal model of the generic lithium-ion cells to investigate the electrical and thermal behavior under different driving cycles at different ambient temperatures.

Keywords Electric vehicles (EV) · Lithium-ion cell · State of charge · Electrothermal · Urban driving pattern (UDP) · Highway driving pattern (HDP)

1 Introduction

Nowadays, with the sharp increase in carbon gases (CO_2) emission has raised the concern towards environmental pollution and global warming. The automotive industry is the prime cause towards this environmental pollution. The fossil fuel based internal combustion engines (ICE) vehicles, industries and manufacturing units are the main source of carbon dioxide emission. The change in the climatic pattern and day by day increment in the temperature at global level has raised the risks and the concern in the usage of ICE vehicles [1]. This alarming factor has made a potential shift towards the development and usage of electric vehicles. Electric vehicles are the most efficient as well as sustainable transport compared with the ICE vehicles.

V. Verma (✉) · B. Kumar
Netaji Subhas University of Technology, Dwarka, New Delhi, India
e-mail: vikas.verma.phd22@nsut.ac.in

J. Kumar et al. (eds.), *Recent Advances in Power Systems*, Lecture Notes in Electrical Engineering 1192, https://doi.org/10.1007/978-981-97-2788-9_21

The maximum projected growth of electric vehicle in the future is dependent on the successful development of the economical and powerful battery [2, 3].

The revolutionary growth in the electric vehicle sector is due to rapid development of the rechargeable battery that is mainly the lithium-ion battery. The usage of lithium-ion battery is initialized from cell phones & laptops and now its high-power rating specifications are used in electric vehicles as well as in military and space. Lithium-ion battery has many positive attributes like less self-discharge rate, large energy density, high power density, maximum cycle life, less per volume cost and negligible memory effect which make it more adaptable for the long run electric vehicles [4, 5]. Currently in the research and development area of electric vehicles lithium battery is quite significant because of its carbon anode and usage of many materials at the cathode creating the possibility of its large co-variants. The market share of lithium battery usage is very large at the global level. Mostly in Electric vehicle applications (two-wheeler & four-wheeler vehicles) lithium-ion battery is mostly employed because of fast charging and discharging within the permissible operational range. It is the most common portable battery because of its high energy density and light weight [6].

For the efficient testing of the battery cell and whole stack modelling of many batteries is done earlier like numerical modelling which is also known as operational modelling that provides the internal parameters of the operational functionality of the battery. Another one is the electrochemical modelling in which chemical reactions occur to extract out the empirical values as well as the basic behavior of the cell [7, 8]. The mathematical model is composed of both empirical and internal operational models which shows the complete behavior and operation of the basic lithium-ion battery. This gives the deep insight investigation of the parameters and generic behavior of the battery. In the case of battery modelling of the electric vehicle many lithium-ion battery models are being proposed earlier like the internal resistance based equivalent circuit model, Thevenin based equivalent circuit model, polarization-based model, resistance-capacitor based model, lumped resistance-capacitor based model [9, 10] etc. Among all the corrected ones, Shepherd model is the basic effective model which is considered for the behavioral analysis of the lithium-ion battery.

In this paper the basic lithium-ion battery modelling is done to analyze the correlation as well as correlative effect of various electrical and thermal parameters on one another of the battery. To monitor the critical analysis of the system and its behavior this complete model is simulated under two different operating cycles in different ambient working conditions [11, 12]. This basic model gives the basic insight understanding of the factors that are responsible for the inappropriate variations which needed to be managed effectively.

2 Mathematical Modelling of General Battery Model

In this paper the mathematical model of the basic lithium-ion battery is modelled to analyze the cell dynamics and the transient behavior of the battery. The whole model is analyzed in two counterparts; first the electric modelling is done to understand the electrical parameters and second the thermal modelling is also incorporated to see the heating effects in the battery as shown in Fig. 1. The prime focus of this modelling is to see the behavioral pattern of the voltage, current, state of charge (SOC) and discharging current (i^*t) with respect to time under the three instances that is no-load (exponential and nominal voltage region), discharging region and the charging region at different cycles [11, 12].

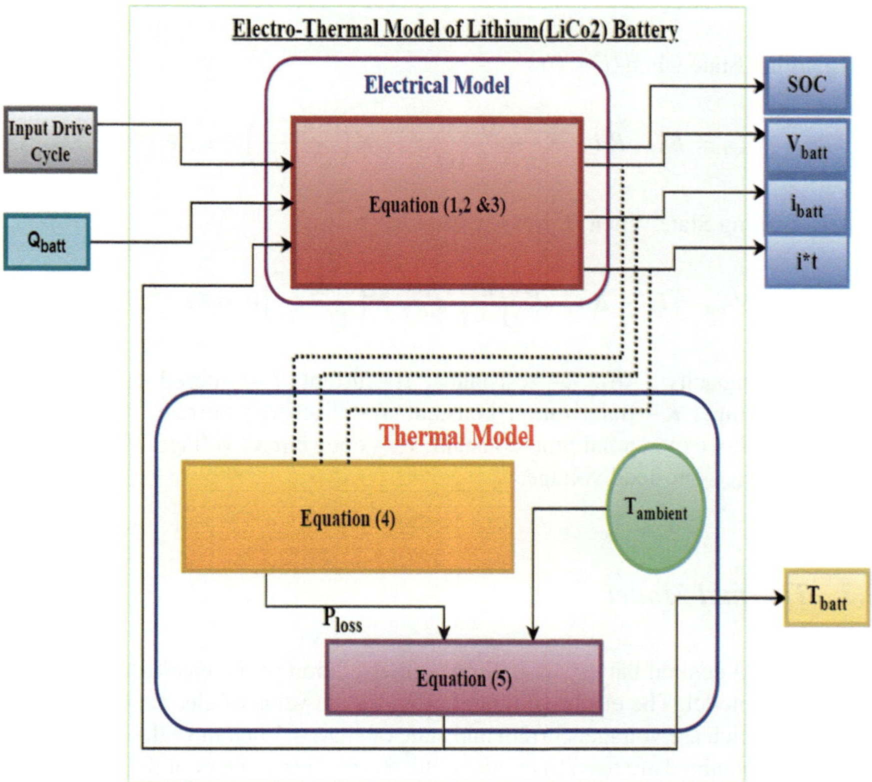

Fig. 1 Electro-thermal model of lithium-ion battery

2.1 Electrical Model

The behavior of the battery cell through electrical model is modelled and evaluated using its basic electrical equivalent characteristic equation. The power is input to the model in terms of two different driving cycles and its capacity in terms of charge. The corresponding voltage, current and state of charge are computed under various battery states that are no-load, charging and discharging states. These different states' equation used is shown below [13, 14]. The output obtained from this model is used as an input to the thermal model. The Basic mathematical equations used for this modelling are

$$V_{noload} = V_o - K\frac{Q}{Q - it}.\left(it + i^*\right). \tag{1}$$

Charging State when ($i^* < 0$)

$$V_{batt} = E_0 - R.i - K\frac{Q}{it - 0.1Q}.i^* - K\left(\frac{Q}{Q - it}\right)it + Ae^{(-B.it)} \tag{2}$$

Discharging State When ($i^* > 0$)

$$V_{batt} = E_0 - R.i - K\frac{Q}{Q - it}.i^* - K\left(\frac{Q}{Q - it}\right)it + Ae^{(-B.it)} \tag{3}$$

where Q—capacity (Ah), R—resistance, i—current, i^*—filtered current, E_0—constant voltage, K—polarization constant, it—discharge current, A—exponential voltage, B—exponential time-constant, V_o—open circuit voltage, V_{batt}—battery voltage, V_{noload}—no-load voltage.

2.2 Thermal Model

The proposed general battery modelling is interrelation of the electrical model and the thermal model. The electrical model provides the value of electrical parameters of battery which are voltage, current and state of charge. Then the value of electrical parameters obtained are used to compute the thermal parameters of the battery. The thermal model of the battery comprises of two counterparts; in the first part the losses are evaluated and in the second part measurement of the cell temperature occurs. The heat is generated inside the battery is due to power losses which is comprised of three losses that are ohmic losses, losses due to change of entropy and losses due to the internal resistance of the battery. The ohmic losses are mainly due to the different current driving cycle, the entropy losses are computed with the change in voltage with respect to temperature and the losses are due to internal resistance which varies with the material specificity. The amount of power losses in the battery is computed through the below equation which includes the combined effect of all the respective losses in the thermal model using the electrical model parameters that are current, voltage and discharging current.

$$P_{loss} = iE_o - \left(i.V_{batt} + T\frac{dV}{dT}\right) + i^2R_{int} \tag{4}$$

where i—battery current, E_o—constant voltage, T—temperature, V—maximum voltage, R_{int}—internal resistance and P_{loss}—power losses.

The battery temperature is calculated in the thermal model which is based on the summation of the heat generated within the thermal model and the ambient temperature in which the operation is being done.

$$\text{Cell Temperature}_{\text{int}} = \text{Heat generated due to } P_{loss} + \text{Ambient Temperature} \tag{5}$$

The resultant of this summation provides the exact operational battery temperature generated during that instantaneous working conditions. The variations in the electrical and thermal parameters of this electrothermal battery model shows the correlation and dependency on each other. The change in the value in any sole model directly affects the other one which would affect the overall performance of the battery. The output through the thermal model is taken as the feedback signal to the electrical model after which all electrical parameters are updated with the correlative effect of internal battery temperature.

3 Results and Discussion

This electro-thermal battery model is simulated under two different driving patterns that are the urban driving pattern (UDP) and highway driving pattern (HDP) for the three ambient temperature working conditions that are low (−10 °C), moderate (10 Degree Celsius) and high (40 °C) as shown in Figs. 2 and 3. In this driving cycle the battery is operated in a sampling cycle of 2500 s of various driving states of continuous charging and discharging. With the change of the ambient temperature, the battery temperature and its dynamic behavior significantly varies because of different amount of heat generation and dissipation in the low, moderate and high temperature conditions (according to conditions respectively. The change in ambient conditions also affects the corresponding electrical parameter which results in overall change in dynamic performance of the battery.

The two different input driving patterns that are urban driving patterns (UDP) and highway driving patterns (HDP) which is given to the battery model in the form of current for the sample cycle of 2500 s.

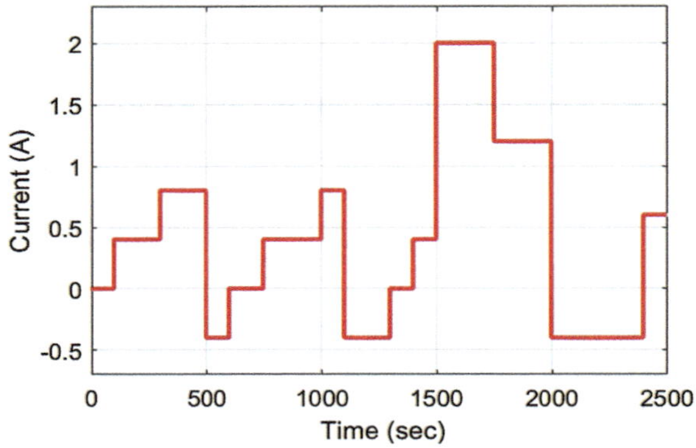

Fig. 2 Input current urban driving pattern (UDP cycles)

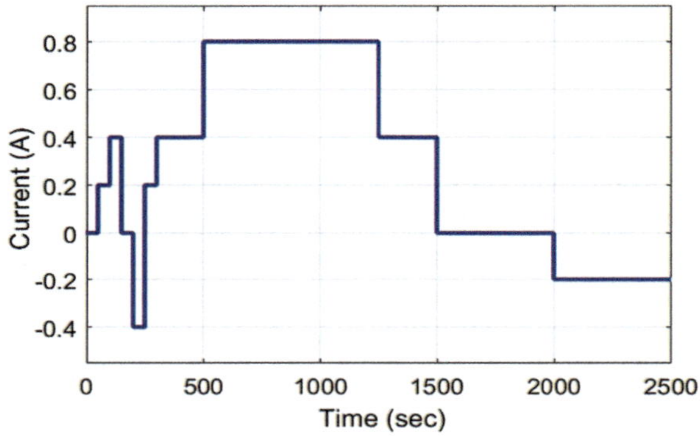

Fig. 3 Input current highway driving pattern (HDP cycles)

The variation in the state of charge percentage of the battery under different ambient temperatures and different driving cycles. Figures 4 and 5 shows the change in behavior of SOC in the urban driving pattern (UDP) and highway driving pattern (HDP). Initially it was at full but during working in their corresponding cycles there is continuous variation in the driving speed that makes the non-uniform change of the charging state. The variation is also affected according to different ambient temperatures. The input driving variation is maximum in UDP as compared to HDP so the corresponding SOC curve shows the non-uniform change in UDP and smooth change in HDP cycles. Moreover, the loss of charge is maximum at high temperature and minimum at low temperature.

Figure 6 shows the dynamic behavior internal cell temperature accumulated due to losses in different driving cycles as well as working conditions at (−10 °C) low ambient temperature. Figure 7 shows the corresponding change in the voltage response of the battery due to the actual cell temperature generated at low ambient operating conditions. There is continuous variation throughout the plot which shows the internal charging and discharging process of the battery in the context of its respective negative and the positive current cycles. The voltage of the battery is initialized with the discharging process and proceeds further. At the negative instances of the

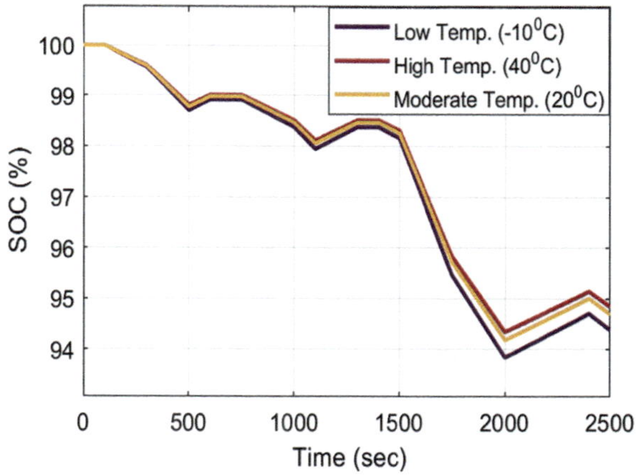

Fig. 4 State of charge (SOC) of battery in UDP cycle

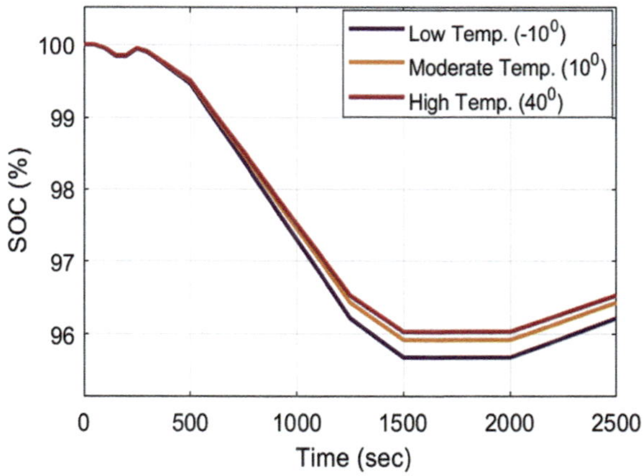

Fig. 5 State of charge (SOC) of battery in HDP cycle

current the voltage value rises show the charging process of the battery and at the positive instances of the current there is drop in voltage which marks the discharging process.

Figure 8 shows the cell temperature at (10 °C) moderate ambient temperature where the cell temperature is nominal in the HDP driving cycle as compared to the UDP cycle. Figure 9 shows the corresponding change in the voltage response of the battery in which the charging and discharging is uniform in the HDP cycle as compared to the UDP cycle due to moderate ambient operating conditions.

The cell temperature is maximum at (40 °C) high ambient temperature as compared to the other. The high ambient temperature response is shown in Fig. 10. The corresponding change in the voltage response of the battery at different working cycles is shown in Fig. 11.

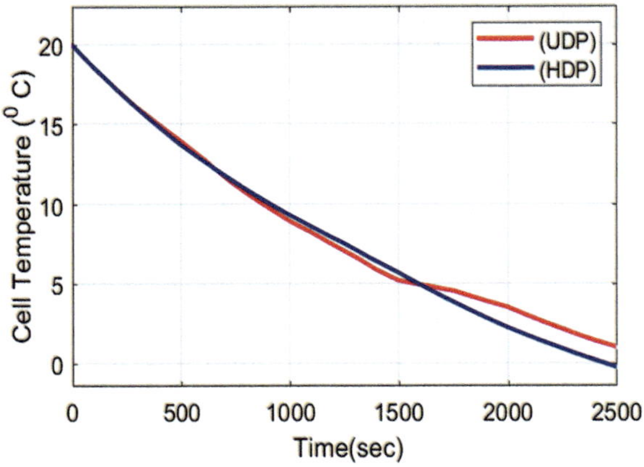

Fig. 6 Cell temperature at low ambient temperature (−10 °C)

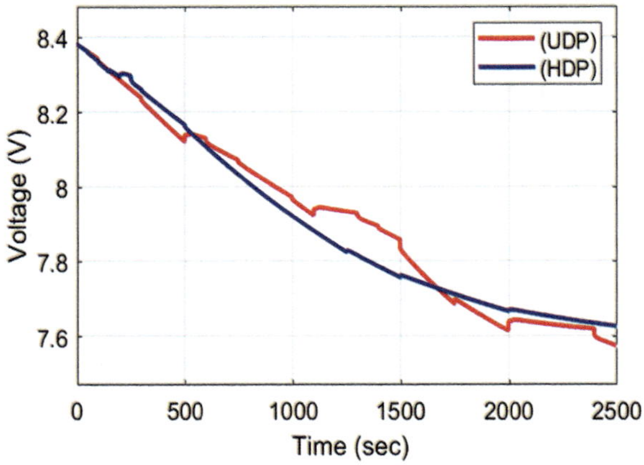

Fig. 7 Voltage at low ambient temperature (−10 °C)

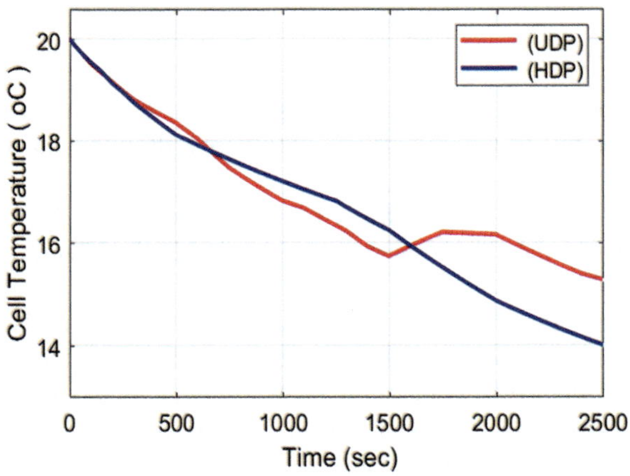

Fig. 8 Cell temperature at moderate ambient temperature (10 °C)

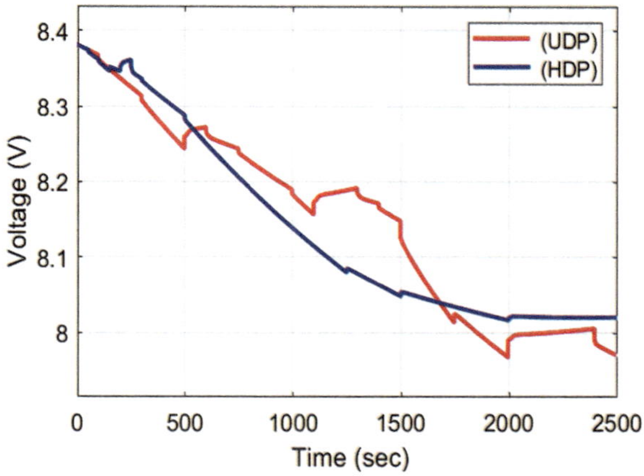

Fig. 9 Voltage at moderate ambient temperature (10 °C)

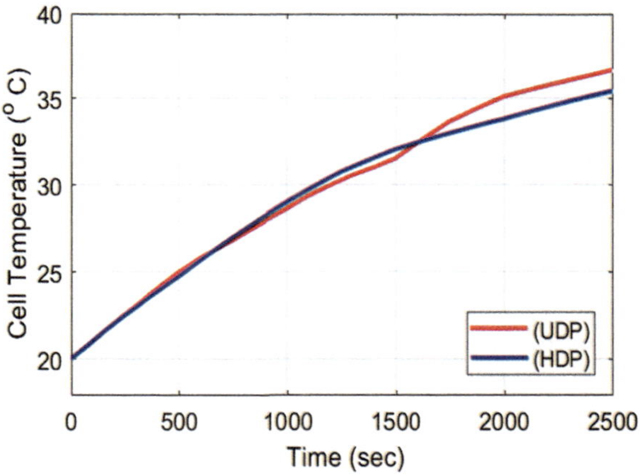

Fig. 10 Cell temperature at high ambient temperature (40 °C)

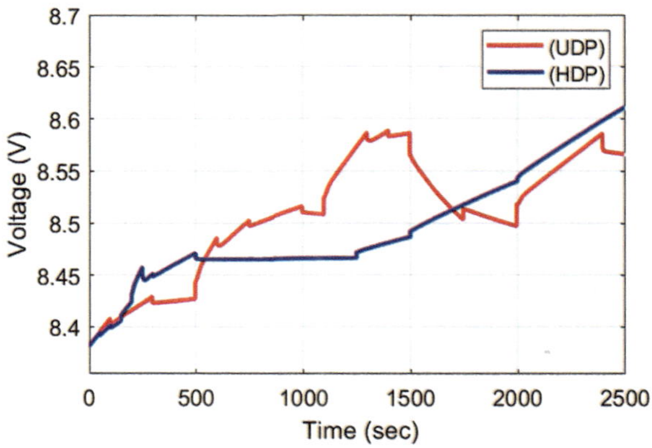

Fig. 11 Voltage at high ambient temperature (40 °C)

4 Conclusion

An electro-thermal model of the Li-Ion battery is presented to study the behavior under varying temperature and load. The effects of thermal changes on electrical behavior have been presented using the developed model. The battery response with two input driving cycles provided in different ambient working conditions shows the corresponding different variations in each case. The highway driving pattern (HDP) shows the smooth variations in the cell temperature, state of charge, charging and discharging voltages but the urban driving pattern (UDP) shows the abrupt variations. Both the driving pattern gives the critical parametric analysis of the battery. The variation of cell temperature largely varies with change in the working conditions at low, moderate and high ambient temperature as well as the internal losses which together affect its dynamic behavior.

References

1. Hasan MK, Mahmud M, Habib AKM, Motakabber SMA, Islam S (2021) Review of electric vehicle energy storage and management system: standards, issues, and challenges. J Energy Storage 41:102940. ISSN: 2352-152X
2. Faraj M, Basir O (2016) Range anxiety reduction in battery-powered vehicles. In: 2016 IEEE transportation electrification conference and expo (ITEC), Dearborn, MI, USA
3. Rimpas D, Kaminaris SD, Aldarraji I, Piromalis D, Vokas G, Papageorgas PG, Tsaramirsis G (2022) Energy management and storage systems on electric vehicles: a comprehensive review. Mater Today: Proc 61(Part 3):813–819. ISSN: 2214-7853
4. Aishwarya M, Sailaja M, Brisilla RM (2021) Modeling of lithium-ion batteries: an overview. In: 2021 Innovations in power and advanced computing technologies (i-PACT), Kuala Lumpur, Malaysia, pp 1–9
5. Olanyi K, Olyungiue A (2020) Review of strategies for hybrid energy storage management system in electric vehicle application. Int J Electr Electr Comput Eng
6. Horiba T (2014) Lithium-ion battery systems. Proc IEEE 102(6):939–950
7. Khanum F, Louback E, Duperly F, Jenkins C, Kollmeyer PJ, Emadi A (2021) A Kalman filter based battery state of charge estimation MATLAB function. In: Proceedings of the 2021 IEEE transportation electrification conference and expo (ITEC), Chicago, IL, USA, pp 484–489
8. Feng T, Yang L, Zhao X, Zhang H, Qiang J (2015) Online identification of lithium-ion battery parameters based on an improved equivalent-circuit model and its implementation on battery state-of-power prediction. J Power Sources 281:192–203
9. Zhu C, Li X, Song L, Xiang L (2013) Development of a theoretically based thermal model for lithium-ion battery pack. J Power Sources 223:155–164. ISSN: 0378-7753
10. Chen X, Shen W, Vo TT, Cao Z, Kapoor A (2012) An overview of lithium-ion batteries for electric vehicles. In: 2012 10th IEEE international power and energy conference (IPEC), Ho Chi Minh City, Vietnam, pp 230–235
11. Piller S, Perrin M, Jossen A (2001) Methods for state-of-charge determination and their applications. J Power Sources 96(1):113–120. ISSN: 0378-7753

12. Rojas OE, Khan MA (2022) A review on electrical and mechanical performance parameters in lithium-ion battery packs. J Clean Prod 378:134381. ISSN: 0959-6526
13. Tremblay O, Dessaint L-A, Dekkiche A-I (2007) A generic battery model for the dynamic simulation of hybrid electric vehicles. In: VPPC 2007—proceedings of the 2007 IEEE vehicle power and propulsion conference, pp 284–289. https://doi.org/10.1109/VPPC.2007.4544139
14. Saw LH, Somasundaram K, Ye Y, Tay AAO (2014) Electro-thermal analysis of lithium iron phosphate battery for electric vehicles. J Power Sources 249:231–238. ISSN: 0378-7753

Batch number: 10583315

Printed by Printforce, the Netherlands